NORTH ATLANTIC
1943 - 1945

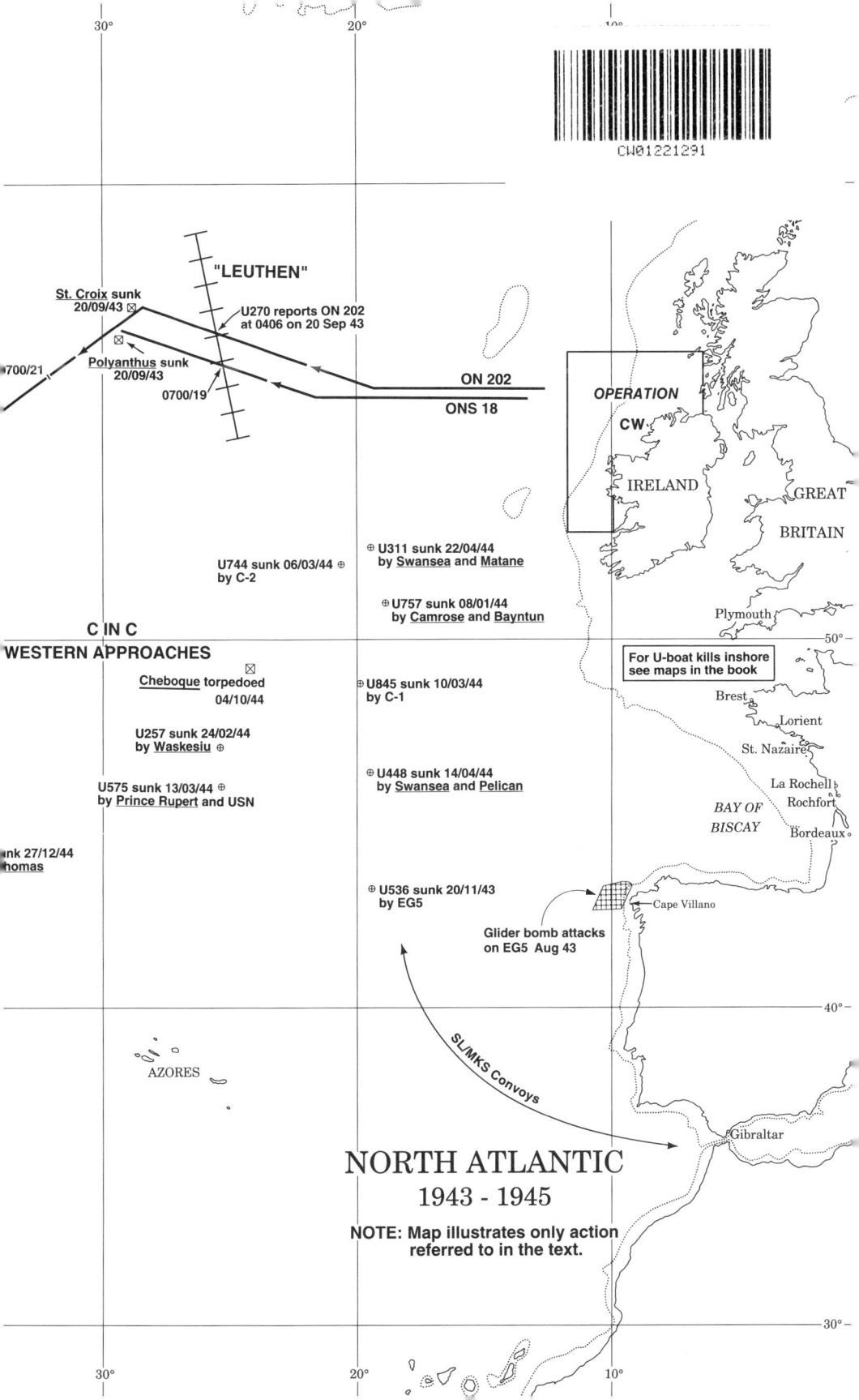

NOTE: Map illustrates only action referred to in the text.

"LEUTHEN"

St. Croix sunk 20/09/43
U270 reports ON 202 at 0406 on 20 Sep 43
Polyanthus sunk 20/09/43
0700/19
700/21

ON 202
ONS 18

OPERATION CW

IRELAND
GREAT BRITAIN
Plymouth

For U-boat kills inshore see maps in the book

C IN C WESTERN APPROACHES

Cheboque torpedoed 04/10/44

U257 sunk 24/02/44 by Waskesiu

U575 sunk 13/03/44 by Prince Rupert and USN

nk 27/12/44
homas

U744 sunk 06/03/44 by C-2

U311 sunk 22/04/44 by Swansea and Matane

U757 sunk 08/01/44 by Camrose and Bayntun

U845 sunk 10/03/44 by C-1

U448 sunk 14/04/44 by Swansea and Pelican

U536 sunk 20/11/43 by EG5

Glider bomb attacks on EG5 Aug 43

Cape Villano

Brest
Lorient
St. Nazaire
La Rochelle
Rochfort
Bordeaux

BAY OF BISCAY

AZORES

SL/MKS Convoys

Gibraltar

The U-Boat Hunters

The Royal Canadian Navy and the
Offensive against Germany's Submarines

The U-Boat Hunters

The Royal Canadian Navy and the Offensive against Germany's Submarines

MARC MILNER

NAVAL INSTITUTE PRESS
Annapolis, Maryland

© Marc Milner 1994

First published in the United States of America by the
Naval Institute Press
118 Maryland Avenue
Annapolis, Maryland 21402-5035

Library of Congress Catalog Card Number 94-66600

ISBN 1-55750-854-2

Printed in Canada on acid-free paper ∞

Cataloging-in-Publication Data is available from the
Library of Congress.

This book has been published with the help of a grant from the Social Science Federation of Canada, using funds provided by the Social Sciences and Humanities Research Council of Canada.

For Bob

'You are not far removed from it [RN tradition] yourselves, you know. You are a part of the Empire and much of your stock is British.'

'That's so, sir,' I acknowledged. 'But ... many of us feel that we have no direct right to your traditions ...'

A book lying on his desk, which I had several times seen him reading, rather demonstrated my point, I thought. It was Southey's *Life of Nelson*. 'Our tradition,' I suggested, 'is possibly being made now.'

Alan Easton, 50 North

Contents

ABBREVIATIONS / ix
PREFACE / xiii
ACKNOWLEDGMENTS / xvii

   Prologue / 3
1 The Fortunes of War / 21
2 Cinderella: Act II / 59
3 Triumph without Celebration / 96
4 The Summer Inshore / 134
5 A Sea of Troubles / 175
6 Victory ... of a Sort / 217
   Epilogue: Loch Eriboll / 256

APPENDIXES
I Escort Building and Modernization, 1941–3 / 269
II U-Boat Kills by the RCN, May 1943 – May 1945 / 280

NOTES / 283
BIBLIOGRAPHY / 313
INDEX / 319
PICTURE CREDITS / 327

Maps

The north Atlantic, 1943–4 / endpapers
The Channel and the Bay of Biscay, June–September 1944 / 136
United Kingdom waters, September 1944 – May 1945 / 196

Figures

1 Theoretical asdic patterns / 90
2 Variations in temperature gradient of sea water / 180

Abbreviations

ACI	*Atlantic Convoy Instructions*
ACNS	Assistant chief of Naval Staff
A/S	Anti-submarine
Asdic	British term for SONAR
ASW	Anti-submarine warfare
B group	British escort group of MOEF
BdU	Befelshaber der U-boote (C-in-C, U-boats)
BT	Bathythermography
C group	Canadian escort group of MOEF
CAFO	Confidential Admiralty fleet order
CAT	Canadian Anti-Acoustic Torpedo gear
Cdr	Commander
CHOP	Change of Operational Control
C-in-C	Commander-in-chief
CNA	Canadian northwest Atlantic
CNMO	Canadian Naval Mission Overseas (London)
CNS	Chief of Naval Staff
CVE	Escort or auxiliary aircraft carrier
CWC	Cabinet War Committee
DA/SW	Director of Anti-Submarine Warfare, Admiralty
DAUD	Director, Anti-U-Boat Division, Admiralty
DF	Direction finding (of radio transmitters)
DHist	Directorate of History, NDHQ, Ottawa
DNO	Director of Naval Operations, NDHQ
DNOS	Director of Naval Operational Studies, RN

x Abbreviations

DOD	Director of Operations Division, NSHQ
DOR	Director of Operational Research, Ottawa
DWT	Director of Warfare and Training, NSHQ
EAC	Eastern Air Command, RCAF
EG	Escort group
FONF	Flag Officer, Newfoundland
FOXER	British and American anti-acoustic torpedo gear
FSL	First Sea Lord
GNAT	German navy acoustic torpedo
HE	Hydrophone effect (listening on asdic)
HF/DF	High-frequency direction finding
HMCS	His Majesty's Canadian Ship
HX	Fast eastbound convoy
Lt Cdr	Lieutenant commander
MF/DF	Medium-frequency direction finding
MOEF	Mid-Ocean Escort Force
NAC	National Archives of Canada
NB	Naval Board, RCN
NEF	Newfoundland Escort Force
NHB	Naval Historical Branch, London
NHC	Naval Historical Center, Washington, DC
NRC	National Research Council, Canada
NS	Naval Staff, RCN
NSHQ	Naval Service Headquarters, Ottawa
OIC	Operational Intelligence Center
ON	Fast westbound convoy
ONS	Slow westbound convoy
PNM	Pipe noise-maker
PRO	Public Record Office, Kew, England
RAF	Royal Air Force
RCAF	Royal Canadian Air Force
RCN	Royal Canadian Navy
RCNMR	*Royal Canadian Navy Monthly Review*
RCNR	Royal Canadian Navy (Reserve)
RCNVR	Royal Canadian Navy (Volunteer Reserve)
RDF	Radio direction finding, British term for radar
RN	Royal Navy

RX/C	Canadian-developed 10-cm radar
Salmon	Joint naval and air hunt to exhaustion for a U-boat
SC	Slow eastbound convoy
SNWE	'Summary of Naval War Effort,' RCN
SOE	Senior officer of the escort
SW1C/2C	Metric wavelength Canadian shipborne radar
TU	Task unit (USN designation for escort group)
USN	United States Navy
VCNS	Vice-chief of Naval Staff
VLR	Very long-range aircraft
WAC	Western Approaches Command, RN
WACI	*Western Approaches Command Instructions*
WLEF	Western Local Escort Force (Halifax, to April 1943)
WEF	Western Escort Force (Halifax, from April 1943)

Preface

The Royal Canadian Navy came of age during the Battle of the Atlantic, 1939–45. In the minds and memoirs of former RCN officers, this struggle for command of the sea against the German submarine marked a rite of passage. It was for them a time of adventure and high achievement. Not until the 1980s, however, was their experience subjected to rigorous analysis, and the result was not always flattering. My *North Atlantic Run: The Royal Canadian Navy and the Battle for the Convoys*, published in 1985, took a fairly hard-nosed look at Canadian convoy operations in the mid-Atlantic from 1941 to 1943. That account ended at what was, unquestionably, the low point of the war for the RCN. As one reviewer of *North Atlantic Run* complained, the book ended on such a negative note, he wondered if there was anything positive to say about the RCN's operational efficiency during the war. The same thought crossed my mind many times. Much of the impetus behind my continuing the study of the RCN's anti-submarine war after mid-1943 was to determine if and when the fleet became truly proficient. *The U-boat Hunters* is the fruit of that enquiry.

The present book seeks to follow the action by pursuing the RCN's role in the offensive against the U-boat which dominated the last two years of the Atlantic war. In part, my approach is based on the simple assumption that the point of contact between opposing forces provides the clearest measure of their operational effectiveness. I am aware that such an assumption is not entirely fair, nor is it an accurate measure of the value of RCN operations. From 1943 to 1945 the vast majority of RCN anti-submarine vessels remained engaged in close escort duty. Their work formed the bedrock – as it had done in 1941–3 – of the western alliance. Without close escorts there would have been no offensive: no Combined Bomber Offensive, no assault landings in North Africa and Italy, no D-Day, and certainly no offensive against the U-boats themselves. The purpose here is to follow the action, however, and to assess how the RCN performed at the point of contact with its old enemy from 1943 to 1945.

In doing so, this book follows a less well-defined path than my earlier, tightly focused analysis of the mid-Atlantic theatre. In particular, anti-submarine operations in 1944–5 were a dreary contrast to the dramatic days of 1940–3. Few merchant ships were sunk, Allied plans and operations went ahead on every front without hindrance from the Kriegsmarine, and few U-boats were found and destroyed. Because the late-war battle with the U-boats has been perceived – with some justification – as dull, uneventful, and irrelevant to the outcome of the war, accounts of the Atlantic war usually stop abruptly in the late fall of 1943 after the failure of the German acoustic torpedo campaign.

The appeal of the earlier period is evident in the documents themselves, which offer a distinctly traditional, almost Nelsonic, view of the Atlantic war. 'Anti-submarine warfare,' especially the great convoy battles, took place largely on the surface and were easily explained in familiar language. Reports of proceedings read like action narratives, relying heavily on visual imagery. U-boats were usually spotted on the surface, engaged by gunfire, and typically driven down to face a hasty and very shallow depth-charge attack. If the charges were close enough, the U-boat was forced back to the surface, where it endured a hail of gunfire and, quite often, a ramming by the attacking vessel. By 1944, however, submarines were forced to operate wholly submerged, and anti-submarine warfare became, in Roger Sarty's description, more like guerilla warfare. The Allies tried to locate or pre-empt attacks by individual submarines operating almost entirely submerged, while the U-boats employed stealth and the ocean environment to their advantage in hit-and-run attacks. This cat-and-mouse game in the sea forced the

Allies to devise methods of discriminating between false and true echoes from what might be a submarine. Reports of proceedings after 1943 lose their narrative flavour and become highly technical, filled with the jargon of classification procedures for unseen underwater contacts, references to operations explained only in long-forgotten – or discarded – Confidential Books, and garnished with extensive appendices of cryptic signal traffic and asdic traces. Frequently, the only visceral contact between the hunter and the hunted was sound and high explosives. Indeed, in the absence of debris or bodies following an A/S attack, one never really knew if a U-boat was present or if all that fury was spent on an anomaly in the water. And if the escort destroyed a submerged submarine, its crew perished – taking with them their own story and much of the human drama of the event.

The dull, highly technical, and inconsequential nature of late-war ASW may, of itself, have been enough to keep historians of the Second World War away from it. But there was another, more prosaic, reason. Late-war U-boats had become very hard to find and sink. Schnorkels, improved batteries, radar detectors, and the like transformed the old 'diving boats' into complete submarines able to operate submerged for prolonged periods and at great ranges. By 1945 all aspects of Allied anti-submarine warfare, ships, equipment, tactics, and doctrine, developed over six years of war, were thrown into a state of uncertainty. Moreover, the Russians had captured U-boat building yards and design personnel, as well as working examples of the very latest German submarines for which the Allies had no clear solution.[1] The story of 1944–5 could hardly be told in detail during the Cold War, even assuming anyone was interested.

The central story told here is the development in anti-submarine warfare from 1943 to 1945 and how the Royal Canadian Navy responded to the challenge. It is a natural corollary to the earlier book, but ironically the challenge was not one that the wartime RCN took to easily or with much enthusiasm. For most of the war the RCN proper – the professional institution – concentrated its attention on acquiring carriers, cruisers, and modern destroyers for the postwar fleet. For only a brief period, from late 1942 to late summer 1943, did the interests of the escort fleet and the professional navy coincide. That brief convergence of efforts was a result of the collapse of the escort fleet in late 1942, as described at length in *North Atlantic Run* and reviewed in passing here. The consequences of that collapse for RCN participation in the great Allied offensive against the U-boats that followed in 1943 are explored further in this book. During that traumatic and painful year, the Allies won a resounding victory in the Atlantic and sank nearly 200 submarines in waters frequented by Canadian ships. Throughout the RCN struggled almost in vain to sink a single U-boat. The story of Canadian naval participation in this, the greatest of all offensives against the U-boats, forms the first portion of the book.

By late 1943 the major problems with the escort fleet were resolved, the Atlantic was secure beyond any doubt, and the professional RCN turned its attention back to long-term fleet building. In that sense, the RCN's A/S story after 1943 had to be extracted from the records of an institution that was preoccupied with other interests. The Canadian navy tinkered with solving the practical problems of late-war ASW, but the threat in Canadian waters remained so small – and the navy's interest so focused on large ships – that the A/S fleet in home waters was left to finish the war mainly on sheer momentum. Anti-submarine forces sent overseas to serve under British command in the eastern Atlantic fared much better and earned an enviable reputation for excellence. The connection between the two experiences – that of the fleet in Canadian waters and that of those in Europe – was, it seems, a tenuous one. But they were tied together by the common problems of an emergent form of modern war. For that reason considerable effort has been made here to set the Canadian experience into the broader context of developments in anti-submarine warfare.

The bulk of this book, therefore, is devoted to an analysis of the disparate story of the RCN's A/S forces during 1944–5 within the wider settings of both the Atlantic war and the development of ASW. In doing so, *The U-boat Hunters* does not attempt to offer a definitive history of the Canadian experience in the last two years of the Atlantic war. Instead, it seeks to answer a simple question: How did the small-ship navy fare in the battle against the U-boats from 1943 to 1945?

Acknowledgments

Although this book is a logical sequel to *North Atlantic Run*, it would not have come about without the encouragement and help of a great many people. The most important push came in the fall of 1985, when it was clear that I was leaving the Directorate of History, NDHQ, to assume a teaching post at the University of New Brunswick. Dr Norm Hillmer, the senior historian, realized that my mind was miles away and asked simply what *I* would like to do for the next six months. The result was the DHist narrative that formed the basis of this book.

Since 1986 this project has been encouraged and nurtured by friends and colleagues, particularly those at DHist. Dr Roger Sarty and Mike Whitby have given unfailingly of their time, and we have shared the fruits of our own research over the years. I can only hope that they got as much from me as I have from them; for without them this project would have taken many more years to complete. Special thanks also go to Mike Whitby for help with the photos.

Support and encouragement have come from the veterans of the Atlantic war. The following endured at least one interview: Nelson Adams, Louis C. Audette, Ian H. Bell, R/Adm P.W. Burnett, RN, Cdr E.M. Chadwick, RCN, Cdr Peter G. Chance, RCN, Jim S. Davis, Lloyd Dawson, L.P. Denny, William Evans, Adm John Grant, RN, R/Adm Dan Hannington, RCN, Capt. G.H. Hayes, C.K. Hurst, Gordon A. Kennedy, A.F.C. Layard, R/Adm R. Murdoch, RCN, Capt. Pat Nixon, RCN, Cmdr Howard Quinn, RCN, A.B. Sanderson,

xviii Acknowledgments

R/Adm R. Timbrell, RCN, R/Adm R. Welland, RCN, and Capt. W. Willson, RCN. Many others kindly replied in writing to my requests for help: R. Erskine, Don Brooke, Cully Lancaster, Jack Bathurst, Capt. Denis Jermain, RN, Alastair Carrick Smith, W.T. Borror, Ernie Doctor, Cdr J.S. Filleul, RN, J.E. O'Neil, R.J. Kennedy, Cdr R. Phillimore, RN, George P. Pickard, Lt Cdr George Moore, RCN, S.G. Hawkes, George Devonshire, George W. Bate, W.P. Chipman, Lloyd Fryer, R. Shales, Lt Cdr C.J. Scott, RCN, J.H. Dunn, Lt Cdr Ted Burke, RCN, David Howett, J.B. Lamb, Gault Findley, Sir Edward Fennessy, and Claude Powell. I owe a special debt of thanks to the late Capt. J.A.M. Lynch, RCN, of Ottawa, who put me in touch with so many Old Salts. If I have missed someone from the list of correspondents, please accept my apologies.

My debt to historians and archivists is, of course, substantial. J. Rohwer and E. Rössler, both of Germany, answered my queries on late-war U-boats. Dr Rohwer and Thomas Weis, both of the Bibliothek für Zeitgeschichte, Stuttgart, provided some U-boat photos. Dr Mike Palmer (now at the University of North Carolina) and Dr Gary Weir of the Naval Historical Center, Washington Navy Yard, provided help from U.S. sources. In the United Kingdom I would like to thank David Brown, head of the Naval Historical Branch, MOD, and his associate Arnold Hague for their particular help and encouragement, and Andrew Lambert. My thanks as well to the staff of the British Museum; Rod Suddaby and his staff at the Imperial War Museum; the staff of the Public Record Office, Kew; and the archivists of Churchill College, Cambridge. The staff at the National Archives of Canada, especially Glenn Wright and Carl Vincent, were a pleasure to work with. So, too, was the staff at the Directorate of History, NDHQ. In addition to support from Sarty and Whitby, I am particularly grateful to Don Graves and Shawn Cafferky for allowing me to read their narratives on the development of naval aviation in the RCN, to Helen Desjardins for pitching in when Roger Sarty was away and I needed something quickly, and to Anne Martin for her superb work in copying documents. John Burgess very kindly dug through his mountain of information on convoys to answer my questions on escort group assignments, and Ken Macpherson gave readily – as ever – from his photo collection. I would also like to thank Lt Cdr Doug McLean – a fellow traveller in the field of late-war ASW – for permission to use his MA thesis, and for easing a novice into the intricacies of sound propagation in water. That task was taken up by Lt Cdr Mark Tunnicliffe, the command oceanographer of the Canadian navy, whom I thank heartily for setting aside a day from his busy schedule to demonstrate how Second-World-War-vintage asdics functioned in different water masses. Oscar Sandoz, one of the scientists in Halifax who worked on the solution to the acoustic homing torpedo, provided much-needed insights into its operation and the development of countermeasures. Information on and photos of the Canadian anti-acoustic torpedo gear were very kindly provided by Walter Ellis, the historian of Defence Research Establishment Atlantic. Thanks also go to Bill Constable for the superb maps and drawings, and to the Social Science and Humanities Research Council of Canada and the Department of National Defence through the UNB Military and Strategic Studies Program for funding the research.

I owe special thanks to those who read the draft. Louis Audette, Michael Hadley, Mike Hennessey, and Roger Sarty endured the early versions and forced me to cut, shape, and focus. Howard Quinn, Bill Willson, Pat Nixon, and Bob Timbrell read portions of the text, as did my colleague Steve Turner, and I am especially grateful for their comments. Thanks to Gerry Hallowell of the University of Toronto Press and Paul Wilderson of the U.S. Naval Institute, both of whom had faith in the project from the outset, to Catherine Frost for her thorough and very effective copy editing, and to Rob Ferguson. Despite the efforts of fellow historians, veterans, and editors, I am responsible for the errors, omissions, and judgments that remain.

No list of acknowledgments is complete without a special word of thanks to those who supported and suffered on a daily basis while I struggled to bring this book to completion. Carole Hines and Elizabeth Hetherington, the secretaries in my office, were unfailing in their help, and I owe them an enormous debt. Families typically bear the brunt of a historian's passion for writing and mine has been no exception. No amount of thanks can replace the lost hours and days, the grumpiness, and the frequent total distraction. Thanks to Matthew and William for their understanding and support. If there is to be a reward for forbearance, however, it must go once again to my sidekick Barbara Jean (known to most as Bobbi), and it is to her that I dedicate this book. Thanks Bob.

The U-Boat Hunters

The Royal Canadian Navy and the
Offensive against Germany's Submarines

Prologue

> Ralston said if the war ended now we would have to hang our heads in shame ... What Ralston meant was that our effort in fighting may have been less than that of other Dominions ...
>
> Angus L. Macdonald, *Diary*, 23 December 1942

On 1 February 1943 William Lyon Mackenzie King rose in the House of Commons to begin debate on the throne speech that had opened the latest sitting of Canada's parliament. It was to be a stormy session. Canada was war weary and impatient with King's policy of maximum effort and minimum fighting. After nearly four years of war the army, smarting from the disasters at Hong Kong and Dieppe, was still sitting in garrisons throughout Canada and Britain. Canada's sustained military effort was limited almost exclusively to bombing Germany and escorting convoys in the north Atlantic. These vital activities had so far produced no heavy casualties – just what King had aimed for.

But Canadians were unimpressed with King's war. The big air force and navy his government had built had not captured the public's imagination and had not produced the political dividends King desperately needed by early 1943. It was true that Canada was busy with war work, but for many such activity smacked of profiteering. Driven by the myths and memories of the Canadian Corps of a generation earlier, Canadians identified with only one kind of military activity: hard fighting against the enemy's armed forces.

By the end of 1942 that kind of war seemed to include everyone but Canada. The Russians had stopped Germany and its Allies at Stalingrad, while in North Africa the British, Australians, New Zealanders, South Africans, Indians, Free French, and Americans had the remnants of the Axis on the run. The tide had turned in the Pacific, too, most notably at Guadalcanal. In late 1942 many Canadians – among them the minister of national defence – feared that the shoddy edifices of fascism might simply collapse before they had a chance to fight.

When King rose to address parliament on that cold February day, he had only one Canadian victory in his pocket – and a very small one at that. The corvette *Ville de Québec* had recently sunk a U-boat in the Mediterranean. The prime minister noted with particular satisfaction that this victory had been won by a vessel named after the 'ancient capital of Canada.' With this ammunition he ridiculed Quebec's isolationist Bloc Populaire members of parliament for their parochialism. The Bloc members simply jeered at King's attempt to derive credit for his government from the sinking of one submarine thousands of miles away. They were more concerned with what King was going to do about the U-boats in the Gulf of St Lawrence. During the previous summer the dull thud of exploding torpedoes had echoed around the gulf. Twenty ships had been sunk and the wreckage of war littered the shoreline. King's navy had been powerless to stop the attacks, and Canada's most important commercial waterway was closed to oceanic shipping in September 1942.[1] Canadians knew that much, at least. King did not tell parliament, or the nation, that Canadian escort groups operating in the mid-Atlantic had suffered heavy losses to their convoys in 1942 as well. In fact, losses to Canadian-escorted shipping in the mid-ocean were so heavy that the Canadian war cabinet was petitioned by the British government in late December to remove its escorts for retraining. It seems fair to say that as King spoke to parliament on 1 February 1943 the navy's, and perhaps the nation's, fortunes were at their ebb.

Met with laughter, King admitted to the House that *Ville de Québec*'s triumph had indeed been small. Undaunted, he responded with a seemingly gratuitous remark about the Royal

Canadian Navy's victories at sea. 'While these [U-boat sinking] announcements are not made from day to day,' King observed, 'it is a fact that something like 8 or 9 or 10 of our corvettes and other ships have succeeded in sinking U-boats during the course of this war.'[2] King had exaggerated the navy's score. So far only five Axis submarines were known to have fallen victim to the RCN. Clearly, even a big navy engaged in important tasks had not enough political clout by 1943. King's government needed victories; it needed U-boat kills it could announce in the House.

If King was forced to deceive the House with exaggerated claims of RCN success, it was not because the navy was not trying to sink submarines. The RCN was large and growing rapidly: nearly 50,000 strong with just under 300 warships. Sailing largely Canadian-built ships in Canadian formations, the RCN was devoted almost exclusively to the crucial task of protecting merchant shipping in the north Atlantic. More importantly, by early 1943 fully half of all the escorts in the decisive theatre of the Atlantic war were Canadian. King and his cabinet had been assured by the chief of the Naval Staff that the transfer of the best of the RCN from the scene of action in January was not only temporary, but was intended to improve 'the killing efficiency' of the fleet itself. As the prime minister ended his speech amid howls of protest from the Opposition and the thumping of desks on his own side of the House, it was reasonable to assume that there would be more – perhaps many more – Canadian naval victories in 1943.

The RCN was committed to sinking U-boats and always had been. Prior to 1939 Canadians shared the common desire of all navies to control the submarine menace by sinking subs. This had certainly been the case in the First World War,[3] and enthusiasm for destroying submarines had never waned after 1918. The biggest impediment to successful anti-submarine warfare (ASW) between 1914 and 1918 was the inability of searching vessels to pinpoint a sub's position once it had submerged. Accurate underwater location was crucial to the placement of depth charges within lethal range, twenty-five to thirty feet or less. The perfection of an active underwater sound location device, known by the British as 'asdic' (now know by its American acronym sonar) in the early 1920s appeared to give the edge to anti-submarine forces. Consequently, Canadian naval plans in the 1930s called for an aggressive and largely offensive anti-submarine doctrine, which also would be part of any defence of merchant shipping strategy. In a 1937 memorandum on defence of shipping, Rear Admiral Percy Walker Nelles, the chief of the Naval Staff, reflected the widely held belief that the combination of escorted convoys, airpower, and asdic made the submarine virtually obsolete. Nelles concluded that convoys escorted by asdic-equipped warships would result in 'very heavy' losses to attacking submarines and might force them to abandon attack on trade entirely. The RCN planned to provide equipment 'for prosecution of *offensive* [emphasis added] means against submarine attack.'[4]

The RCN's offensive ASW doctrine mirrored that of the RN, and both navies clung to it in the early years of the Second World War. Until the spring of 1941 the British sent ships hither and yon to hunt U-boats in response to a variety of stimuli: sightings, distress calls from ships under attack, and position fixes from U-boat radio transmissions. In the event, this approach simply scattered forces to little effect. The U-boats proved remarkably hard to find, and they displayed an understandable reluctance to stay where they could be killed. The RCN had little opportunity in the early years of the war to take the offensive against real submarines, but the navy remained wedded to the concept of offensive anti-submarine warfare. This view point is clearly reflected in its initial plans for the first major war-built auxiliary class of ships, the Flower class corvettes. By late 1940 the navy intended to organize its corvettes into anti-submarine 'hunting groups' to contain and destroy U-boats in the approaches to Canada's main ports.[5]

Driven by the operational failure of their offensive doctrine, the British officially abandoned it in favour of concentrating on the 'safe and timely arrival of the convoy' in the spring of 1941. From then on the anti-submarine escort's duty was to stick closely to the convoy, shun all but the most promising situations for destroying U-boats, and see the shipping through. The overseer of this new orthodoxy was Western Approaches Command, established at Liverpool, England, in April. 'Safe and timely arrival of the convoy' became the standard British measure of escort efficiency for the balance of the war. The new doctrine was transmitted to the escort fleet in the *Western Approaches Convoy Instructions* (*WACI*).[6]

The RCN adopted *WACI* as a matter of course when the Newfoundland Escort Force (NEF) was established in May 1941. NEF represented a major change in deployment for the RCN's burgeoning corvette fleet. The notion of hunting groups attached to ports was now traded for the idea of anti-submarine escort of convoys between the Grand Banks and Iceland. It proved to be demanding, dreary, and dangerous work. By late 1941 NEF had absorbed virtually all of the RCN's ocean-going escort fleet, as corvettes streamed from Canadian shipyards directly into battle against the best of Germany's submariners. NEF was the RCN's principal task in 1941, its baptism of fire and the formative influence on the navy's escort fleet.[7]

In the summer of 1941 NEF operated under Western Approaches Command. Noticeably absent from NEF both then and much later, however, was any mechanism for actually imposing WAC doctrine, especially its subtle emphasis on defence over offence. NEF's main base at St John's had neither a training establishment nor a tactical analysis unit. In fact, it had little more than a few overworked officers and a lot of new ships. The Canadians' eastern terminus in 1941 was the RN's advance base in Iceland, where contact with WAC was tenuous at best. The poorly trained reservist officers of Canada's escort fleet tended to follow their instincts. During 1941 senior WAC officers were often appalled by the Canadian propensity to neglect safe and timely arrival if an opportunity for offensive action arose. It appears that Canadian sailors thought they had gone to sea to fight the Germans.

Had the NEF remained under WAC operational control, the RN might well have checked the Canadian proclivity for offensive action in 1941. Before the scions of Western Approaches tactical orthodoxy could make any imprint on the NEF, however, it was transferred to American operational control. On 17 September 1941 the United States Navy (USN), still officially neutral, took responsibility for convoy and escort operations west of Iceland. *WACI* still formed the basis of NEF's actions, but the Canadians now operated under a navy whose *Escort-of-Convoy Instructions* placed safe and timely arrival of the convoy dead last on its list of priorities. The primary duty of an American convoy escort was to sink any U-boat it came into contact with.[8] The Americans made no attempt to impose their doctrine on the RCN,[9] but after September 1941 Western Approaches Command was not in a position to do so either. The RCN was left to follow its own predilections under the direction of a navy that shared its enthusiasm for offensive action.

The RN was delighted to have the Americans involved in the shooting war but anxious to get the Canadians back under their wing as soon as possible. Shortly after the official U.S. entry into the war in December 1941 the British tried to resume operational control of NEF, but the Americans refused. The British regained a measure of control over Canadian doctrine in February 1942, when Iceland was abandoned as a relay point and the new Allied 'Mid-Ocean Escort Force' (MOEF) took charge of convoys between Newfoundland and Northern Ireland. This arrangement put at least the eastern operations of Newfoundland-based RCN escorts back under WAC. Since most Canadian operations remained in the American zone, however, the development of the RCN's escort fleet remained largely in Canadian hands throughout 1942.

If there was a driving force behind Canadian anti-submarine and escort doctrine in 1941 and 1942 it was Commander James Douglas 'Chummy' Prentice, RCN(R). In 1941 Prentice was appointed 'Senior Officer, Canadian Corvettes,' a title apparently connected to the hunting-group concept originally envisaged for the corvettes. Prentice arrived in St John's with the first corvettes assigned to NEF and worked closely with NEF's first commander, Commodore (later Rear Admiral) L.W. Murray, RCN, to improve the efficiency of the fleet. In fact, Murray and Prentice were the two central figures in the development of the escort fleet and ASW doctrine from 1941 to 1944. When Murray went to Halifax in September 1942 to become Commanding Officer, Atlantic Coast, Prentice followed as Captain (D), Halifax, in December. That appointment formalized Prentice's responsibility for escort fleet efficiency. The significance of Captain (D), Halifax, increased dramatically in 1943 when Murray became C-in-C, Canadian Northwest Atlantic, and the previously independent Newfoundland command was subordinated to him. Prentice then became the equivalent in substance, if not in form, of Commodore (D), Western Approaches, the doyen of RN escort doctrine.

Other RCN officers, such as Jimmy Hibbard and Bob Welland, made substantial contributions to training the escort fleet, but none left such an indelible mark on doctrinal development as Prentice. He epitomized the view that the ability to sink submarines was the real measure of an escort's efficiency. Prentice believed that the RCN's principle escort vessel, the corvette, was ideally suited to aggressive use and offensive action. As he confided in 1947 to Gilbert Tucker, the RCN's official historian, corvettes were 'the handiest *anti-submarine* [emphasis added] ship that was ever built,' and he developed tactics for the corvette based on its agility.[10] The British ultimately adopted some of his tactical methods and adaptations to the asdic equipment – those developed to ensure a more aggressive counter-attack on submerged U-boats by corvettes.

But Prentice's interest extended beyond simply improving the RCN's skill at depth-charge attacks. From the moment he arrived in St John's, Prentice's enthusiasm for putting escorts in contact with U-boats never wavered. Within days of joining NEF Prentice sailed in his own corvette *Chambly* along with two others to support a threatened convoy. By August 1941 he had built a training command around *Chambly*, the first of two that he established while based in Newfoundland. Murray encouraged Prentice's training schemes not least because they filled an enormous void. No one in the RCN – apart from Prentice – was yet dealing with operational training.[11] His training emphasized offensive ASW and 'support' operations, and he believed that the best place to train A/S ships was at sea against the enemy. Moreover, the pattern of his exercises indicates that sinking U-boats – not defending convoys – was the primary thrust of instruction. On his group's first training cruise in September Prentice took *Chambly* and *Moose Jaw* to the mid-Atlantic in support of the embattled convoy SC 42. The reward was *U501*: the first known RCN U-boat kill of the war.

The idea of providing 'offensive' support to a convoy under

threat was not new. Nor was it limited within the RCN to Prentice. Cdr H.S. Rayner, RCN, captain of the destroyer *St Laurent*, developed a scheme in late 1941 which involved withdrawing all Canadian destroyers from NEF groups and combining them into hunting groups. Rayner wanted to keep two such RCN support groups permanently on station in the mid-Atlantic to hunt U-boats and reinforce threatened convoys. The RN had already tried this tactic in the eastern Atlantic in September 1941, with some success.[12] The British opposed reducing NEF's groups to the point where they consisted of corvettes alone, and so nothing came of Rayner's scheme.[13] The only Canadian 'support' group operating with NEF in 1941 was therefore Prentice's, but even it was short lived. Canadian-escorted convoys suffered heavy losses in the fall, and there was concern about the ability of NEF to get the convoys through. The solution adopted was to increase the size of escort groups, which drew Prentice, *Chambly*, and ships under training into close escort duty.

Prentice's 'training-cum-support group'[14] was re-established in April 1942, when spring weather and new escorts eased the strain on ships and men. This second group enjoyed a more successful and influential period of operation. From May to August approximately twenty-five ships underwent training, and Prentice supported roughly seventeen convoys in the western Atlantic. He typically refused to allow his group to be drawn into the close-escort screen of the convoys, or even to place his training group under the operational command of the senior officer of the escort. Instead, Prentice patrolled around threatened convoys, trying in vain to repeat his success with *U501*.[15] By July his efforts had prompted the ire of the American admiral in command of convoy operations in the northwest Atlantic, who complained of Prentice's 'freelance operations.' The American's pleas appear to have been politely ignored, but in August Prentice's group was again disbanded to fill gaps in close escort groups. The importance of Prentice's support for convoys in the area of the Grand Banks was demonstrated shortly thereafter by the Anglo-American decision to establish a permanent force of short-range destroyers to do exactly the same thing. Canadians did not lack ideas; they lacked only the ships and clout to make them happen.

Whether Prentice actually shaped RCN doctrine or merely reflected it is hard to judge. What is important is that throughout 1941–2 the RCN's anti-submarine escort doctrine remained aggressive. This attitude put them at odds with the RN's Western Approaches Command, where the notion of Canadians charging madly around the ocean looking for U-boats was widely held. Indeed, it was pervasive enough for W.S. Chalmers to use it to characterize the RCN in his biography of Admiral Max Horton, the second C-in-C, Western Approaches.[16] In fairness, Chalmers tarred both the RCN and the USN with the same brush. Senior WAC officers were offended by the North American tendency to judge the effectiveness of escort operations by how aggressive they were in the face of the enemy. Geography exerted an influence here. For both the Canadians and the Americans the Atlantic was the front line: 'battling the enemy' described the job. The British, on the other hand, saw the Atlantic as a rear area: a line of communications behind the front. Their primary object was getting the trade through, and it showed in the way they approached battle assessments. Operations of Canadian escorts in the mid-Atlantic during 1942, for example, demonstrated the North American penchant for aggressive action and the British reaction to it. The British were rather annoyed at the comments of Captain E.R. Mainguy, RCN, Captain (D), Newfoundland, on action around convoy ONS 100 in May 1942. Mainguy observed that a confirmed U-boat sighting was 'well worth leaving the convoy unguarded' to pursue. WAC officers considered this 'a rash generalization.' They carped again when *Skeena* and *Wetaskiwin* abandoned convoy ON 115 to pursue – and sink – *U558* in July. Captain R.W. Ravenhill, RN, the deputy chief of staff (Operations) at Western Approaches, complained that the Canadians had left the convoy 'singularly unprotected by day.' Yet even Ravenhill had to concede that, 'they got their s/m [submarine] and must receive their due for that.'[17] Other remarks were less kind.

To a considerable extent British assessments in 1942 were coloured by the legacy of 1941. They tended to view the Canadian escort fleet as an assembly of wildly misdirected amateurs. Such may have been a reasonable assessment in 1941 when the fleet was expanding rapidly and was new to operations. But by 1942 the RCN was much better than the British were prepared to admit. Where the Canadians saw themselves as a glass of water half filled, the British always saw them as a glass half empty. Western Approaches Command would have been happier if the Canadians had simply circled the wagons and waited for the cavalry when the going got tough. This fundamentally different perspective held tremendously important consequences for the RCN's role in the great offensive unleashed against the U-boat in 1943.

While the RCN failed to meet the British standard for convoy defence in 1942, Canadian escorts in the mid-Atlantic got their fair share of U-boat kills. RCN escorts made up only 35 per cent of the Mid-Ocean Escort Force, but in the last six months of 1942 they claimed five of the ten U-boats sunk by MOEF. Significantly, only three of these kills were known at the time, and all occurred in the summer months. *St Croix* was escorting convoy ON 113 in late July when she sank *U90*. A few days later *Skeena* and *Wetaskiwin* sank *U558*, and in August *Assiniboine* sank *U210*. The fourth kill occurred around convoy SC 97 in September, when the corvette *Morden* sank *U756*. The incident was without clear result at the time, and the kill was not awarded to *Morden* until forty-five years later. The fifth kill, too, was unknown at the time: that of *U356* during the battle for ONS 154 in December. Her loss resulted from one of many depth-charge attacks made by group C 1 and was not credited to the RCN until immediately after the war.[18] It is interesting to speculate on what might have been the fate of the RCN during 1943 if its success in sinking submarines in the mid-Atlantic had been known at the time.

Half of the U-boat kills by MOEF in late 1942 with only 35 per cent of the escorts was a good score, but in fact the Canadian score ought to have been higher. The majority of Wolf Pack operations in late 1942 were directed against slow convoys, most of which were escorted by the RCN. That fact alone accounts for a very much higher loss rate suffered by the Canadians in the mid-ocean groups in late 1942. But it also meant that RCN ships were more frequently in contact with the enemy and had proportionately more opportunity to sink submarines. The reason they did not do so was the same reason German submariners found it so easy to sink ships from RCN convoys: outdated escort ships and equipment.

The vast majority of the RCN's escorts were corvettes, simple designs built to mercantile standards and fitted with rudimentary equipment. Their construction suited King's wish to exploit the opportunities of war in order to develop Canada's industrial base, and throughout 1941 and 1942 basic warships poured from Canadian yards. Their sheer number made the convoy system – the bedrock of the Allied victory – possible. But Canada could not keep pace with the change in the high-tech war. By 1942 the corvette fleet was already obsolete, not only in weapons, but in electronics and sensors. The problem offered no quick fix. Some piecemeal improvements, such as the fitting of modern radar and heavier secondary armaments, were possible. These efforts were pursued with increasing vigour as the year wore on. But modernization required not only a radical improvement in the ships' asdic, navigation, and anti-submarine weapons, but also major structural changes.

The details of this problem are outlined in appendix I, but a few salient points affecting the offensive capabilities of the fleet should be mentioned here. Because of a shortage of gyro compasses, all Canadian corvettes completed before the end of 1942 were equipped with a single magnetic compass and the correspondingly primitive asdic type 123A. Even by 1939 standards the magnetic compass (initially graduated in 'points' instead of 360 degrees) and the type 123A asdic were obsolete. Primitive asdics and compasses adversely affected the fleet's navigation (another source of British criticism), and made coordination of A/S searches with British ships difficult. Primitive asdics and compasses also made asdic searches and depth-charge attacks on submerged targets much less accurate. As Captain (D), Newfoundland, observed in mid-1943 (when the problem was still prevalent), 'The problems of a Corvette Captain when attempting an accurate [depth charge] attack with a swinging magnetic compass are well nigh insoluble.'[19] Destroying U-boats with such basic equipment required very considerable skill – and some educated guesswork.

Up to the end of 1942 the desperate state of asdic and navigational equipment aboard RCN corvettes undoubtedly cost the navy a number of U-boat kills. The same can be said of other shortfalls. Comparable British escorts by 1942 were fitted with the latest shortwave (10-cm) type 271 radar with automatic 360-degree sweeping and a range, on small targets under good conditions, of 5,000 yards or more. This radar could detect a U-boat approaching a convoy on the surface at night, while the 360-degree sweep allowed British groups to maintain a radar barrier around the convoy itself. In several convoy battles of late 1942 British MOEF groups pushed waves of attacking U-boats aside using type 271 radar. In contrast, the most common radar on RCN escorts until 1943 was the Canadian-built SW1C and SW2C, a set based on the earlier British type 286. The Canadians improved the set by developing an antenna that could be pointed by turning a handwheel, but the 1.5-metre wavelength could still locate only rather large targets. It was possible to pick up a U-boat on an SW1C/2C if conditions were good, the submarine was close enough, and the antenna was pointed at it. Not surprisingly, lack of modern radar in RCN ships allowed the Germans to attack at night with impunity, resulting in a higher loss of merchant ships. The lack of 10-cm radar also prevented RCN escorts from sinking more submarines. *Sackville*, for example, detected two U-boats by SW1C radar in calm seas and poor visibility around ON 115 in July 1942. But the radar could not provide enough information to get a fix and a proper classification. All the escort's operator knew was that something was out there. *Sackville* had to close the contacts slowly to confirm their nature, giving the U-boats precious time to react. Both U-boats escaped, as did a third later located on the surface by asdic. As a British staff officer admitted, '*Sackville*'s two U-boats would have been a gift if she had been fitted with RDF [radio direction finding] type 271.'[20] To compensate for the lack of modern radar Canadians developed a night illumination tactic called 'Operation Major Hoople' (named after a popular comic-strip character). Based on an escort commander's estimate of just when the attacking U-boats were close enough to strike, Major Hoople called for the escort to fire illumination around the convoy in hopes of finding the U-boats visually. As innovative as such tactics were, they remained a gamble and were no substitute for modern radar.

Other advances, such as a proper number of destroyers in Canadian MOEF groups and shipborne high-frequency direction finding (HF/DF), would have improved the score in the fall of 1942 as well. Shipborne HF/DF sets could determine the number and location of U-boats making siting reports on the convoy. Once a fix was established, the senior officer of the escort usually dispatched a destroyer to drive the shadower down and force him to break contact. High-speed destroyers were essential for such work and British groups typically had two. That allowed them to conduct HF/DF directed sweeps without fear of completely exhausting the fuel in either ship. The British enjoyed success with destroyer sweeps in MOEF during late 1942, both defending convoys and sinking submarines. Canadian groups of MOEF, on the other hand, often made do with a single destroyer which could seldom be spared for offensive sweeps. More critical still, RCN groups typically had no HF/DF set. Thus C groups found it hard either to break up the pack as it formed or to counter-attack the shadowers. Apart from establishing advanced screens during daylight hours (the ones that accounted for *U90* and *U558* in July 1942), they substituted sweeps based on medium-frequency direction finding (MF/DF), using their own navigational set. This technique provided rough fixes on a shadowing U-boat's homing signals, but MF/DF-directed sweeps were a poor substitute, because MF transmissions gave imprecise bearings.[21] The implications of technological backwardness in the mid-Atlantic were therefore enormous. Moreover, it cut both ways, inhibiting the RCN's ability to defend shipping and also to sink U-boats.

Equipment shortfalls were identified in 1942, but since the key pieces could be only grudgingly obtained from the British or Americans, there was no quick fix. The Naval Staff moved on radar, gyros, HF/DF, and the like on its own, but it took the need for more destroyers directly to the government following the sinking of *Ottawa* in September. Since Canada was no more capable of pulling destroyers from a hat in 1942 than it had been in 1917,[22] the Naval Staff asked that at least some of the River class frigates being built in Canada for the RN be taken over for the RCN. Like corvettes, frigates were a war-emergency class built in haste to mercantile not naval standards. They were, however, designed for deep-water anti-submarine warfare and escort, being 100 feet longer than corvettes, with twin screws and twice the range of the early corvettes. Moreover, their bridges, electronics, and weapons sytems were modern. Indeed, these ships were so unlike their corvette antecedents that Adm Nelles recommended, and the British approved, changing their designation from 'twin-screw corvette' to 'frigate,' a term abandoned by navies at the end of the Age of Sail.

By 1942 (as outlined in appendix I) the RCN had its own frigates on order, and in time these ships would form the basis of the navy's A/S effort. For the moment, however, the ten British frigates on order in Canada had priority and were well advanced. When the RCN asked that these ships be taken over for its own use, the government refused. It depended on the British contracts both for American cash (they were financed under lend-lease) and for British expertise in shipbuilding, and it would not requisition them for its own struggling navy. The RCN saw the first two Canadian-built frigates, retained by the Americans as the USS *Ashville* and *Natchez*, in late 1942 and early 1943 as they stopped in Halifax on their way to Boston. The solution to the problem adopted by the Naval Staff was to switch its priority in destroyer acquisition in October 1942 from fleet types to escorts.[23] The only quick source of escort destroyers was the RN, which meant a direct appeal to the British for the transfer of the needed ships: a candid admission to the British by the Canadian government that the RCN needed help.[24]

Over the fall of 1942 Canadian fortunes in the mid-Atlantic went from bad to worse, reaching a new low in November with the passage of convoy SC 107 – fifteen ships sunk and no attackers destroyed. It was these faltering Canadian transatlantic operations and the legacy of 1941 that weighed heavily on the minds of Western Approaches officers as 1942 drew to a close. However, the situation in Canadian coastal waters during 1942, operations that so profoundly influenced Canada's domestic political scene, looked equally grim. The RCN was unable to deal effectively with U-boats operating in the Gulf of St Lawrence because of increasing operational commitments offshore. In any event, the British preferred not to route oceanic shipping to Montreal, 1,000 miles 'inland,' but rather to work out of the east-coast railheads of the Maritimes. For a number of related reasons, then, the RCN decided to close the gulf to oceanic shipping in September. This appeared to be a clear tactical defeat. Even the navy's own gloomy analysis produced a dismal picture of the inshore campaign of 1942. By their reckoning, the ratio of ships lost to U-boats destroyed in the Canadian coastal area in 1942 was 112 to 1. The global average was 10.3 lost ships to every U-boat sunk. By this standard alone Canadian naval operations in home waters in 1942 were a failure[25] – and the lone kill in the Canadian zone was by the Royal Canadian Air Force. As the navy's operational researchers concluded, 'It will be seen that the quota for last year should have been *at least 10* [emphasis in original] U-Boats [sunk in Canadian waters] ... to compare favourably with losses sustained.'[26] The trade-off between U-boat kills and shipping losses was an erroneous measure of efficacy; the fact that it was raised in early 1943 is an indication of how anxious the RCN was to kill submarines. The real Canadian triumph in 1942 was the establishment of the convoy system, which kept losses much lower than would otherwise have been the case. The majority of the 112 ships sunk in the Canadian zone were independently routed or from mid-ocean convoy battles that spilled over into the fog of the Grand Banks. Coastal convoys in the Atlantic off Nova Scotia and Newfoundland (as distinct from transatlantic convoys in the same waters) were entirely successful: only one of their ships is known to have been sunk in 1942. The only Canadian coastal convoys heavily attacked were those in the gulf itself, and they travelled a predictable and easily intercepted route. Even so, the gulf convoys had a loss rate in 1942 of only 1.2 per cent:[27] sustainable operationally, if not politically.

More escorts with better equipment, especially aircraft with 10-cm radar, would have made the gulf a much less profitable U-boat hunting ground in 1942, but these planes were simply not available. Moreover, what plagued the RCN, particularly inshore, during 1942 was the ocean environment itself. Asdic conditions in the Gulf of St Lawrence and in the mouth of the river were notoriously poor. It was impossible, for example, to obtain asdic contact on U-boats immediately following attacks, even though the general location of the submerged submarine was known. This problem would haunt the RCN to the end of the war, but the impetus to solve it derived from the experiences of 1942. By the end of the year the National Research Council (NRC), which acted as the navy's research establishment, was asked to investigate the problem of asdic sound propagation in east-coast waters. The NRC sent a scientist to Woods Hole Oceanographic Institute in Massachusetts to learn about the new science of bathythermography (the temperature profile of a water mass) and how it affected sound.[28] In December, the director of Anti-Submarine at Naval Service Headquarters, Commander A.R. Pressy, asked that an oceanographic survey vessel be acquired for Halifax: the beginning of a long, and ultimately fruitless, search.[29]

The RCN's decision to concentrate its efforts in 1942 on establishing the basic convoy systems needed was both strategically and operationally sound. Tactically, however, the RCN had been badly mauled by the U-boats in Canadian waters. The navy could neither fully prevent losses inshore nor stop the U-boats by sinking them. Canadian naval officers expected the Germans to draw the same conclusions from the U-boat kill / shipping-loss rates that they had done and to launch a major campaign in Canadian waters in 1943.[30] And although Canadian escorts in the mid-ocean sank submarines, at least in the summer, their convoys suffered heavily. In the last half of 1942 fully 80 per cent of shipping losses in the mid-ocean came from RCN-escorted convoys. Judged by the standards of safe and timely arrival of individual convoys, the RCN appeared to be the weak link in the Allied chain at the end of 1942.

When the Canadian government's request for more destroyers arrived in London in December 1942, the British were already considering asking the Canadians to remove their escorts from the embattled mid-Atlantic. On 17 December, Mackenzie King's birthday, Churchill cabled a personal appeal to the Canadian government to remove its groups from MOEF and transfer them to the RN for retraining. 'I appreciate the grand contribution of the Royal Canadian Navy to the battle of the Atlantic,' Churchill explained to King, 'but the expansion of the RCN has created a training problem which must take some time to solve.'[31]

Between Christmas and the new year the issue of Canadian failure in the mid-ocean was debated at a special conference in Washington, within Naval Service Headquarters, by the Naval Staff with a special British representative, and eventually by the War Committee of cabinet.[32] With a dismal year behind it, this was not the kind of news the cabinet needed. Colonel J.L. Ralston, the minister of defence, confided to his cabinet colleagues at the 23 December meeting that if the war ended now, 'we would have to hang our heads in shame.' King and other members of cabinet were appalled. Angus L. Macdonald, the naval minister, was forced to defend Ralston. 'What Ralston meant,' Macdonald wrote later in his diary, 'was that our effort in fighting may have been less than that of the other dominions.'[33] It seemed that King's 'maximum effort, but minimum casualties' war effort was beginning to unravel. The tide was turning against the Axis and Canada needed to get in the fight.

Whether the RCN or the government ever seriously considered challenging the British request that the RCN abandon – even temporarily – its most prestigious operational commitment remains a mystery. RCN staff officers correctly argued that blame for the heavy losses to Canadian-escorted convoys lay primarily with poor equipment rather than with poor training. In the event, they were undone by developments at sea. The backdrop to the debate over Churchill's request was the tragic passage of the Canadian-escorted convoy, ONS 154. For five days at the end of 1942 it had struggled through the widest portion of the air gap, losing fourteen ships in the process without any known retribution exacted from the enemy. By early January Nelles was forced to recommend that cabinet accept the British request for the transfer. He coated the pill by noting that this action was part of a new plan to improve the 'killing efficiency' of all mid-ocean groups. Cabinet agreed to the transfer on 6 January 1943. Canadian acquiescence was forwarded to Churchill and the First Sea Lord in the days that followed. The British were reminded that Canada saw its involvement in the mid-ocean as central to its naval war effort. As Nelles informed Admiral Sir Dudley Pound on 7 January, 'Public interest in the Canadian Navy is centered on the part it has taken in this task, which is without question one of the highest and enduring priority upon which the outcome of the war depends. We are satisfied that the Canadian Navy can serve no higher purpose than to continue to share this task, which we have come to look upon as a natural responsibility for Canada.'[34] The collapse of the RCN's effort in the mid-ocean could be – and indeed was – hidden from the Canadian public.[35]

King did not inform the House that the cream of the RCN's escort forces was being temporarily withdrawn from the decisive theatre of the Atlantic war for retraining. It seems likely that King and his cabinet colleagues accepted Admiral Nelles's assertion that the Canadians were simply the first to undergo a special program to improve their U-boat-killing efficiency. When King rose to address parliament on 1 February, it was therefore not unreasonable to expect that the RCN would be in the thick of the fighting in 1943, and that there would be many more U-boat kills to report.

Certainly that was the sentiment within the navy itself. Whatever the compelling reasons for the RCN's poor showing during 1942, the navy was not content to let it continue. By early 1943 plans were already underway to modernize the fleet, consolidate and reorganize the naval war effort at home, and prepare forces to take the offensive against the U-boat menace. In many ways, the end of 1942 marked the end of one story and the start of a new one. Clearly, some of that hope and optimism for a better year in 1943 was conveyed to King and his cabinet. Perhaps it was just as well that no one in Ottawa realized how difficult it would be to clear away the wreckage of four years of unbridled expansion. Nor could anyone have realized how strongly the British mistrusted Canadian abilities, how they would influence and manipulate events, or how capricious the fortunes of war would be.

1
The Fortunes of War

> In all the successes against submarines H.M.C. Ships had little part. None of them [was] employed for any length of time in the areas where U-Boats could have attacked ... and only a few Canadian ships have as yet been incorporated in the Killer groups which can seek out and destroy the foe.
>
> 'Summary of Naval War Effort,' third quarter, 1943

By early 1943 there were roughly 100 U-boats, organized into several Wolf Packs, prowling the main north Atlantic convoy routes beyond the limits of Allied air cover. More were on the way. The mid-Atlantic was literally filling up with submarines. Admiral Karl Dönitz, the driving force behind the U-boat fleet and as of January the head of the German navy, actually had little other choice. The extension of Allied convoy systems throughout the Atlantic in 1942 had robbed him of easy hunting grounds. By the winter of 1942-3 the mid-Atlantic was the only place left where the large fleet of short-range type VIIC U-boats could operate with any hope of decisive result. After nearly four years of war the U-boat fleet was committed to a showdown. It was, in the end, a battle that German submariners could not win. In the first five months of 1943 the Allies sank approximately 100 German submarines around the main convoys. Over the summer and into November a further 100 were destroyed in all theatres. A major portion of the Allied forces engaged in these operations were Canadian, yet the RCN played virtually no direct part in this great victory. For the Canadian navy 1943 proved a year of consolidation, frustration, and lost opportunities; it was the offensive that never was.

The realization that the Battle of the Atlantic had reached its moment of decision dawned slowly on the Allies in the early days of 1943. The British and Americans were committed to winning it in January, but for reasons of their own. In that month Roosevelt, Churchill, and their senior staffs met at Casablanca to plot the next moves in the war. This marked the first – and only – time when events in the Atlantic impinged upon the formulation of Allied grand strategy. All the plans for offensives in Europe, particularly those for the invasion of France, now hinged on the undecided contest in the mid-Atlantic. At Casablanca, then, the Atlantic got top priority: that meant greater resources and increased flexibility in their use. The U-boat was to be beaten, finally.

Therefore, just as the RCN's mid-ocean escort groups were forced into exile in early 1943, the larger Allies moved to solve long-standing problems in the mid-Atlantic. 'Support groups' for convoys passing through the air gap were formed after the disastrous passage of convoy SC 118 in February. It was the first British-escorted slow convoy to be heavily attacked in months. Despite its powerful escort and heavy reinforcement, SC 118 lost eleven ships in a bitter battle, while no U-boats were sunk by the escort. Ships for the new support groups were drawn from MOEF groups, but included were destroyers from the Home Fleet and auxiliary aircraft carriers. More very long-range (VLR) patrol aircraft, especially the B-24 Liberator, which could operate in the depths of the mid-Atlantic, were also on the way.

These resources held the key to defeat of the Wolf Packs, but they could not be employed effectively under the existing command structure in the north Atlantic. Control of anti-submarine operations and convoys was divided in mid-Atlantic between the Americans and the British roughly along 35° W longitude. The Mid-Ocean Escort Force sailed through this 'Change of Operational Control' (CHOP) line, but overall control of the action changed in the middle of the area of Wolf Pack attacks – and often at the height of a battle. During 1942 very different command philosophies existed on either side of the CHOP line. The British gave the senior officer of the escort (SOE) as much latitude as possible in the conduct of his battle. The American convoy controlling authority in Washington took positive control, often ordering the escort to take certain measures. This difference had caused problems. In one instance the Americans had ordered a Canadian escort to take action that the senior officer knew would endanger his convoy. The Canadian SOE compromised by taking his convoy in a complete circle in the middle of the ocean, which brought it back onto the waiting Wolf Pack. One ship was lost in the process. The British prevailed upon the Americans to adopt their arm's length approach by late 1942. However, Allied operational control of convoy battles still changed in the middle of the Atlantic. Everyone agreed that that practice contravened the principles of war and common sense. By early 1943 the British, in particular, wanted operational control of the whole mid-ocean battleground in order to use their resources effectively.

The haphazard command structure in the western Atlantic also cried out for change in early 1943. On the British side of the Atlantic, escort and anti-submarine operations were coordinated by a single authority, the RN's C-in-C, Western Approaches Command, at Liverpool. In November 1942 this command was taken over by an aggressive submariner, Admiral Max Horton, who held the post to the end of the war. Integration of naval and air operations in the eastern Atlantic was comparatively easy, since the Royal Air Force's Coastal Command was under the operational control of the Admiralty. At Horton's level cooperation between naval and air forces was facilitated by placing the headquarters for Coastal Command's 15 Group, which did most of the oceanic convoy escort duty, beside WAC's headquarters in Liverpool. Horton also worked closely with the key staff divisions of the Admiralty, such as naval intelligence and trade. By 1943 the British system was the most effective of its kind in the Atlantic.

The contrast between the tight command structure east of 35° and that west of the CHOP line in early 1943 could not have been greater. In the northwest Atlantic two nations overlapped authority in no less than nine separate and poorly coordinated commands.

The United States held strategic and operational control in the area in accordance with agreements reached in 1941. Convoy movements were controlled from Washington, and the USN disseminated naval intelligence. Both of these tasks had earlier been exercised by Naval Service Headquarters (NSHQ) in Ottawa. The Americans assumed responsibility for them in mid-1942, but Canadians maintained the capabilities and continued to develop them. The conduct of naval escort and U.S. naval air operations in the northwest Atlantic fell to two USN authorities: Commander, Eastern Sea Frontier, who controlled the U.S. Atlantic coast from Florida to the Gulf of Maine, and Commander, Task Force-24, at Argentia, Newfoundland, who ran operations between the Gulf of Maine and the CHOP line. U.S. Army air force aircraft also flew seaward patrols. However, because of American legislation governing responsibility for 'naval' and land-based aircraft, army maritime patrols remained strictly under U.S. army control. By 1943 no effective coordination of effort between USN and USAAF had been achieved.

The Anglo-American agreement reached at Argentia in August 1941 also gave the USN control over almost all RCN operations beyond Canada's twelve-mile territorial limit. This delegation was coordinated with the two separate RCN commands: Flag Officer, Newfoundland, at St John's, and Commanding Officer, Atlantic Coast, in Halifax. Both were responsible not only for Canadian naval operations in their respective areas, but also for the provision of escort groups for the oceanic convoys under USN control. The subordination of the RCN to American operational control made good sense in 1941, when large USN forces were present in the northwest Atlantic. By early 1943, however, only a token American naval presence remained, while the U.S. admiral at Argentia retained de facto control over what was now a very large Canadian fleet.

The Royal Canadian Air Force never subordinated itself to American control in the Atlantic war. Like the navy, the RCAF maintained two separate commands in the area; Eastern Air Command (EAC), in Canada, and No. 1 Group in Newfoundland. Unlike the RAF's Coastal Command, which concentrated solely on maritime air operations, EAC was a regional command tasked with a multitude of roles, including air defence, army cooperation, and spotting for coast artillery. Finding U-boats, protecting convoys, and helping the RCN comprised only part of its mandate. Air operations on the east coast therefore remained the prerogative of the RCAF, and they reflected the current air force reluctance to cooperate too closely with other services. Canadian airmen 'entertained' the USN Admiral's daily 'requests' for anti-submarine and escort operations, none of which seems to have been refused. RCAF patrol aircraft not assigned that day to USN-controlled operations remained under RCAF control. Coordination with the RCN was on an ad hoc basis. In early 1943 no system yet existed for RCN-RCAF coordination in the mutual tasks of escort and ASW.

Clearly the muddled command and control arrangement west of 35° was a serious impediment to the Allied war effort. The British had realized this difficulty soon after the American entry into the war in 1941; but so long as the crises lay elsewhere, they could not force the issue. The return of the war to the mid-Atlantic in late 1942 thrust command and control problems to the fore again. Canadians, too, were restless with the command structure by late 1942 and threatened to become obstructionist. By early 1943 both the RCN and the RCAF wanted the Americans out of the northwest Atlantic and operational control transferred to a Canadian. They began to restructure the Canadian command system, both to make the system itself more effective and as a way of pointing out the absurdity of the USN's position. Canadian lobbying, the threat of obstructionism, and the pressures of the war itself finally forced the Americans to convene the 'Atlantic Convoy Conference' in Washington in early March 1943. This meeting settled existing command and control problems in the Atlantic and, more importantly here, erected the framework within which the last two years of the Atlantic war was fought.[1]

The backdrop to the Washington Conference was the high point of Germany's mid-Atlantic war. As Allied naval and air planners gathered to sort out their campaign, a gap in special intelligence coupled with the enormous number of U-boats swirling around in the mid-ocean produced the highest interception and loss rates for convoys during the war. In the first weeks of March every transatlantic convoy was located, over 50 per cent were attacked, and fully 22 per cent of shipping that sailed failed to arrive. These staggering losses bred despondency in some circles and served to focus debate wonderfully in Washington. The spectre of crisis in the mid-Atlantic, therefore, hovers over the discussion of command arrangements that follows.

Many on both sides of the Atlantic had argued for an Allied 'Supreme Commander': a single mastermind to counteract the unity of Dönitz's command. The concept was sound (one NATO later adopted), but in 1943 the British could not accept an American, and the growing power of the USN would not permit a Briton. Both parties resigned themselves to continued Anglo-American strategic partition of the Atlantic. Coordination of the Atlantic effort was passed to a special 'Allied Anti-Submarine Survey Board,' which was given broad powers to review and make recommendations on standardizing training, tactics, and doctrine throughout the north Atlantic.

While the western Atlantic remained an American strategic responsibility, a more practical distribution of roles was possible at the operational level. The British came to Washington determined to push their operational control of the anti-U-boat war as far west as possible, perhaps to the Gulf of Maine. The Americans were equally determined to limit their concessions to either the British or the Canadians. The Canadian challenge was to carve out something for themselves between these two ambitious and suspicious rivals. The initial Canadian claim was bold. As Rear Admiral Victor Brodeur, the naval member of the Canadian Joint Staff in Washington, informed the conference, what the Canadians wanted was operational control of convoys and anti-submarine warfare from just south of New York northward and east – presumably – to the existing CHOP line. The RCN had already unified its own command in the area under Admiral L.W. Murray, Commanding Officer, Atlantic Coast, in Halifax by making Flag Officer, Newfoundland, a subordinate command. Brodeur now recommended that Murray be elevated to the status of Allied 'Commander in Chief, North West Atlantic' and be given 'general direction of all surface and Air Force employed in A/S warfare in the N.W. Atlantic.'[2]

The Canadian case was ambitious, but not without merit. The RCN's larger partners jealously guarded their own primacy, however, and worked to trim those ambitions and compress both the size and the effectiveness of the proposed Canadian zone. While the British were sympathetic to the Canadian position, they could hardly argue against sharing control of the mid-ocean with the Americans because it contravened a principle of war, only to hand the western portion over the RCN. The RN's primary objective at Washington was to secure operational control of Mid-Ocean Escort Force and all operations in the mid-Atlantic. It was agreed, therefore, to move the CHOP line for naval escorts to 47° W, from the mid-ocean to roughly the tip of the Grand Banks. That move gave Horton complete control over the current battle against the Wolf Packs. There is no indication that Canadians were displeased with the extension of British naval control so far westward, perhaps because the conference had conceded to Admiral Murray operational control of Allied ASW aircraft 'to the limit of aircraft range from bases in Labrador, Newfoundland and Canada.'[3] The Canadians would clearly have preferred to exercise control of naval forces further to seaward as well. Indeed, over the last two years of the war the RCN would, for purposes of planning and operations, consider that its own zone extended to 40° W.

If the new Canadian zone was squeezed from the east by the British, it was also pressed from the south by the Americans. The Canadians argued that their provision of escorts for the main transatlantic convoys between the Grand Banks and New York gave them proprietary interest in control of such operations to 40° N (just south of New York). This claim constituted a major intrusion into U.S. waters. Not surprisingly, the USN did not budge an inch. The southern edge of the new Canadian zone was restricted by the Washington Conference to the northern limits of the USN's Eastern Sea Frontier, roughly 42° N. In addition, instead of running along the 42nd parallel eastward to the new CHOP line, the Canadian zone was pared away on its seaward edges until it conformed roughly with the line of the continental shelf. This arrangement left a pie-shaped piece of territory southeast of Nova Scotia under U.S. operational control, giving the Americans ample room in which to route their U.S.-Mediterranean (and in 1944–5 U.S.-Europe) military convoys without fear of entering Canadian-controlled waters. The new Canadian theatre was, in fact, nothing more than the old American-designated 'Canadian Coastal Zone,' which had included Newfoundland. Having thus boxed the Canadians in, the senior alliance partners put paid to RCN-RCAF ambitions by designating the zone pejoratively as the '*Canadian* Northwest Atlantic' (CNA). Operational control passed to Admiral L.W. Murray in Halifax as C-in-C, CNA, effective 30 April 1943. The USN admiral at Argentia remained the American authority in the northwest Atlantic, and Argentia became an important base for USN 'Hunter-killer' groups.

Murray's new powers mirrored those of Horton on the other side. He was fully responsible for Allied convoy and escort operations within his area, and air patrols to the limits of their range. To facilitate coordination and control of operations, a new Area Combined Headquarters was built in Halifax to house the naval and air operations staff. It opened on 17 July, when daily naval and air coordination meetings began.[4]

What really seems to have tipped the balance in favour of this new Canadian command was the excellence of RCN communications, intelligence, and convoy-routing staffs in Ottawa. The Americans' initial objection to a separate Canadian theatre had stemmed from their belief that the RCN lacked the expertise to exercise all the command and control functions required of independent command. In fact, the RCN was well up to the challenge. As the Americans were gently reminded, the RCN had controlled shipping in the western Atlantic north of the equator for the first six months of 1942 while the United States made the transition to war. Well connected with the British naval control of shipping and naval intelligence network, Ottawa also provided the USN with daily intelligence summaries over the same period.[5] The RCN continued to develop its own naval intelligence system during 1942. By early 1943 it had ten direction-finding stations in operation and more building, and in June an Operational Intelligence Centre was established at NSHQ. In March 1943 the Americans were persuaded, somewhat reluctantly, that Canada could manage escort and convoy operations off its own coast.[6]

The Washington Convoy Conference corrected much of what had been wrong with the defensive system in the Atlantic. In that sense, its spirit was in keeping with the directives that emanated from the January meeting in Casablanca. In order for planning to go ahead for a cross-channel invasion of France in 1944, more materials needed to be brought – safely – across the Atlantic in 1943 than was the case early in the year. The issue of the moment, then, was securing and increasing the flow of war materials to Britain. The tough, aggressive, and anglophobe head of the USN, Admiral Ernest J. King, cautioned the delegates in his opening remarks that the purpose of Atlantic operations in 1943 was 'to maintain the flow of shipping that is essential to the general aims of the war.' This was a remarkable position for King, given his previous rejection of convoys in favour of offensive ASW. Pointing out that only limited resources were available for actually 'fighting the submarines,' King emphasized 'that anti-submarine warfare for the remainder of 1943, at least, must concern itself primarily with the escort of convoys.'[7] Thus, at Washington in March 1943 King endorsed the long-held British position emphasizing safe and timely arrival. The British, in contrast, were ready to adopt an offensive posture. Vice-Admiral Sir Henry R. Moore, the vice-chief of Naval Staff at the Admiralty, alluded to that changed stance in his response to King's remarks, which advocated flexibility and a less dogmatic approach to present problems. By early 1943 the British clearly saw what the Americans did not: that Dönitz's submariners were committed to a fight they could not win and that it was time to crush them. The Canadians, anxious to take the offensive themselves but wary of the consequences of failed action, seem to have hoisted King's message wholly inboard, but it was Moore's approach that would prevail in the Atlantic during 1943.

Increasing and securing transatlantic convoys required the return of Canadian groups to the mid-Atlantic. Since their departure in February, MOEF had run with only eight groups, seven British and the nominally American A 3. They bore the brunt of savage Wolf Pack attacks and, with fewer groups, of a brutal convoy schedule and brutal winter weather. If the convoy cycle was shortened to intervals of five days for fast convoys and eight for slow ones and the Canadians were brought back, more convoys could be pushed across the Atlantic in the spring of 1943. In addition, it was agreed at Washington to increase the number of both British and Canadian groups in MOEF, to ten B groups and the four C groups. The token American presence in the mid-ocean, A 3, was to be withdrawn.[8]

The increase in transatlantic traffic made it necessary to increase the number of Canadian escorts at Halifax as well. Consolidation of RCN escorts within the new Canadian northwest Atlantic therefore became an Allied priority in March 1943. The USN had already agreed to return Canadian escorts operating between New York and Cuba, and at Washington the British agreed to return those operating in support of the North African campaign. Significantly, the ships recalled from foreign service were among the RCN's best, and they had demonstrated skill in sinking U-boats.[9] *Oakville*, for example, had sunk *U94* off Cuba the previous summer while operating with the USN. *Ville de Québec*'s sinking of *U224*, as announced by Mackenzie King in the House on 1 February, was followed by the sinking of two Italian submarines in the Mediterranean: *Tritone* by *Port Arthur* and *Avorio* by *Regina*.[10] This much was confirmed at the time. But we now know that another of the TORCH corvettes, *Prescott*, had sunk *U163*,[11] and the MOEF ship *Shediac* had destroyed *U87* off Portugal. Five U-boats sunk in a few weeks in early 1943 was an enviable record by any standard. Only three were known at the time and all were in the Mediterranean. It would have helped the RCN to know in the spring of 1943 that Canadian corvettes – despite their important equipment shortfalls – were more proficient in ASW than their RN counterparts. Ironically, pulling the best Canadian corvettes back into the decisive theatre of the Atlantic war did not mean that their chances of sinking more U-boats would improve.

To a considerable extent the tenor of Canadian naval operations during 1943 was set by the Washington Conference. Still smarting from the disasters of 1942, the navy's plan was to concentrate – as Admiral King had urged – on securing the convoys first and foremost, and, secondly, modernizing the fleet as quickly as possible. The British, after all, were clearly set on retaining the largest responsibility for *close escort* of convoys along the main trade routes over which they had just won operational control. No one could be sure how long the battle in the mid-ocean would last. It was incumbent upon the RCN to ensure that its MOEF component, now due to return to the fray, was up to the mark. Moreover, in view of the anticipated German offensive in Canadian waters during 1943 the Naval Staff wanted no repetition of the disastrous losses of 1942 there either.

That being said, the Washington Conference also laid the groundwork for the offensive against the U-boats that followed. The establishment of support groups for the mid-ocean, an idea 'strongly approved by Admiralty and NSHQ,' was debated at length in Washington.[12] In order to establish five support groups already authorized for the Atlantic in the near future, the British planned to draw escorts from the Russian convoys, MOEF groups, and the Home Fleet. They also expected to have two escort carriers in the mid-ocean by the end of April. At the Washington Conference the USN agreed to field a 'Hunter-killer' group built around the auxiliary carrier *Bogue*, while the RCN provided by mid-May a 'Canadian support group' composed largely of corvettes. The Canadian group would operate east of Newfoundland, exactly where Prentice's long-defunct training group had patrolled. Inside the new Canadian zone a special force of short-range destroyers known as 'W 10' would provide support.[13] In addition, the RCN had its own plans for offensive forces in the Gulf of St Lawrence (discussed below).

Operational control of support groups was a contentious issue at Washington, but in the end command of these forces broke along the lines of the new command arrangements. Admiral Horton was given control of support groups to the new CHOP line, while the USN maintained control of offensive forces in the western Atlantic outside the Canadian zone. Murray's operational control of support groups was officially confined to his own small area. In practice, Murray ignored these limits on control over Canadian support groups based in the western Atlantic. British and American support groups operating on the fringes of Canada's tiny zone remained outside Murray's control, however, which gave Murray's new command little depth and restricted his ability to conduct operations against U-boats destined for Canadian waters. In March 1943 the British wanted American assurances that the USN would reinforce the Canadian zone as need be. The Americans were not interested in propping up the Canadians. They made it clear to the British that they refused to allow the Canadians to have operational control over their auxiliary carrier A/S forces, even within the Canadian zone. The issue was contentious enough for Admiral Moore to report to the First Sea Lord from Washington that the Americans had threatened 'to step out and leave us to deal with the Canadians.'[14] The threat was not serious enough to scuttle the command changes. In 1944-5 the USN gave Murray the loan of several destroyer-escort groups, but he never got command of any USN carrier groups. The Americans, as it turned out, would play an increasingly important role in defending the Canadian zone from U-boat intrusions as the war went on.

The Washington Conference has been seen by Canadians as a watershed in national and naval history, and as the culmination of a long battle for autonomy in home waters. And so it was. But the arrangements made in March 1943 also reflected a passing phase of the war in the Atlantic. The command structure erected to fight the 1942-3 war, primarily in the mid-ocean, was not well suited to the 1944-5 period. None the less, the establishment of a separate Canadian theatre and the return of its best escorts were important accomplishments in the spring of 1943. Under Murray's command the northwest Atlantic and the naval operations at home soon settled in to a semblance of order and system. Equally important changes were under way in Ottawa. The most significant was the establishment of an assistant chief of Naval Staff directly responsible for operations, plans, signals (communications and radar), trade (shipping and convoys), intelligence, naval air, hydrography, and the newly formed Directorate of Warfare and Training (DWT). Previously all these crucial staffs had come directly under the vice-chief of the Naval Staff who, as deputy to the CNS, was also responsible for personnel, supply, engineering, and the routine administration of NSHQ. The new organization separated the key staff functions under the ACNS. He also became primarily responsible 'for the conduct of anti-submarine warfare,'[15] a task formerly left to a junior 'staff officer A/S' who was on equal footing with other, less vital, functions such as minesweeping and coastal forces. The new Directorate of Warfare and Training was responsible for the establishment and maintenance of fighting efficiency standards, development of tactics, assessments of operational reports, research, and advice on staff organization. Like the changes on the east coast, these were long overdue. They also meant much temporary dislocation of effort through mid-1943 as the new organizations settled in.[16]

The RCN, as an institution, was bent on substantive change in 1943, of that there can be no doubt. This determination derived in part from an earnest desire to strike hard at the U-boat fleet. On the opening day of the Washington Conference Brodeur had spoken of the RCN's support for any scheme that would 'destroy the present submarine menace' in the Atlantic. Brodeur's words were likely chosen with care; the RCN wanted something more concrete done about U-boats in early 1943. The navy made clear in March its intention of keeping the Gulf of St Lawrence closed to oceanic shipping for the 1943 season unless the area could be made untenable to U-boats. It was no easy decision. As one historian has observed, closing the gulf in September 1942, 'With little more than two months left in the shipping season ... seemed ... prudent.'[17] Closing it for the whole of the 1943 shipping season was political dynamite, and the government came under bitter attack in parliament when the announcement was made.[18] Clearly, the inability – or unwillingness – of the Canadian navy to keep open the nation's principal maritime artery in the face of the mere *threat* of enemy action was a public admission of defeat.

In fact, the RCN's decision to close the Gulf of St. Lawrence was a calculated risk, one tied closely to its own plans for offensive action. Based on the 1942 experience, the Canadians expected a major German inshore campaign in 1943; the gulf and the river were particularly vulnerable.[19] Trade still had to be routed where it was easy prey for U-boats, and the difficult asdic conditions admitted of no easy and effective A/S measures. Not surprisingly, the RCN wanted to strike at the U-boats with as little hindrance from escort operations as possible. Hence, to meet the threat in the Gulf of St Lawrence it prepared elaborate plans designed for 'the maximum offensive action against U-Boats.'[20] When needed, close escort was to be limited to trawlers on loan from the RN, but they were to be supported by at least four other formations. Four corvettes equipped with modern surface search radar (type 271) were to be assigned to Quebec Force, to operate in the river where sonar conditions were poor. Six short-range Bangor class minesweepers were based at Sydney to support convoys in the gulf. Four flotillas of motor launches formed 'Gulf Strike Force,' and they were backed up in turn by six Bangors of Gulf Support Force. The task of the latter two formations was to prosecute U-boat contacts to a conclusion.[21] The choice of Bangors as the primary A/S search force in the gulf was deliberate. Because these ships had been built as minesweepers, it was imperative that they have excellent navigational capabilities. Thus, while the early corvettes fitted magnetic compasses, the Bangors were fitted from the outset with gyro compasses. This fact allowed them also to fit a more advanced asdic, the type 128, which made them more efficient at underwater searches than contemporary Canadian corvettes. The choice of vessels for the gulf offensive therefore was a sound combination of numbers and capabilities. Despite a severe shortage of training submarines, and little time available on the two at hand, by the end of April the staff officer (Operations) reported that the gulf forces were ready.[22]

The RCAF, too, had been preparing itself to handle another inshore U-boat campaign by adopting the new concept of 'hunts to exhaustion' in February. They were based on the simple notion that it was possible to hound a diesel-electric-powered submarine to death. U-boats had limited endurance on battery power and had to surface regularly to recharge their batteries. By conducting a systematic and expanding search with radar-equipped aircraft in the area of a U-boat sighting, it was possible to relocate and attack the U-boat when it surfaced again. The process could be repeated until the U-boat was destroyed. The concept was already in place in the eastern Atlantic, and Canadians adopted it, code-naming these hunts 'Salmons.' Salmons, however, never got much beyond the planning stage in 1943. No operational orders were issued and no joint training was undertaken. Not until late 1943 could the RCN and RCAF be said to be marching to the same drummer on the east coast.[23] Naval-air cooperation in 1943 had to be worked out the hard way: on the spot by trial and error. Bad weather, either on the ground or in operational areas, affected the availability of RCAF aircraft in any event. And until the first RCAF Liberators became operational in May, Murray's land-based aircraft had a maximum effective range of about 400 miles.

In early 1943 the lack of a 'very long range' capability excluded the RCAF from supporting RCN operations in the mid-ocean, where it was needed most. This issue was wrestled with at Washington in early March, but without resolution. Neither the British nor the Americans were prepared to assign Liberators from their own allocations to Canadian squadrons with the north Atlantic expertise to use them. Finally at the end of March the British agreed to transfer ten (later fifteen) of their own B-24 allocation to the RCAF for Atlantic operations.[24] While the airmen haggled over Liberators, the RCN preferred its own fleet of auxiliary carriers and in early 1943 was pushing the idea of acquiring four – one for each of the C groups in MOEF. As the directors of Operations and of Plans at NSHQ reported jointly on 6 April 1943, 'It is probable that Air will be the decisive factor in the "Battle of the Atlantic." ' Basing their plans on the need for auxiliary carriers in the war against the U-boats, they recommended the formation of a Canadian naval air service. The Naval Staff concurred and passed the recommendation on to the Naval Board, which concluded that 'all anti-U-boat phases of air operations' should be pursued.[25]

Participation in the air offensive against the U-boat, especially by RCN carriers, lay beyond reach in early 1943, but the navy was not content with a merely passive role. Chummy Prentice, now Captain (D), Halifax, fundamentally altered the tactical doctrine of his Western Local Escort Force – that varied assembly of old destroyers, minesweepers, and corvettes – which escorted convoys between the Grand Banks and New York. Their operations, too, were governed by what were now called *Atlantic Convoy Instructions (ACI)*, which set the safe and timely arrival of the convoy as the escort's primary duty. At the end of March 1943 Prentice issued the first in a series of tactical memoranda called 'Hints on Escort Work.' Revolutionary in intent and design, they turned *ACI* on its head.

Prentice recognized that air patrols made inshore waters unsuitable for pack tactics. As a result, U-boats operated individually in assigned patrol areas, so that no more than a single U-boat was likely to be in contact with an inshore convoy at any given time. In these circumstances Prentice altered the tactical doctrine of his escorts to a primarily offensive one. As he outlined in the first issue of 'Hints,': 'Always think in terms of the destruction of submarines. If you or one of your ships have your teeth into a submarine don't let him go for anything else until he's dead ... Decide whether a continued search in the area where the submarine is known to have been, or the ship's return to the area near the convoy into which that submarine or more of his friends will come, will give you the greatest chance of a kill. Make your decision on that, rather than in terms of the safety of the convoy. All convoys are safe from a submarine which you've sunk.'[26] Later in June, when part IV of 'Hints' was issued, Prentice stated his preference even more emphatically: 'The "safe and timely arrival of our convoys" is the object of all escorts. History has shown that the only sure way of achieving this object is to ensure the destruction of any enemy forces which approach these convoys.'[27] The Canadian shift to an offensive posture inshore in 1943 was well in keeping with the Allied trend, and in some cases anticipated it. Plans for the various strike forces in the gulf and the change of doctrine in WLEF displayed an innovative use of local resources, and for that reason they were well received by the Allies.[28]

While Canadians waited for the U-boat onslaught inshore, the mid-Atlantic war moved apace. By the end of March the worst of the German depredations was over, and signals intelligence revealed serious morale problems in the U-boat fleet. With special intelligence itself re-established in late March, fairer spring weather, support groups, and VLR aircraft, the mid-Atlantic became a killing ground for the U-boat fleet. That process began slowly in April, the month Canadian groups returned to the mid-ocean. As the Newfoundland Command War Dairy observed, it was probably 'the most unique month of the war in the Battle of the Atlantic.' The U-boats were 'extremely active,' but losses to shipping were 'extremely small,' and 'The innovation of the Commander-in-Chief Western Approaches' five support groups,' the Newfoundland War Diary concluded, 'has saved convoy after convoy threatened with pack attacks.'[29]

By early April senior Canadian naval officers and politicians were aware that the counter-offensive had begun in the mid-ocean. The cabinet knew that the RCN planned to get involved by fielding a support group in the spring of 1943.[30] The anticipation of Canadian U-boat kills is reflected in the drafts of Mackenzie King's speech to the Canadian Club of Toronto, which launched the Fourth Victory Loan Campaign in mid-April. 'The menace of the U-boat is in 1943 so grave and so constant,' the draft speech read, 'that it can only be met by the strongest offensive we can make against it.' The final version, delivered in Toronto on 19 April, omitted reference to the offensive, staying on solid ground by pointing out that the RCN provided half of the escorts on duty in the North Atlantic.[31] Presumably the change in emphasis was the work of those at NSHQ who vetted the draft. Canadian groups were back in the mid-ocean, but British groups were doing the fighting. J.J. Connolly, the naval minister's executive assistant, probably reflected the mood at NSHQ when he opined to his diary on 7 April that 'If we were to start over again, we would build convoy destroyers.'[32]

None of the eighteen U-boats sunk around North Atlantic convoys during April fell to the RCN. This pattern continued into May, when the British took operational control of the mid-ocean and the RCN fell even further behind the pace. Convoys that could not be routed clear of danger were strongly reinforced and driven through waiting packs. The first great victory came, appropriately enough, around the westbound convoy ONS 5 in the first days of May in what had until 30 April been the U.S. side of the mid-ocean. The storm-battered convoy was pursued by forty U-boats and lost eleven ships before the weather abated and the escort could take the offensive. As U-boats groped through the fog around ONS 5 on the night of 6–7 May, the British escorts of group B 7 under Cdr Peter Gretton counter-attacked. On that night alone Gretton's escorts located fifteen submarines, chased four away, rammed and damaged two, and sank no less than four.[33] In a single night the mystique of the Wolf Packs was broken.

The pace and significance of the battle by late April left little scope for niceties. With special intelligence providing U-boat positions and evidence of the submariners' shaky morale, the British used their new operational control to crush an old and resourceful enemy. To do so they cast aside previous agreements and adopted the flexibility to which Admiral Moore had alluded on the opening day of the Washington Conference. Indeed, the ink was hardly dry on the minutes of that meeting when major changes in escort dispositions were put in train to take advantage of the U-boats' weakened position. In late April the British petitioned their Allies for a reduction of MOEF from fourteen to eleven groups. The RN wanted to shift forces to the eastern Atlantic as part of the offensive. This 'reduction' was in RN groups promised, but not yet committed, to the mid-ocean. Other changes followed in May with the demise of the token American force A 3. Its Canadian corvette component was coupled with the first of the second-generation RCN River class destroyers, *Ottawa II*, to produce a fifth C group in MOEF.

In May, therefore, five of the twelve close escort groups in the decisive theatre were Canadian, yet they fared little better than the four of April. Five support groups, including the USN's group led by the auxiliary carrier *Bogue*, prowled in the mid-ocean. Also a wave of Liberator aircraft, some of them from 10 Squadron RCAF, flooded the mid-Atlantic, effectively eliminating the air gap. The impact on the Wolf Packs was devastating. In May the Allies sank no less than forty-seven U-boats in the north Atlantic. These were unprecedented losses, but none – apart from one-third of a kill awarded to *Drumheller* – fell to the RCN. Convoys escorted by C groups were supported in every case during that month by an auxiliary aircraft carrier group, which arrived a day before contact was imminent and left when the danger was well past.[34] Under these conditions it was all but impossible for Canadian ships to gain contact with U-boats. *Drumheller*'s portion of a kill was shared by HMS *Lagan*, also of group C 2 (which had a large British component), and an RCAF aircraft in the battle around HX 237. That convoy was the only one under C group escort to fight its way through a Wolf Pack in the spring of 1943.

From the RCN's perspective, all the forces around them – naval and air – were sinking submarines. Mackenzie King was still looking for good news in May when he attended the 'Trident' conference of senior Anglo-American leaders in Washington. He took the opportunity to express to Roosevelt and Churchill his 'dissatisfaction with the [lack of] recognition ... given Canada's war effort' in the press releases issued by the larger powers.[35] By the spring of 1943 he had secured a place for the First Canadian Division in the forthcoming invasion of Sicily, but nothing could be said in advance. Unless the RCN came through with dramatic results, King would now have to wait until the Sicilian landings in July for some high-profile news on Canada's participation in the fighting.

King would get no help from his navy. Despite the heavy toll of U-boat kills in the north Atlantic in May, the RCN ended the month without a victory. In fact, by the end of May unsustainable losses forced Dönitz to withdraw his U-boats from the main north Atlantic routes.[36] The legacy of 1941 and 1942, which left the RN cautious about the efficiency of the RCN, coupled with British resolve to destroy the Wolf Packs with their own forces, had denied the RCN participation in the great victory of early 1943. The pattern for 1943 was set: only support groups – or the very best escorts – were allowed to deal directly with the U-boat fleet. The problem for the RCN now, therefore, was how to gain access to that elite club.

As the spring rolled on, the British 'found' escorts for the offensive through further reductions in MOEF, facilitated by changes in the routing, size, and scheduling of convoys. Starting in May, convoys sailing northern tracks followed a standard path along the Great Circle Route, thus reducing by about three days the average time on passage for westbound convoys, by about one day for those eastbound.[37] The fixing of southern routes followed in June, so that by high summer north Atlantic convoys sailed 'tram lines.'

These changes made the routing of convoys much simpler and more efficient – and predictable. The Allies hoped that predictability would facilitate their own offensive against the U-boats. Further reductions in the number of groups needed were achieved by opening the interval between convoy sailings and reducing their frequency, while allowing the size of convoys to swell. By mid-1943 convoys in excess of eighty ships became common;[38] the volume of transatlantic shipping remained high, but it took fewer close escorts to protect it.

The RCN was swept along in the flow of these events with little opportunity either to affect them or to alter its own role. The best Canadian ASW vessels – even the newly commissioned destroyers acquired from the RN (*Ottawa II* and *Kootenay*, both of which went to the new C 5) – were committed to close escort duty with MOEF. This much had been agreed at Washington. It had also been agreed that the RCN would field a 'Canadian support group' by mid-May. In mid-April Admiral Sir Dudley Pound, the First Sea Lord, petitioned Nelles to get the Canadian support group into action quickly, and he offered two newly commissioned RN River class frigates to bolster the group.[39] But Pound's prodding and the offer of HM ships *Nene* and *Tweed* could do little to hurry the pace of the group's formation. Finding four Canadian corvettes modern and competent enough to act with credibility in a purely offensive role in the spring of 1943 was no mean feat.

The RCN's problem in the spring of 1943 and for the next full year was how to find ships with enough range and the right equipment to get into the support group business. No one in Ottawa was prepared to remove destroyers from mid-ocean groups. The first River class frigate for the RCN would not hit the water until June, and it would be late 1943 or early 1944 before these vessels arrived in numbers. The only type of ship with the endurance needed to operate in support groups was the RCN's corvette. Over the previous winter many corvettes had been re-equipped with type 271 radar and 20-mm guns. None of them, however, was fully modernized.

The Naval Staff had discussed the modernization of its fleet of some seventy corvettes in 1942 and had concluded it would be a waste of effort. It was clearly better to concentrate on the production of more modern ships. The most the staff would authorize in 1942 was the updating of a single corvette as a trial to work out problems and costs. It took until early 1943 to find the dockyard space, material, and the time to take *Edmundston* in hand for the trial. The Naval Staff also took the opportunity in early 1943 to have the revised corvette *Calgary* modernized at Cardiff, Wales, where she had gone to have nagging mechanical problems fixed. The virtual collapse of the navy's effort finally spurred the Naval Staff in February 1943 to commit itself to the complete revamping of all corvettes before waiting details of *Edmundston*'s conversion.

With the benefit of hindsight, we can see clearly that the Naval Staff made the wrong choice in 1942. When asked about the prospects of modernizing the fleet quickly in 1943, the supervising naval engineer in Halifax was pessimistic. New construction, including merchant ships, had priority, the available work force was too small, and capacity on the east coast was inadequate. Work at the Halifax dockyard was up 300 per cent over the previous year, and it was all that three shifts working seven days a week could handle. Severe winter weather, the worst of the war so far, choked Canadian ports with damaged shipping. Shunned by busy British and American yards, ships of marginal value in need of repair gravitated to Canada, and the government was under pressure from the British to do more repair work.[40] In the first quarter of 1943 850 vessels needed repairs in Atlantic Canada, and 125 required docking. In addition, most east-coast yards were small, undermanned, lacked experience in major work, and did not have the political clout to obtain supplies. They suffered from indifferent management and in some cases from strained labour relations.[41] Officers in Halifax estimated that perhaps half of the corvette fleet might be modernized in Canadian yards during 1943, an estimate the chief of Naval Engineering and Construction at NSHQ rightly considered optimistic. In any event, the idea was not uniformly popular, because refitting took ships out of service for a long time. When asked about the prospects for the refitting schedule, the chief engineer in Newfoundland complained that the idea of losing ships from operations for ten weeks was 'appalling'. In the event, he would lose his ships for much longer than that.[42] Early 1943 was a poor time to think about modernizing the fleet.

The refit situation was chaotic when the Allied Anti-Submarine Survey Board toured the east coast in May. Canadian officers protested that their hands were tied, because ship repair, including access to resources, 'rests in civilian hands.' After listening to personnel in Halifax grumble over their inability to obtain any priority for naval repairs and modernization, Commodore J.R. Mansfield, RN, the senior British member of the board, put his finger on the nub of the problem. He agreed that modernization called for Canadian government involvement, but he 'doubted whether the case had really been represented to them as a long term policy.'[43] In fairness, the RCN had had the foresight early in the war to order the construction of a number of marine railways throughout Atlantic Canada large enough to lift ships the size of destroyers out of the water. By the spring of 1943 two of these hoists – one at Sydney and the other at Pictou – were in operation; four others were under construction in Nova Scotia and another was being built at Bay Bulls, Newfoundland.[44] They ultimately eased maintenance problems, but they could do little to help in early 1943.

Beyond congested Halifax and Saint John, there was not a single centre where all the necessary work could be done simultaneously. Ships under refit ricocheted from one port to another until the work was completed. Thus *Sackville* went to refit in Liverpool, NS, in mid-January and stayed alongside until late March. She then sailed to Halifax for docking before continuing on to Pictou to have her minesweeping gear removed; while this work might logically have been done at Liverpool, the Pictou yard – in Angus Macdonald's home county – had won the government contract for removal of the 'sweep gear. Work-ups followed and the corvette returned to operations in early May. *Sackville*'s refit lasted about twenty weeks, and even then she still had the old-style short forecastle, outdated bridge, and none of the modern navigation and asdic equipment needed. Small wonder Mansfield confided to Admiral Sir Dudley Pound that the Allied A/S Survey Board was 'undoubtedly shocked with conditions as they found them on visiting Canada.'[45]

By the spring of 1943 modernization was moving, but only slowly.

By the end of May, *Calgary* and *Edmundston* were refitted and a further ten were in hand. Thus, only two out of the RCN's seventy corvettes were suitable for support groups in June 1943. Figures from the RN for later in the summer (received by the RCN in August) revealed the backwardness of the Canadian fleet. Of the sixty British corvettes in Western Approaches Command, only two had not yet been at least partially modernized. By mid-1943 the dearth of modern equipment in the Canadian escort fleet helped to keep the RCN out of the U-boat killing business, with grave consequences for the chief of the Naval Staff.

The Admiralty relieved some of the pressure of modernization by arranging to have its eight corvettes on loan to the RCN upgraded in the United States under lend-lease. The British also promised to modernize a further six RCN corvettes in the United Kingdom during 1943. There were still about fifty ships without yards, even assuming that east-coast yards could do the work expected. Shortages of material and equipment also plagued the modernization effort. When the program was getting under way in earnest in June, electrical equipment was in short supply. Much of the vital equipment came from Britain and the United States and could not be obtained quickly. The War Diary of the Canadian Northwest Atlantic Command complained in June that, 'delay in supply of raw materials, especially brass and electrical gear, has again delayed the completion date of many of the ships under refit.'[46] Government regulations caused further delays. Despite the press of the war, tenders had to be called for all work, a lengthy and bureaucratic process.[47] Not surprisingly, the modernization program fell into arrears so quickly that by the end of June Admiral Murray's chief of staff, Captain Roger Bidwell, complained that no sensible planning was possible for work-up, training, and refit schedules.[48] The staff at Halifax even considered not attempting any more modernizations until those already begun had been completed. In all, twenty-six Canadian corvettes were modernized in 1943. This figure was well short of early, optimistic projections, and few ships were available for operations before the end of the year.

It was not surprising, therefore, that all the first choices for the Canadian support group in April were recently returned TORCH corvettes, *Algoma, Kitchener, Port Arthur,* and *Woodstock.* They carried the latest in radar and a heavy secondary armament of six 20-mm oerlikon guns. All except *Algoma* were revised corvettes with improved hulls and with bridges modified by the addition of a second compass binnacle and loudspeakers from the asdic position (the type 123D standard). But all had to 'make do' with magnetic compasses and without the latest anti-submarine weapon, 'hedgehog.' Magnetic compasses reduced most depth-charge attacks to little more than educated guesswork and made it all but impossible to coordinate an A/S search with British escorts (which were fitted with gyro compasses). Hedgehog, too, was considered essential for efficient U-boat killing in 1943. Depth-charge attacks were delivered over the stern, making it necesary to run over the target, losing it in the process, before pounding an area with high explosives in hopes of crushing the U-boat. In contrast, hedgehog was designed to be a precision weapon fired ahead of the A/S ship while the escort maintained asdic contact with the U-boat. Its gyrostabilization held the weapon on target in anything up to a 20-degree roll. The hedgehog's twenty-four small 'bombs,' each carrying 32 lbs of Torpex, were fired from a spigot mortar to a range of 200 yards, where they fell in an oval 120 ft long by 140 ft wide. The bombs were contact fused, so explosions occurred only when the submarine was hit, and they were good to a depth of 1,300 ft. The hedgehog promised a much higher U-boat kill rate. The success rate in depth-charge attacks during the war hovered around 6 per cent, while hedgehog success reached approximately 30 per cent by 1944.

In the absence of such modern equipment, the initial choices for the Canadian support group were perfectly sound. However, although this group appears in the records as 'Western Support Force' in early May, its members never undertook operations as a group.[49] A more permanent Canadian support group composition was promulgated on 2 June, by which time the RCN had something better to work with. *Calgary* and *Edmundston* were assigned, as was the British-owned corvette *Snowberry,* undergoing modernization in Charleston, NC, and ready for operations by July. *Camrose, Lunenburg,* and *Prescott,* all undergoing major rebuilding, were promised for later in the year. The British frigates *Nene* and *Tweed* joined in mid-June to round out the group. With *Nene* came the new group senior officer, Cdr J.D. Birch, RNR. Birch's appointment followed the current practice of bolstering Canadian operations by providing experienced RN senior officers, a trend that continued into 1944. By June the RCN finally had a support group, albeit with a core of British ships under an RN officer, based at St John's. Unfortunately, by the time the Canadian support group sailed on its first operation east of the CHOP line, the Wolf Packs were long gone.

The closing of the air gap and the collapse of the Wolf Pack campaign in the mid-ocean in May radically altered the tactical situation around transatlantic convoys. Among the first forces released because of these changes were the British Home Fleet destroyers of support groups EG 3, EG 4, and EG 5.[50] They were never very successful at ASW, but they allowed other, more skilled, north Atlantic veterans to take offensive action. Their departure left four support groups currently available for the north Atlantic: two British (EG 1 composed of sloops and frigates and EG 2 composed exclusively of sloops), one American (EG 6 built around the escort carrier *Bogue*), and the Canadian support group. By June the Admiralty hoped to maintain five support groups on station in the north Atlantic. After the withdrawal of the Home Fleet destroyers and eventually the American EG 6, this task could be accomplished only by rearranging A/S forces. At the end of May the number of close escort groups in MOEF went from twelve to nine, since two of the B groups released were shifted to offensive operations.[51] On 7 June the Canadian support group was redesignated 'EG 5,' a number that lapsed when HMS *Biter*'s group dissolved at the end of May.[52] Two new British groups, EGs 7 and 8, were forming.[53] When they became operational and the American EG 6 was withdrawn, the Admiralty would have enough groups to keep five at sea.

The reallocation of forces reduced MOEF to nine close escort groups by July. Significantly, five were Canadian (groups C 1 - C 5) with a smattering of RN ships, while the remaining four, B 2, B 3, B 6, and B 7, were nominally British – although the corvette component of B 3 was Free French and that of B 6 largely Norwegian.[54]

In the end, not all the British ships released from close escort with MOEF went U-boat hunting. Many of them were switched to Gibraltar convoys or to escort duty in the western Mediterranean.[55] These assignments had more to do with the Allied landings in Sicily in July and the Italian mainland in early September than with the counter offensive against the U-boats then under way in the eastern Atlantic. It is none the less significant that the British role in close escort in the mid-ocean was much diminished by mid-1943; well below that agreed upon at Washington. Moreover, the shift of British forces to better hunting grounds and the consequent shifting onto the RCN of the burden of close escort in the mid-Atlantic by July 1943 did not go unnoticed in the Canadian fleet.

Meanwhile, in an attempt to deceive the Allies of the true extent of the collapse of his campaign, Dönitz continued to keep a few U-boats on station in the mid-Atlantic throughout June. The ruse was an utter failure. The British concentrated their efforts over the summer on the so-called 'Biscay Offensive,' an all-out assault on U-boats in transit through the Bay of Biscay. It began as an air attack, to which the Germans responded by equipping their submarines with much heavier anti-aircraft armament. In June Dönitz tried sailing U-boats through the bay on the surface in groups, relying on concentrated anti-aircraft fire to get them through. The tactic was not altogether successful. Such gun battles produced heavy casualties among key bridge personnel, even when U-boats were not severely damaged or sunk. Moreover, the Allies responded by filling the bay with aircraft, supported by surface ships, and overwhelming the submariners. Between 20 July and 2 August ten out of seventeen outbound U-boats were sunk. Over the whole month, from 1 July until 2 August, when outward-bound sailings from Biscay ports were temporarily stopped, the Allies sank seventeen of the eighty-six U-boats transiting the bay.[56]

The bay offensive continued into August, growing in intensity as the Germans moved in forces to help their beleaguered U-boat fleet. Meanwhile, they shifted operations to supposedly less dangerous theatres. In July they attempted to operate a Wolf Pack against American convoys southwest of the Azores. The pack failed to find a single convoy and lost five of its fifteen boats. In August U-boats concentrated in an arc from the Caribbean, along the north coast of South America, east to Africa, and up towards Gibraltar. This was an attempt to maintain some presence in what were believed to be less well-defended Allied theatres, but this strategy, too, failed to stem the tide of U-boat losses. Of the thirty-nine U-boats deployed in these waters, fourteen were sunk – only two of the seven sent to operate off Brazil ever returned. These operations were crushed primarily by American auxiliary aircraft carrier hunter-killer groups. Operating along the U.S.-Mediterranean route and south of the Azores, these groups, directed by special intelligence, took a dreadful toll of Germany's submarine fleet. In July and August they accounted for thirteen U-boats, including virtually all of Dönitz's small, but crucial, fleet of tanker U-boats.[57]

The extent of the punishment inflicted on the U-boat fleet in July and August is evident when losses in those months are compared with those of the spring. The victory of May was achieved with a 22.2 per cent loss rate of U-boats at sea. In July the loss rate peaked at 39.3 per cent and in August it dropped only slightly to 35.6 per cent – in that month fully 64 per cent of U-boats in transit were sunk.[58] Altogether in June, July, and August Germany lost seventy-four U-boats in exchange for fifty-eight Allied merchant ships (nearly half of which were lost in the thinly defended Indian and South Atlantic oceans). Aircraft, both shore-based and from aircraft carriers, accounted for fifty-seven of the seventy-four U-boats destroyed.[59] Surface ships therefore accounted for only seventeen of the U-boats sunk from June to August. None was claimed by the RCN.

Almost without exception, these summer sinkings took place far away from the concentration of RCN ships. Even the feared German assault in Canadian waters failed to materialize. The few submarines to arrive confined themselves largely to clandestine operations – landing spies, establishing a weather station on the Labrador coast, attempting rendezvous with escaped prisoners, or – in one of the most aggressive acts – mining the approaches to Halifax harbour.[60] All of these U-boats scrupulously avoided prolonged surface runs and any aggressive action that might expose them to defending forces. Neither the elaborate gulf plans nor the altered doctrine of Prentice's groups was put to the test. The gulf was gradually reopened to oceanic shipping. In the summer of 1943 not a single U-boat was sunk inshore by the RCN.

The failure of the RCN to participate in the great destruction of the U-boat fleet in early 1943 did not go unnoticed. When in early January Admiral Nelles, the chief of the Naval Staff, recommended that the Canadian groups of MOEF be transferred to the British for additional training, he sold the plan to the war cabinet as part of a larger effort to increase the killing efficiency of mid-ocean groups. It was clear by the summer who was doing the killing, and it was not the RCN. All Nelles could offer to the cabinet in his briefing of 2 July was the promise that modern corvettes were on the way. This promise was almost too little, too late. Nelles was already the subject of a whispering campaign at NSHQ intended to discredit him. Politicians were certainly dissatisfied with the performance of his fleet. By June so, too, were the men at sea.

The hotbed of RCN discontent at sea was centred in Mid- Ocean Escort Force. C groups had spent the winter in the bosom of Western Approaches Command, where the RN made it clear that their splendid effort was being wasted by incompetence at home. Lt J. George, RCNVR, a Historical Records officer posted in the United Kingdom, reported on the wardroom grumblers of MOEF in the spring of 1943: he had found them a surly lot. They were deeply annoyed at what appeared to be general mismanagement – if not plain incompetence – on the part of NSHQ in Ottawa. They were aware of the inadequacies of Canadian maintenance facilities but could not understand why the navy's current priority on modernization was not abandoned in favour of manning new ships. They were mystified by the assignment of corvettes with extended forecastles to theatres other than the mid-ocean, why they lacked gyro compasses, oerlikons, and even modern depth-charge throwers, which were being produced in Canada for the RN and new RCN construction. The general supply of equipment, George reported, was currently managed by a government department, and the men at sea thought it should be taken over by the navy. After visiting one or two ships in each group and meeting all commanding officers, George concluded, 'My impression is that the failure to equip these ships with modern material is keenly felt by all their Officers and widely discussed, and is tending to discourage them.'[61]

George's sentiments were echoed in the June report of the Canadian liaison officer aboard the Western Approaches training ship *Philante*, and summarized again in a lengthy memo by Cdr Desmond Piers, captain of the destroyer *Restigouche*.[62] The Naval Staff was well aware of the mood at sea but could not prevent it rising to a boil. Canadian sailors were embarrassed enough about the situation, but it was given wider currency in early July when the lack of modern equipment aboard ships proposed for a second RCN support group was noted in a general signal by Admiral Horton. The purpose of the signal was to acquaint staffs and officers with the limits of the Canadian group, but the public airing of RCN weakness struck hard at the morale of the fleet.[63]

The lid finally came off in August as a result of two of the most notorious memos in RCN history. In July Captain W. Strange, the dour assistant director of Canadian Naval Intelligence, took passage to Britain in the RN destroyer *Duncan*. Her captain, Cdr Peter Gretton, confided in Strange that it was a pity that Canada's excellent effort at sea was crippled by the lack of modern equipment. Strange was vaguely aware of the situation from his discussions with Canadian officers, and he found Gretton's concerns echoed by senior RN officers in the United Kingdom, including the irascible Commodore (D), Western Approaches, G.W. Simpson. Rather like Saul on the road to Damascus, Strange was converted to the belief that something was seriously amiss and that he would have to serve as the prophet of change. His conversion and the justice of his cause were confirmed during the return passage to Canada aboard HMCS *Assiniboine*. Her captain, Cdr K.F. Adams, had returned to sea in February and by July was nonplussed by the RCN's passive role. Adams and Strange conspired to write two memoranda on the matter: Adams's was to pass through proper channels, Strange's was to circumvent all protocol and pass directly to the naval minister himself.

Adams' sent his memo to Captain (D), Newfoundland, on 9 August. It reviewed the equipment situation in the ships of group C 1, noting that the three RN ships assigned were up to date, but that the five Canadian ships were not. Even the two Canadian destroyers, *Assiniboine* and *St Laurent*, could not compare. Under those circumstances, Adams admitted, 'I have found myself again and again forced to the reluctant conclusion that the R.N. ships would – in the interest of safety of the convoy and destruction of the enemy – have to form the striking units [of the group].'[64] Strange's memo found the minister on 20 August at the Quebec Conference. As of early August, Strange noted, only five of the sixty RN corvettes in Western Approaches Command were *without* gyro compasses and hedgehog, while only two of seventy RCN corvettes (presumably *Edmundston* and *Calgary*) had so far been fitted with them. The disparity was staggering. Putting the point in terms a politician was better able to comprehend, Strange charged that lack of modern equipment had 'prevented our ships from making a good showing' and as a result the RCN had been relegated to minor roles in the Atlantic.[65]

It was, of course, not only Canadians who measured victory in the Atlantic in terms of U-boat kills. On 20 August, the day Strange's memo arrived in Macdonald's hands, Churchill met with the Canadian war cabinet. Both Mackenzie King and Macdonald complained to Churchill that nothing had been said in British or American press releases about the Canadian contribution to the recent Atlantic victory. King was still steaming over the omission of the Canadians from the initial press releases on the landings in Sicily in July. He now pressed Churchill about why so little was being said about the role of the RCN in the Atlantic, even though half of the escorts were Canadian. King and his cabinet were dumbfounded by Churchill's bald admission: he had no idea that Canadian naval forces were so large or so important. Yet Churchill clearly knew how many U-boats had been sunk and by whom. Both he and Roosevelt issued the monthly press releases on the subject. The Canadians had half of the escorts, but of the 100 U-boats destroyed in the north Atlantic from January to May 1943 *none* had clearly fallen to a Canadian warship.[66]

Churchill's shocking revelation came just as the Canadian government's fortunes reached their nadir. Even as the Quebec Conference was getting under way – itself staged to lend support to

Mackenzie King's faltering popularity – a war-weary electorate displayed its anger. In early August four federal by-elections went against the government, and the Liberal party of Ontario was swept from power. If Macdonald needed any further evidence of why the navy had failed to salvage the fortunes of his party it was probably provided by Admiral Sir Dudley Pound, the First Sea Lord. Pound was sent the official reports of the Allied A/S Survey Board visit to Canada and a confidential – and critical – brief by Mansfield. Moreover, Strange's memo suggested why the navy had faltered. The next day, 21 August, Macdonald ordered Nelles to report on the state of equipment in the RCN, especially asdic, radar, gyros, and hedgehog. Nelles was instructed to spare no effort in getting the information.[67]

At the best of times Percy Nelles could have been described as a quiet, competent, but uninspired man, one promoted well beyond his limits. By 1943 he had been chief of the Naval Staff for ten years. Until 1939 he had run the RCN like a small family franchise of a larger global corporation – the Imperial Navy. The war, however, took its toll on Nelles's strength and tested his abilities. Like so many others, his health had collapsed in that frantic year of 1942. He ought to have been replaced then, but no one was available. In any event, such decisive and far-sighted action was not likely to come from Macdonald. He, too, perhaps even more than Nelles, had power well beyond his ability to wield it. There is little evidence that the minister ever took much interest in the navy outside the routine administration of accounts and liaison with other government departments. All Macdonald, and more importantly the cabinet, seemed to know by August 1943 was that there was something wrong with the navy.

What the naval minister got from Nelles at the end of August, in response to his specific request for comparative data on the state of equipment in comparable RCN and RN ships, was obfuscation, avoidance, and delay. Nelles's response was probably typical of what he and his staff had fed Macdonald over the previous three years. This time, however, Nelles judged his minister incorrectly: the nature of the game, its rules, and the pitch all had changed. Nelles's totally inadequate response nourished Macdonald's growing conviction that the Naval Staff's weak grip had let down the men at sea, the Canadian people, and the federal government. Thus began a protracted campaign on the part of the naval minister to find the cause of this abject failure. There was nothing grand in Macdonald's subsequent action. Clearly, the failure of the RCN to help deliver the government from crisis put him under pressure in cabinet and he had to find a scapegoat. He soon fixed his sights on Percy Nelles.

Nelles was not without friends within the Naval Staff in late 1943, but even his vice-chief, Vice-Admiral G.C. Jones, was already talking behind his back to Macdonald. Nelles was also at odds with many of his senior staff by August 1943 over the plan for a Canadian naval air service. They, Macdonald, and even Mackenzie King were supportive of the immediate acquisition of small carriers. Nelles supported the British view that the RCN's effort was better spent elsewhere during the war, and without British support it would be impossible to build a Canadian naval air service quickly. At Quebec Nelles – and the British – won their case. They deflected the issue away from the immediate acquisition of escort carriers into a long-term plan to acquire light fleet carriers for the postwar navy.[68] The extent to which Nelles himself obstructed the carrier plan is open to question. But his opposition to the hasty establishment of a naval air service was entirely consistent with the policy of consolidation, modernization, and maintenance of the best ships in close escort groups which characterized RCN priorities in 1943.

The Quebec Conference was also important because it brought to an end the brief period when the fortunes of the navy's wartime escort fleet coincided with the ambitions of the professional service itself. The RCN as an institution had always conceived of the war on two quite distinct levels: the hastily built 'sheep dog' navy manned by reservists doing dreary but essential escort duty, and the professional navy's struggle to acquire the basic components of a balanced postwar fleet. When the RCN was called to task in late 1942 for inefficiency in the north Atlantic, the professional navy reoriented its priorities to escort destroyers and auxiliary carriers for the A/S war. That brief enthusiasm lasted until August 1943, when the British offered the RCN the fulfillment of its long-cherished dream: modern cruisers and fleet class destroyers, and the promise of light fleet carriers and a naval air arm. The combination of victory in the Atlantic in 1943 and the offer from the RN to man some of its latest ships set the professional navy's course for the balance of the war. The best of the young RCN officer corps would be poured into the new ships, leaving the reservists – and the old – to run the war against the U-boats.

The impact of this new concentration in effort would not be felt until 1944. In the meantime, while the sound of knives being sharpened echoed through the corridors of NSHQ, the A/S fleet finally joined the shooting war again. The opportunity to do so originated in July, when the Admiralty proposed that EG 5 be transferred to Western Approaches Command[69] and that another Canadian support group be established by reducing the size of each MOEF group. The Admiralty proposed that this second RCN support group also be allocated to Western Approaches Command for 'special duties.' On 14 August NSHQ signalled its plan to assign its two long-range Town class destroyers, *St Francis* and *St Croix*, and the corvettes *Chambly*, *Morden*, and *Sackville* to the new EG 9.[70] The choice of ships could be described as the best of the second line. *St Francis* was never a consistent long-range performer: she might be expected to do better if not tied too closely to a convoy. *St Croix* was among the most successful of her class: the only Town – British or Canadian – to claim two U-boats. None of the corvettes was modernized, but all were recently refitted. *Chambly* and *Sackville* had particularly noteworthy mid-Atlantic careers, and there were important continuities in the crews of all the corvettes.

It may be that the new group was selected more for the abilities of the personnel than for the hardware on the ships. If so, that reasoning escaped Captain Frank Houghton, the senior Canadian naval officer (London). Houghton was probably instrumental in the formation of EG 9 and certainly was well placed to lobby for greater Canadian participation in the offensive. That might explain his bitter complaint to Murray's chief of staff about the make-up of the group. 'When the Admiralty asks us for some ships, over and above the 6 per [MOEF] group to form a "capital hunting group," ' Houghton wrote Bidwell on 18 August, 'I think we all felt a bit – well embarrassed to say the least – when they were offered two four stackers and three corvettes without hedgehogs – two of which are still short fxles [forecastles], all of which still [have] 123A [asdics].'[71] Presumably Houghton was looking for a little boldness and imagination – a few of the RCN's River class destroyers, perhaps. Such a change would not happen under Nelles, and the staff in Halifax displayed much of the same conservatism throughout the war. EG 9 assembled in the United Kingdom by the end of August.[72] Its composition was completed with the transfer of the British frigate *Itchen* from C 1, along with Cdr C.E. Bridgeman, RNR, former senior officer of C 1, to command the new group.

Meanwhile, EG 5, with the few available modernized corvettes, arrived in the United Kingdom. From 16 to 18 August it trained under HMS *Philante*, Western Approaches Command's sea training ship. *Philante*'s report on EG 5 was rather typical of all assessments of groups that transferred to Britain from Canada during the war. EG 5 was found to be less than skilled at its principal task, a hardly surprising evaluation, since the RCN still lacked an operational training establishment and its own submarines to train on. EG 5's main weakness was in the use of its asdic. 'All ships require much more individual A/S practice,' the report observed, and 'There is much work to be done by the Group A/S Officer in instructing the A/S teams in the individual ships.'[73] There is, in fact, no evidence that EG 5 had ever trained as a group on a tame submarine at any time prior to its transfer to the United Kingdom.

The lack of submarines for ASW training in the western Atlantic, and in the Canadian zone in particular, was noted at the Washington Conference in March 1943 and was singled out for attention. But the provision of training submarines for Canadian-based escort remained a problem that the British and Americans had to solve on Canada's behalf, a process that took until 1944. Even then, the provision of target submarines for operational ships based in Canada remained an intractable problem to the end of the war.

Nothing in *Philante*'s report on EG 5 marked the group as exceptional. The ships fresh from modernization, *Edmundston* and *Snowberry* (*Calgary* missed the group exercises), were unfamiliar with their new equipment. But these and other problems could be remedied with time and attention.

EG 5 sailed for its first operation in the eastern Atlantic on 22 August. By then the Biscay Offensive was nearly over, but this time it was not the Germans who quit. The battle in the Biscay had escalated throughout the summer of 1943. Air patrols needed surface A/S forces to make them more effective. Anti-submarine ships needed support by heavier naval forces because of the threat from German destroyers in the Biscay ports. The Germans responded, in turn, with increasingly powerful air support for their U-boats and by long-range bombers to attack the warships. By August the work of Allied A/S forces was routinely supported by a light cruiser or several fleet class destroyers and by fighter patrols. The first RCN Tribal class destroyers to commission, *Athabaskan* and *Iroquois*, participated in these operations during July and August. These long-awaited fleet class destroyers represented the cutting edge of the professional RCN's ambitions for a major postwar navy. Designed primarily for surface gunnery battles, these Canadian destroyers had no hand in sinking U-boats. The A/S work grew more dangerous in August as the U-boats moved close to the Spanish shoreline, slipping along inside territorial waters and hiding themselves from radar along the rugged coastal mountains. By the time EG 5 arrived on patrol off Cape Villano, Spain, considerable forces were engaged on both sides, and the action had shifted to an area that favoured German aircraft.

The group rendezvoused off Cape Villano with the sloops of EG 40 on 22 August, supported by the destroyers HMS *Grenville* and *Athabaskan*. There was little U-boat activity, and the group contented itself with searching Spanish shipping and keeping an eye on patrolling German aircraft. *Nene* joined on the 24th, the same day that a prowling Luftwaffe FW 200 Condor pushed in close enough for the group to engage it with gunfire. The constant air surveillance heightened tensions later that day when the arrival of three Condors was followed shortly by the appearance of three very large ships on the horizon. The Canadians were aware that the powerful Narvik class destroyers based on the French coast would make very short work of EG 5. Fortunately all these ships, a cruiser and two destroyers, were Spanish: a subtle warning, perhaps, about the harassment of their coastal shipping.

German air reconnaissance was indeed a portent. Three Condors appeared the next morning, as usual, but shortly after noon an American Liberator 'going hell for leather towards Gibraltar' passed EG 5 a curt warning, 'Twenty-one enemy planes heading this way!' and then promptly dove into cloud cover. Twenty minutes later the men of EG 5 counted fourteen Dornier Do 217s and seven Ju 88s on the eastern horizon, and the group braced for a bombing attack. As the ships waited, the aircraft broke formation, divided into three groups, and then circled EGs 5 and 40 at a considerable distance. Eventually several small black objects trailing jets of white smoke fell from the wings of the Dorniers. They travelled parallel to the warships, then turned sharply and followed each move of the target ship. It quickly became evident that this was something new: a remotely controlled missile called erroneously a 'glider bomb.' Two ships of EG 40 suffered near misses, but as the captain of *Snowberry* wrote later, 'this being their first attempt their homing technique was poor.' The ships of EG 5 took individual evasive action and went to maximum speed. Under these conditions chief engineers found reserves of power they never had suspected. When *Snowberry* was queried by HMS *Tweed* about her speed, the response was '15 knots,' which drew a sharp, 'Don't give us that, we're doing 18 knots and can't shake you!' *Snowberry*'s chief engine room articifer later admitted that he was told there were *fifty* German aircraft overhead and with that incentive he found considerably more reserve power in the corvette's machinery.[74]

The German 'glider bombs' were quickly dubbed 'Chase me Charlies.' They were actually small radio-controlled, jet-propelled aircraft, with a speed of 3–400 knots and 1,100-lb warheads.[75] Their accuracy improved dramatically the next day, when the Do 217s reappeared over Cape Villano. EG 5 was still there, but EG 40 had been replaced by EG 1; *Athabaskan* and *Grenville* remained in support. One of the bombs pursued *Egret*, the senior officer's ship of EG 1. As she turned to avoid, the weapon passed astern of the sloop, swung around, and struck her amidships. *Egret* disappeared in a huge explosion which claimed all but twenty-five of her 225 crewmen; the senior officer was recovered by *Grenville*.

During the pause that followed, the bow of *Egret*, pointing skyward out of the water, was seen from *Athabaskan*, where one of her crew mistook it for a sailboat. Another case of mis-identification occurred at the same time in the German aircraft, where it was decided that *Athabaskan* was a cruiser. Fuses for the next attack were switched to a coarser setting to allow for heavier plating, perhaps some armour, and the greater breadth of a larger ship. Five aircraft then turned their attention to the Canadian destroyer, and three launched a simultaneous attack on the destroyer's port side. *Athabaskan* could not avoid all of them. At 1315Z a glider bomb struck the ship directly under the bridge. Its coarse setting allowed the bomb to pass through the plotting room and the chief petty officers mess, and out the starboard side before exploding. The blast crushed and scorched *Athabaskan*'s hull and superstructure and riddled the ship with splinters. Cdr G.R. Miles, the destroyer's captain, and other personnel on the bridge were knocked down, one crewman from B gun was killed instantly, another was blown overboard, and several men were severely wounded – three of them fatally. Men at other guns also were wounded, but all guns kept firing, and the destroyer sailed on despite heavy flooding forward that brought her down by the bows, loss of fire control, fires amidship, and a fifteen-degree list.[76]

Athabaskan steamed back to Britain under her own power; it took two months to repair the damage. The 'glider bomb' attack of 27 August also effectively ended the Bay of Biscay Offensive. Since U-boat kills were diminishing in number and given the pressure of German air attacks, there was little point in pushing surface ships into dangerous waters. As the British official history observed, 'After the first few days the month of August thus produced few successes in our sea and air patrols, and considerable losses were suffered by the latter.'[77] It is hard to disagree with German historians, who see the abandonment of the Biscay Offensive as a victory for Germany. It was unfortunate – if rather exciting – for the RCN to arrive on the scene just as the battle was tipping in the enemy's favour. The fortunes of war had conspired to keep the RCN out of the main action from January to mid-August, and now the introduction of a radical new weapon checked the RCN's first serious attempt to carry the war to the enemy.

EG 5 completed one more Biscay patrol before that offensive was abandoned. The battle soon shifted further to seaward in any event, as the Germans moved to re-establish themselves along the main convoy routes. Their hopes for a mid-Atlantic comeback rested on another dramatic new weapon. It proved cruelly ironic that the first use of this weapon virtually destroyed the RCN's modest efforts to participate in the 1943 offensive.

2
Cinderella: Act II

> D.A.U.D. feels that as far as the Royal Canadian Navy is concerned, a bad impression would be created with R.C.N. authorities ... if a comparative statement concerning RN successes [at sinking U-boats] were now to be promulgated.
>
> Director Anti-U-Boat Division Minute to the
> First Sea Lord, 20 February 1944[1]

Through the summer of 1943 the battle against the U-boats swirled and eddied around the eastern and central Atlantic as Dönitz sought to maintain a semblance of a threat. At the same time he planned a return to the main trade routes of the north Atlantic. The Allies also knew nothing decisive could be gained from the disparate U-boat attacks in the far reaches of the Atlantic. If the U-boats were to accomplish anything, they had to return to the main trade routes in force. 'Knowing that this is so,' Roger Winn, director of the Admiralty's Operational Intelligence Centre, wrote in his appreciation at the end of July, 'the enemy in withdrawing from the North Atlantic must have intended an ultimate return to this area.' That return, Winn noted, would depend on the development of new weapons, and could be expected in September or October. While it might occasion 'heavy losses of merchant shipping and escort forces on the North Atlantic convoy routes,' Winn warned, 'no fear need be felt as to the ultimate outcome.'[2]

Sailors and airmen knew the U-boats were not yet beaten, and the mood of expectancy grew as the summer wore on. In early September the Admiralty's assistant chief of Naval Staff (U-Boats and Trade), Rear Admiral J.H. Edelsten, warned Admiral Sir Andrew Cunningham, the incumbent First Sea Lord (Pound was dying from a tumour, which claimed him on Trafalgar Day), that they had not seen the last of the U-boat fleet. 'The Politicians are firmly convinced,' Edelsten wrote, 'a) that the U/B war is over, b) that they won it. I endeavour to enlighten them by pointing out that the Germans could have 300 operational boats in the Atlantic tomorrow if they choose and that one day, when they are re-armed with A/A and better radar they will choose.'[3] By mid-September the Germans were ready to renew the battle in the mid-ocean.

It was precisely because of this latent danger that the strength of the RCN remained with the Mid-Ocean Escort Force during the summer of 1943. MOEF's nine groups, five of which were RCN, defended the only target of decisive strategic importance in the Atlantic. Any new German offensive in the mid-Atlantic would put the best of the RCN, including two Canadian support groups, in contact with the enemy. That the result would be another Allied victory there seemed little doubt. But even the most pessimistic would not have predicted that the Canadians would be shut out again. This time it cost the chief of Naval Staff his job.

With virtually all of the precious replenishment submarines sunk and increasingly oppressive Allied air patrols, the operational effectiveness and range of the U-boat fleet was sharply reduced. The technology that had carried the burden of two major German submarine campaigns in two great wars had reached its limit. Radically new weapons were needed to carry on the war in the Atlantic. Two mutually supportive avenues of development were open in the wake of the 1943 defeat: retro-fitting the existing fleet with enough improvements to maintain a viable threat, and producing totally new submarines with little dependence on surface manoeuvrability. Germany pursued both, but the best solution was a totally new submarine.

The limitations of existing U-boats were well known, and by 1942 considerable efforts had been made by Professor Helmuth Walter to develop an 'air independent' submarine using hydrogen-peroxide-driven engines. Walter's project suffered from serious technological problems and offered no quick solutions. In November 1942, however, a group of U-boat constructors recommended that the efficiency of existing diesel-electric technology could be radically improved by streamlining hulls and increasing battery capacity. By June 1943 the preliminary designs for this new 'Electro boote' were complete. In order to carry the additional batteries the submarine was large, at 1,600 tons more than twice the size of the type VII, and the hull was built entirely for submerged travel. The designed maximum submerged speed of the new submarine was eighteen knots for one and a half hours, twelve to fourteen knots for up to ten hours, or five knots for sixty hours. It was also designed to dive deeper, fire its torpedoes faster, and carry new, highly sophisticated acoustic sensors, accurate enough to provide target data to the U-boat's twenty programmable torpedoes.[4]

Here indeed was a radical solution to the U-boat fleet's problem. Hitler approved the new type XXI submarine and a smaller version for inshore work, the type XXIII, on 8 July 1943. In mid-August all construction on types VII and IX U-boats not already under way was cancelled, and priority shifted to the new types. Dwindling production of older types would cover losses until the first type XXI submarines entered service in the summer of 1944.[5]

Meanwhile the Germans sought ways of keeping the old submarines in the fight. Until the end of July, when the idea of transiting the Bay of Biscay in groups providing mutual anti-aircraft fire was abandoned, it looked as though simply providing heavier A/A weaponry was a workable solution. The ability of Allied air patrols to swamp groups of U-boats on the surface in the bay ended that experiment. But conditions in the mid-ocean were likely to be more favourable. Long-range and very long-range maritime patrol aircraft were slow and vulnerable and unlikely to be present in enough numbers to overwhelm easily a well-armed U-boat.

The introduction of a new radar warning device, *Wanze*, also promised to improve reaction times in the presence of aircraft. *Wanze*, like its predecessor *Metox*, was designed to detect radar signals in the metre range, but it did so automatically, covered a wider wave band, and could detect intermittent radiation, which *Metox* could not.[6] The new radar detector was credited – erroneously – with making transits of the Bay of Biscay safer after August.[7] Thus it appeared that the problem of radar warning was solved, for the moment at least. Remarkably, it was not until January 1944 that the Germans confirmed the Allied use of centimetric radar, their real enemy in 1943.

The centrepiece of Dönitz's plan was a new acoustic homing torpedo, the 'T-5,' known to the Germans as *Zaunkonig* and to the Allies as GNAT, the acronym for German Naval Acoustic Torpedo. The idea of an acoustic homing torpedo was not uniquely German. The Allies were working on versions of their own, and an air-launched homing torpedo entered service in 1943. The Allies had also been aware of German developments for some time. In April 1943 the Admiralty warned that a German acoustic homing torpedo, known as 'Taffy,' had entered limited service.[8] The Taffy had a contact pistol; since it still had to hit something solid to detonate, it was difficult to distinguish from conventional torpedoes. Both the Taffy and the GNAT were designed to home on the propeller noises of a ship travelling at the rate of ten to eighteen knots. The self-steering acoustic homing mechanism required a comparatively slow speed, twenty-four knots as opposed to the thirty-to-forty-knot speeds of other torpedoes. However, this rate suited it well for use against most war-built escorts, which were slow (in the sixteen-to-twenty-knot range) and typically responded to the threat of torpedo attack with evasive action at high speed – just what the GNAT required for accurate homing. The homing device could not detect vessels moving at less than eight knots, while those travelling away at very high speeds might escape the torpedo entirely or find it detonating harmlessly in their wake. The great advantage of the GNAT over the Taffy was its magnetic pistol. The addition of a proximity magnetic detonator overcame the U-boat's enduring problem of hitting a fast, shallow-draught, and highly manoeuvrable warship.[9] The GNAT's slow speed and proximity fuse meant that it was best fired at escorts approaching the U-boat at fairly high speed: the torpedo would be drawn directly towards the target and would detonate under the keel.

The T-5 was not scheduled for release to the U-boat fleet until

1944. Dönitz forced the pace, and the first GNATs were ready for operational use in early September 1943.[10] By then his plans were set. Using *Wanze* to warn of air attack, U-boats were to operate in groups of two to provide mutually supporting anti-aircraft fire, and, travelling submerged whenever possible, a large Wolf Pack was to be deployed in the eastern Atlantic. Once enough U-boats were ready, they were to locate a westbound transatlantic convoy and blast their way through the naval escort with the new torpedoes. Dönitz was anxious that the initial attack prove a resounding success, and he counted on a combination of surprise and shock to reestablish the U-boats' dominance over the main trade routes. In early September twenty-nine U-boats sailed to renew the battle. What followed was the last major campaign of the war in the open ocean.

The new deployments along the main convoy routes initially provided much-needed intelligence on Allied movements, since the Germans were having trouble reconstructing the convoy cycle. In June 1943 the Allies had changed their convoy cypher, the source of much German intelligence on shipping movements, and the new cypher remained secure for the balance of the war.[11] It was not until mid-September then that Dönitz was ready to launch his new offensive. On 15 September he ordered twenty U-boats of group 'Leuthen' to form a patrol line 500 miles long across the anticipated paths of convoys ONS 18 and ON 202. Allied intelligence lagged behind German movements, and it was not until 18 September that Leuthen's patrol line was finally plotted – but further westward than was really the case. The initial threat appeared to be to HX 256, escorted by C 5. EG 9, the RCN's second support group, which had sailed from Plymouth on the 15th on its first operational cruise to the Biscay, was diverted westward to assist. Meanwhile, ONS 18 and ON 202 were given course alterations which, it was hoped, would carry them away from danger, and supporting air patrols were increased. Decrypts of German signals then revealed that Leuthen was to attack only westbound convoys. HX 256 was safe; the battle would be joined around ONS 18 and ON 202.[12]

When the head of the Admiralty's Operational Intelligence Centre had anticipated a favourable outcome to any future confrontation in the mid-Atlantic, his optimism was based on a clear understanding of how the Allied tactical situation had improved over the previous year. Airpower was now omnipresent in the Atlantic, and its influence would be felt in the first battle of the new campaign. Not only were land-based aircraft now flying routinely in the mid-Atlantic on extended air patrols, but auxiliary aircraft carriers were readily available for support. The American carriers *Bogue*, *Santee*, and *Card* attacked the growing U-boat concentration in the eastern Atlantic in early September, albeit with little effect. In addition, a number of bulk carriers (oil tankers and grain ships) had been converted to 'merchant aircraft carriers' (MAC ships). Fitted with a tiny flight deck and a handful of aged Swordfish biplane aircraft equipped with radar, rockets, and depth charges, MAC ships provided integral air support to transatlantic convoys. The MAC ship *Empire MacAlpine*, with eight Swordfish embarked, was part of ONS 18.[13] The air gap, which earlier German campaigns had exploited so successfully, no longer existed.

Dönitz was none the less prepared to fight, and it was a fight the Allies welcomed. With convoy routing now tied closely to predictable tracks along the Great Circle Route, confrontations were easily arranged. To limit steaming for support groups, and to allow for easy movement of forces from one convoy to another, it was now practice to time departures of fast and slow convoys so that they reached the danger area at roughly the same time. So it was with ONS 18 and ON 202, which were on close, parallel tracks with the faster convoy ON 202, closing on the slow convoy as they headed westward into the mid-Atlantic. Aircraft flew in support even before they came within striking distance of the U-boats. During one of these patrols an RCAF Liberator opened the battle by sinking *U341* 160 miles ahead of ONS 18 on 19 September.

When the first U-boat sighted ONS 18 just before midnight on the same day, the two convoys were about forty miles apart. The leading convoy, ONS 18, was escorted by B 3: the destroyers *Keppel*, with Cdr M.B. Evans, RN, the senior officer, aboard, and *Escapade*; the frigate *Towy* (which was delayed in joining); the RN corvettes *Orchis* and *Narcissus*; and the Free French corvette division, *Lobelia*, *Renoncule*, and *Roselys*. The fast convoy ON 202 coming up astern was escorted by C 2; Commander P.W. Burnett, RN, the group's senior officer, was in the RCN destroyer *Gatineau*, accompanied by the RN destroyer *Icarus*, the RN frigate *Lagan*, the RN corvette *Polyanthus* and the RCN corvettes *Drumheller* and *Kamloops*. The strong British component was typical of C 2, but Burnett was no typical RN officer. Command of C 2 was his first stint at sea after several years at the RN's anti-submarine school HMS *Osprey*. Few escort groups enjoyed such qualified and capable leadership.

About the time Leuthen made contact with ONS 18, EG 9 (*St Croix*, *Itchen* (RN), *Chambly*, *Morden*, and *Sackville*) was converging on the convoy from the southwest. Direction fixes during the day indicated U-boat activity on the port quarter and astern (south and southeast) of the convoy, so Commander C.E. Bridgeman, RNR, EG 9's senior officer in the frigate *Itchen*, apparently planned to sweep around the threatened quarter and astern of ONS 18. Acting Lt Cdr A.F. Pickard, RCNR, the captain of *Chambly*, later recalled the approach to battle in rather melodramatic style: 'That evening of the 19th September, as the sun set beneath a saffron hued horizon, we had in the fading light our last sight for all time of the five ships together.' With *Chambly* leading and the rest strung out behind her at 7,000-yard intervals, EG 9 sailed on to what 'a fickle and inscrutable fate' held in store.[14]

As Pickard and his colleagues closed the convoy, they monitored the developing battle. In fact, although B 3 and EG 9 had firm contacts with shadowers around the leading convoy, the battle on the first night concentrated largely on ON 202 coming up behind. The contact report on ON 202, by *U270*, was detected by C 2's HF/DF operators, and *Lagan* went to investigate. Her radar contact twenty minutes later brought *Gatineau* out in support. As the destroyer approached *Lagan*, the frigate was struck by an acoustic torpedo from *U270*, which severed thirty feet of the stern: GNAT's first victim. *Gatineau* damaged *U270* in a counter-attack and then screened *Lagan* while the tug *Destiny* took her in tow. (The shattered frigate made port in Britain but was declared a 'constructive total loss'). Meanwhile *U238* slipped into the convoy between *Polyanthus* and *Kamloops* and torpedoed two ships from the portside of ON 202. One ship sank promptly, but the *Frederick Douglas* – abandoned by her crew – drifted astern of the convoy. *Gatineau* and *Polyanthus* searched the scene of the attack, and the corvette made several unsuccessful attacks on *U238*.

The first round of the battle for ONS 18 / ON 202 therefore went to Leuthen: two merchant ships picked off from ON 202 and one crippled escort with little damage to themselves. But the battle had only just begun. The morning of 20 September dawned bright and clear, and with daylight came the first RAF Liberators from Iceland, which supported both convoys all day. The aircraft also helped bring the two convoys together, since it was decided to combine them and thus obtain more efficient use of the available escorts. Burnett and Evans also agreed to draw EG 9's corvettes into a close escort role, and to use the destroyers and frigates as striking forces. In this instance, at least, the relegation of most of the RCN ships present to close escort duty did not preclude contact with the enemy: there were enough U-boats to go around.

Bringing the convoys together was easier said than done. Their positions relative to each other were not established until the afternoon. With help from an RAF Liberator Evans found ON 202 to the southeast, and by late afternoon Burnett in *Gatineau* had sighted ONS 18. By dusk the convoys still lay some distance apart, and no effective junction was made until later the next day. Meanwhile, according to Evans's now famous comment, 'the two convoys gyrated majestically about the ocean, never appearing to get much closer, and watched appreciatively by a growing swarm of U-boats.'[15]

As the two convoys pushed westward, the battle around them intensified. RAF Liberators repeatedly attacked U-boats circling the convoys during the 20th. On occasion, when their depth charges were expended, airmen were reduced to strafing U-boats with machine-gun fire and calling up escort vessels to prosecute the contacts. The naval escort, meanwhile, found no shortage of U-boats to pursue. A contact made by *Keppel* on *U386* in the late afternoon was passed along to *Itchen*, *St Croix*, and *Narcissus* to prosecute. *Drumeller*, attempting to rejoin the convoys from astern, drove *U338* away from the wreck of the *Frederick Douglas* and became involved in *Icarus*'s pursuit of *U731*. *U305* began to shadow the res-

Cinderella: Act II   67

cue ship *Rathlin* and the corvette *Polyanthus* as they attempted to rejoin. *Drumheller* was joined in her chase of *U731* by an RAF Liberator which had spotted the fall of the corvette's 4-inch shells; it exhorted the escort to 'give the ——— one for us.' Meanwhile another Liberator attacked *U338*, forced it to submerge, and then sank it – ironically – with an acoustic homing torpedo. The pace of the action on 20 September was frantic, with U-boats popping up everywhere and the escorts hard pressed to handle them. The urgency of putting down one after another left little time for prolonged hunts, but there seemed to be U-boats for the taking. In the excitement to get at *U731* the British destroyer *Icarus* managed to ram *Drumheller*, causing slight damage. The corvette's captain, A.H.G. Storrs, RCNR, got his own back at *Icarus* by signalling, 'Having no Submarines?'[16]

Late in the afternoon of 20 September *U305*, the U-boat shadowing *Polyanthus* and the rescue ship *Rathlin*, was detected by an aircraft, and *St Croix* set off in pursuit. Unfortunately, *U305* saw the destroyer first and her approach was a perfect set-up for a GNAT: bows-on, steaming at about twenty knots. At approximately 1935Z a single acoustic torpedo, which ran nearly the length of the destroyer before its magnetic pistol activated, blew off *St Croix*'s stern, leaving her dead in the water. As Cdr C.A. Rutherford, the captain, surveyed the damage from his vantage point on the bridge, the ship's motor boat and whaler were lowered and many of the injured taken off. *St Croix*'s remaining crew then made a brief and futile attempt to get her under way. Smoke was pouring from two of *St Croix*'s four funnels as *Itchen* appeared over the horizon. *U305*, meanwhile, had moved in for a killing shot. Rutherford had time to signal *Itchen* that he was 'leaving the office' before *U305*'s second acoustic torpedo struck the destroyer. Onlookers reported a huge pillar of flame as the explosion cut the remainder of *St Croix* in half. *Itchen* signalled '*St Croix* blown up!;' she was the first ship destroyed by a GNAT. *Itchen* closed the scene to find the destroyer's bows still afloat and the sea littered with survivors. But Bridgeman was given no time to linger before another GNAT from *U305* detonated in *Itchen*'s wake. Not surprisingly, Bridgeman drew off to await support from *Polyanthus* before attempting to rescue what remained of *St Croix*'s crew, most of whom appear to have abandoned ship.

It was thirteen hours before *Itchen* returned for *St Croix*'s survivors, and she would get no help from *Polyanthus*. The corvette was on her way to assist *St Croix* and *Itchen* when she reported a radar contact. As she closed to investigate, *U952* fired a single acoustic torpedo which shattered the little escort and sent her plunging to the bottom. *Polyanthus*'s lone survivor was found some time later and taken aboard *Itchen*. The battle continued throughout the night. *Itchen* and *Narcissus* chased *U270* and *U641* at different stages, the U-boats firing acoustic torpedoes without effect and the escorts having similar luck with their counter attacks. *Sackville*, now screening the convoy, foiled the only serious threat of the night, when she picked up a U-boat transmission (probably an MF homing beacon) ahead of the convoy and put *U378* down. Again the U-boat responded by firing a GNAT, but the torpedo missed entirely, and the crew of *Sackville* remained blithely unaware of their brush with extinction.

By dawn on 21 September the battle became shrouded in dense fog, which slowed the pace of the action for a day. Despite the conditions, Swordfish aircraft flew from the *Empire MacAlpine* and RAF Liberators probed the mist around the convoys. When a brief break in the fog occurred early in the afternoon, Evans, who was now overall commander of the escort, was surprised to discover that ONS 18 was quite nicely in position abeam of ON 202. Those less well informed were impressed with the skill of Evans's handling of the two convoys. Evans himself thought the junction had been affected – without his consent – by the commodore of ON 202, who, upon being asked, professed his innocence and his belief that ONS 18 had actually been astern of him. The commodore could only conclude that 'this most desirable change in formation had been ordered by a higher authority.'[17]

Fortunately the deteriorating visibility did not prevent *Itchen* from returning to rescue the survivors of *St Croix*. They had spent a miserable night. Men died from wounds and exposure. Rutherford worked diligently to save his crew. Able-bodied men were shifted from the motor boat to the whaler, which was packed with injured and was leaking badly. Among those transferred was Stoker William Fisher, who sat for hours with his feet pressed against the plugs needed to keep the whaler afloat. Those in the motor-boat were less fortunate. It filled with water during the night, and in panic several men jumped overboard and drowned; by morning only the first lieutenant was still aboard. Shortly after dawn, *Itchen* arrived, with *Narcissus* in support. In the excitement of imminent rescue eight men, numb with cold, slipped off one of the carley floats and drowned. When the final count was made aboard *Itchen*, seventy-six ratings and five officers from *St Croix*'s complement of 147 had been saved. They now joined the lone survivor of *Polyanthus* aboard the frigate as she headed back into the battle.

Poor visibility persisted throughout 21 September and on into the late afternoon of the 22nd. During those two fog-shrouded days the U-boats of group Leuthen probed around the two convoys, attempting to keep contact. Under such conditions the advantage lay with the radar-equipped escorts. Despite aggressive sweeps and numerous contacts, however, only *Keppel* managed a kill. Acting on an HF/DF fix, she swept astern of the convoys in the early hours of the 22nd, surprised *U229* on the surface, then rammed and sank it.

When the fog lifted in the afternoon the two convoys were still abeam of one another, roughly four miles apart, with a frontage of eighteen columns and spread over some thirty square miles of ocean. That daunting sight was counterbalanced by the presence of Canadian aircraft around the convoys the very moment visibility improved. As Evans wrote in his report, 'it was very nice to come into the open air and find it filled with Liberators.'[18] They had been there for some time, in fact, and had risked much to operate from fog-shrouded runways, but their efforts were not in vain. During the balance of the day RCAF Liberators picked up where those of the RAF had left off, attacking contacts and suffering from heavy anti-aircraft fire in the process. Swordfish from *Empire MacAlpine* joined the air patrols, but the slow biplanes were no match for the U-boats' heavy flak and their crews had to content themselves with directing naval vessels on to contacts.

Despite the best efforts of the aircraft, ten U-boats remained undetected and were converging on the convoys at dusk on 22 September. In terms of losses the worst was now to befall ONS 18 / ON 202. Just before midnight four U-boats attacked, *U952* from the port quarter and *U731*, *U260*, and *U666* from directly ahead. A confused action followed in which the explosion of acoustic torpedoes in wakes was punctuated by detonating depth charges. As this action unfolded, *Itchen* and *Morden* pushed ahead of the convoys to investigate a radar contact, finding *U666* on the surface. A chase began as the U-boat, illuminated by the frigate's searchlights, ran across *Itchen*'s bows not 300 yards away. In the glare of the searchlights *Itchen*'s forward guns opened fire, just as *U666* fired her own salvo of two acoustic torpedoes. The one directed at *Morden* exploded in her wake as the corvette turned away. *Itchen* was not so lucky. The frigate was enveloped in a huge explosion which threw her over on her side and sank her in less than a minute. *Itchen* disappeared so quickly no one was sure what had happened, and it was some time before she was found to be missing. In the meantime, men and debris littered the cold sea. As the survivors collected themselves, the convoy surged through the scene, choking and smothering them in oil-laden wakes and pulling men and wreckage alike down into the wash of passing hulls. No one knows how many survived *Itchen*'s sinking only to be run down by the convoy. Three hours later, when a motor boat from the Polish ship *Waleha* steered its way through the wreckage, three were rescued: Stoker William Fisher from *St Croix* and two men from *Itchen* – all that remained of nearly 400 men from three ships' companies.[19]

In the confusion following *Itchen*'s loss *U238* very coolly penetrated the screen and sank three ships from ON 202; another ship was lost from ONS 18 later that night. At dawn RCAF Liberators arrived again and began another day of air assault on Leuthen. With the battle running out of sea room and a victory in hand, on the morning of 23 September Dönitz called his U-boats back; the battle was over. Leuthen claimed twelve escorts and seven merchant ships sunk, damage to three escorts and three merchant ships. Against this tally two U-boats (later found to be three) were believed lost. It looked to Dönitz like the victory he sought.

The reality was somewhat different. At best the battle for ONS 18 / ON 202 was a qualified tactical victory for the U-boat fleet. Evans, the senior officer for the combined escort, quite rightly noted that only three attacks reached the convoys, and in the end only six merchant ships and three escorts were lost – considerably fewer than the Germans claimed. The situation easily could have been worse, however, since the convoys were saved from heavier attack by some fortuitous alterations of course and two days of heavy fog. Perhaps more significantly, only one U-boat was sunk by the fourteen surface escorts around ONS 18 / ON 202 during nearly four days of continuous contact. As Rohwer and Douglas observed, 'Radar constantly failed to detect U-Boats well within radar range.'[20] This failure, in contrast to the fog-shrouded battle for ONS 5 in the spring of 1943, when B 7 and EG 1 destroyed four U-boats in a single night, may well have been due to rough seas' obscuring radar signals on low-lying targets. It also owed something to size of the battle area, the frequency of sightings, which constantly pulled escorts to some new crisis, and the obvious success of the U-boats in attacking the escorts.[21]

The novelty of the battle for ONS 18 / ON 202 and the heavy loss of life among the escorting forces mitigated criticism of the escort's conduct. However, it is fair to say that it was not up to the standard recently in evidence in the Atlantic. Over thirty confirmed contacts were made on U-boats by the naval escort, but only *Keppel* scored a kill. Worse still, from the RCN's perspective, its ships and groups had ample opportunity to sink submarines but could not take advantage of it. In fact, the vast majority of U-boat contacts during the battle were made – fittingly enough – by ships of EG 9 (at least twenty).[22] The use of the GNAT and the early loss of *Lagan* and *St Croix* unquestionably affected the success of warships in hunting U-boats. Escorts had always been under the threat of counter-attack, but never had the Germans scored so effectively. *Itchen*'s reluctance to linger around the wreck of *St Croix* without support and *Morden*'s abandonment of the scene of *Itchen*'s loss is graphic evidence of anxiety about the new tactical situation. The fact remains, none the less, that seven of the RCN's best escorts spent four days in close contact with twenty U-boats and failed to record a kill.

Fortunately, scientists on both sides of the Atlantic were at work on the problem of acoustic homing torpedoes. By coincidence, on the day *St Croix* was sunk, the RCN's director of Operational Research, J.H.L. Johnstone, reported to Operations Division in Ottawa on possible countermeasures to the Taffy. As Johnstone noted, the Admiralty recently had proposed using the 'pipe noise-maker' (PNM), developed by the RCN in Halifax in 1940 to sweep acoustic mines, as a decoy astern of warships. Like most great ideas, the concept of a pipe noise-maker was a simple one. A Halifax-based scientist, Dr Anesley, was pondering the problem one warm day when the wind rattled his venetian blind.[23] The PNM was a variation of that concept. Two pipes were loosely fitted, one above the other, in a metal frame. A yoke was attached to pull the pipes and their frame sideways through the water. The passage of water around the loosely fitting pipes caused them to rattle against one another. The pitch of the noise could be varied by changing the size of the PNM or the material used for the pipes. The British had adopted the Canadian PNM in 1940 for sweeping acoustic mines and dubbed it 'FOXER.' In response to the threat from Taffy, the Admiralty suggested towing two sets of this FOXER gear 250 yards astern, held 100 yards apart by paravanes (small torpedo-shaped floats). Johnstone thought that this idea was basically sound, but Canadians had their own ideas about how PNMs should be used to counter the torpedo problem.[24]

The suspicion that the Germans were employing acoustic torpedoes in the battle for ONS 18 / ON 202 prompted action. On the evening of 21 September, the day after *St Croix* was sunk, the RCN began experimental work in Halifax on the use of pipe noise makers as decoys for the new weapon. On the 22nd sea trials were conducted with PNMs of various lengths to determine if they could be towed at greater speeds than those used in minesweeping. An optimum size of five-feet-long, 1⅜-inch pipes was settled on. These pipes could be towed 250 yards astern at speeds between 8.5 and 17.5 knots. The trials convinced Canadian scientists that a single PNM produced sufficient noise to drown the sounds of the ship's propellers. The equipment, dubbed the 'Canadian Anti-Acoustic Torpedo' (CAT) gear, was tested again the next day, when it was

Cinderella: Act II 73

learned that the 1$^3/_8$-inch pipe was in short supply. A 1¼-inch pipe was substituted without loss of effectiveness and production began immediately in the dockyard.[25]

While the battle around the convoys still raged (in fact, even before *Itchen* was sunk) the RCN staff in Halifax signalled their estimation of the problem and the fix to interested parties. 'The P.N.M. acoustic sweep may provide a counter measure for the new U-Boat weapon,' Murray advised Flag Officer, Newfoundland, on 22 September, 'which is thought to be an acoustic torpedo.' Orders to construct the sweeps locally and instructions on how to employ them were included in Murray's signal.[26] Once the trials were completed in Halifax, production of CAT went apace. Fifty sets were produced per day, the first being sent to St John's on the 24th – the day after the battle for ONS 18 / ON 202 ended. Thirty more were sent to Newfoundland the next day, where local staff produced twenty of their own. Production stopped when 400 sets of CAT Mk I were available.[27]

The counter to the weapon that sank *St Croix* and two other escorts and damaged a third was therefore waiting for the remnants of EG 9 and C 2 when they arrived at St John's following the battle on 25 September.[28] The speed and effectiveness of the Canadian response were remarkable and contrast favourably with the response of the RN. Richard Phillimore, a staff officer at Western Approaches Command, recalls the first meeting at Liverpool to discuss the GNAT. Horton prodded his staff for answers and was met with a stony silence. According to Phillimore, no one at WAC had anticipated the GNAT; the staff had not worked on countermeasures and they only grudgingly accepted Horton's view that a towed noise-maker was the solution.[29] In fact, the Admiralty was already working on a solution but preferred a twin FOXER because of the initial fear that the new German torpedo had a rudder mechanism capable of continuous and graduated changes in course in response to varying intensity and direction of sound. If that was the case, then the Taffy (as they still suspected it was), instead of being drawn directly to the single loudest sound, might alter course to strike the ship if it passed close enough. In that case, it was necessary to protect both sides of the ship with decoys and two FOXERs were needed. Canadian scientists assumed a much less sophisticated control mechanism with only three rudder positions, centre, left, and right, and that the torpedo would be drawn straight to the loudest sound source without the likelihood of subtle changes at the last moment. The British preference for twin FOXERs presented them with more difficulties of launching, operation, and recovery than the single, light gear proposed from the outset by the RCN. The maximum towing speed for FOXER was fifteen knots – 2.5 knots slower than for CAT gear – and whereas the rods of the CAT gear lasted over fourteen hours in initial trials, the FOXER's rods wore out after only a few hours. There was also no need to use paravanes with single CAT gear, and the equipment itself was so light it could be launched by one man.[30]

For the moment the Admiralty's more conservative estimates held sway. There was still uncertainty, in fact, over whether acoustic homing torpedoes had been used and if the present crisis concerned a new version or was simply a successful application of the Taffy. The first batch of CATs sent to St John's was accompanied by the officer in charge of the RCN's Experimental Section in Halifax, Cdr A.F. Peers, RCN. His task was to interview the officers of escorts and torpedoed merchant ships to determine what had happened. Peers concluded that there was 'no definite indications of the use of this torpedo in the attacks' on ONS 18 / ON 202.[31] Not all ships were clearly hit in the stern – indeed some reported explosions amidships – and the total pattern of hits was not inconsistent with that expected from a spread of conventional torpedoes. The Allies remained ignorant of GNAT's magnetic pistol, which would have explained the pattern of hits. Peers's second interim report about CAT gear, submitted at the end of October, speculated on the possibility of a non-contact pistol. Until the end of November, at least, however, there remained a strong speculation that the mysterious explosions *around* ships were the result of a new German tactic: floating influence mines sown in the paths of convoys.[32]

A major effort of trials and scrutiny of prisoner-of-war evidence followed through the fall in an attempt to resolve the riddle. By November the Allies had sorted the GNAT from the earlier Taffy, and the issue of rudder control was decided in favour of a simple three-position system. By then American trials with an even smaller CAT, as short as eighteen inches, confirmed that the RCN was on the right track. The Admiralty's FOXER system, with its two large noisemakers and complex paravanes, was reckoned to be little more security against a GNAT hit than a single, lightweight CAT and involved far too much effort.[33] Senior RCN officers were cautious about dismissing the British system outright, but they were persuaded by their scientists that CAT was the way to go. Some risks were acceptable. J.H. Johnstone, the director of Operational Research informed the director of the Operations Division at NSHQ on 1 December, 'The Canadian PNM unit is superior to the British PNM or FOXER Unit.'[34] The Naval Staff finally approved CAT gear as the RCN's anti-GNAT decoy system on 13 December. The older, 5-foot-long gear remained briefly in service as CAT Mk I, and it was gradually replaced by a 30-inch CAT gear, which entered service in early 1944 as the Mk II. The only significant modification to CAT gear before the end of the war was a tripping mechanism, which allowed the gear, once streamed astern, to be turned off and on. The British remained unconvinced. They considered CAT gear insufficient protection and its method of streaming a danger to the ship because of the possibility of fouling propellers at slow speeds. Their scepticism lasted well into 1944, probably because the nature of the GNAT's triggering mechanism remained in doubt until one was recovered from a sunken U-boat in June.[35]

Uncertainty over the technical qualities of the GNAT delayed CAT's introduction in the immediate wake of ONS 18 / ON 202. The Admiralty warning about the possibility of fouling propellers with unbuoyed CATs prompted the RCN to order that the initial issue of Mk I gear not be used until the whole question of PNMs was resolved. Thus it appears that the Admiralty's heavy FOXER gear was the first to see operational use, and it was provided to a number of Canadian ships.

The response to the GNAT problem was therefore less straightforward than historians have generally believed. Certainly towed noisemakers were not universally accepted. The British system in particular involved considerable work in launching and recovering, requiring heavy davits on the quarterdeck to handle the paravanes and strong winches to manage the stout cables.[36] The noise-makers failed after a few hours and recovery of the whole affair was laborious and time consuming. The PNMs also tended to drown out the asdic and reduced the A/S effectiveness of ships using them. The alternative was to work at slow speeds that generated less noise than the approximately 20kHz minimum that allowed the homing mechanism to detect the vessel. Such a method always entailed risks. Some bold commanders, such as Captain Johnny Walker of EG 2, preferred simply to operate without FOXERs, trusting in their asdic operator's ability to hear a torpedo coming. The debate over whether to use FOXER or slow speed when in a danger area continued to the end of the war. What did become standard practice was a 'side-step' approach to a contact: a long dog-leg turn, wide to either side of the direct line approach, intended to draw the GNAT well astern. The combination of side step, slow speeds, and PNMs proved an effective counter to the GNAT. According to wartime estimates, from the fall of 1943 until June 1944 only two of the twenty-seven ships attacked by GNATs suffered damage.[37] Jurgen Rohwer has worked out the figures for the whole period of GNAT use. Of the 464 GNAT firings he could analyse in detail, only seventy-seven were hits and only eight of these hits were against ships with some form of PNM streamed.[38] It was not a perfect system, but, as the Canadians rightly concluded in the fall of 1943, the risks were acceptable.[39]

The RCN's superb reaction to the GNAT problem had little impact on the role of Canadian escorts in the balance of the fall campaign. In fact, after ONS 18 / ON 202 the pattern of the earlier mid-ocean victory reasserted itself. *St Croix* had hardly come to rest on the bottom before Admiral Horton recommended that EG 9 be disbanded. The loss of *Polyanthus* and the damage to *Lagan* now left C 2 seriously short-handed. Horton wanted *Itchen* and one corvette transferred to fill the gaps. The other two corvettes were to go to C 5. With the loss of *Itchen* the future of EG 9 was sealed. The support group was officially disbanded when the three remaining corvettes arrived in St John's on 25 September.[40] *Morden* and *Sackville* went to C 2 and *Chambly* to C 5. EG 9's first operation was its last. For the second time in a month a fickle and inscrutable fate intervened to check Canadian participation in the anti-U-boat offensive.

Vice-Admiral P.W. Nelles and Angus L. Macdonald (centre), the naval minister, pondering the shape of the fleet's newest escorts, in this case a model of the revised corvette *Halifax*, in December 1943

The corvette *Saskatoon*, probably in early 1942. By 1943 most Canadian corvettes had extended bridge wings to carry heavier machine-guns and had fit 10-cm radar, but they sailed on with the short forecastle and inadequate original bridge seen here. The antenna atop the foremast is for the SW2C radar.

Control equipment for the type 123A asdic, the most common set in the RCN until 1944. The handwheel on the compass binnacle controlled the aiming of the asdic transducer, while the chemical trace recorder on the left produced range. Much of the crucial data for conducting attacks came through the headphones worn by the operators; note the headset jacks on the recorder stand.

The RCN's most ardent U-boat hunter, James Douglas Prentice, seen here as Captain (D), Halifax, in April 1943

'Data processing' with the type 123A asdic. The asdic rating, on the left, has his eyes on the compass and his hands on the asdic transducer control wheel. The dreamy stare of the unidentified RCNVR lieutenant in front of the range recorder indicates just where the interface between visual and auditory information was.

As good as the short forecastle corvettes got without modernization: *Moncton* in November 1943. The mast is behind the bridge – as is the type 271 radar – and the bridge itself has been extended forward to fit an additional compass binnacle and on either side for 20-mm oerlikon guns.

Ottawa II, first of the six badly needed second-generation River class destroyers acquired from the RN in 1943, seen here at Halifax in October. Her forward gun has been replaced by a hedgehog mounting, the lantern just behind the bridge holds the 10-cm type 271 radar, and the birdcage antenna on the short mast aft is for the HF/DF.

Control equipment for the type 127DV asdic of *Rimouski*, November 1943: range recorder on the left; transducer control, gyro compass repeater, and sending position in the middle; and bearing indicator on the right

The RCN's anti-submarine weapon of choice in 1943–4: hedgehog mounting and bombs aboard *Moose Jaw* in 1944. The bombs are still fitted with safety caps over their firing pins.

The last known photograph of the Town class destroyer *St Croix*, taken from *Arvida* in August 1943

The original short forecastle corvette in its modernized form: rebuilt bridge, extended forecastle, resited mast, 10-cm radar, and heavier secondary armament all are visible in this shot of *Camrose* in early 1944, about the time she shared in the sinking of *U757*.

The object of the professional navy's desire: the Tribal class destroyer *Athabaskan* at Plymouth in early 1944, shortly before her loss in action off the French coast. The glider bomb that crippled her in August 1943 passed through the superstructure just above the pennant numbers on her side.

British FOXER gear aboard a corvette at St John's, October 1943; even the noise-maker, in the foreground, is more complex than CAT gear. The torpedo-shaped 'paravane' in the background was used to pull the noise-maker out to either side of the ship as it was towed: hence the need for heavy cables, davits, and winches to deploy FOXER.

The remarkably simple Canadian solution to the acoustic torpedo: CAT Mk II, seen here in April 1944. Its size is easily judged by the leather mitten lying on the deck just behind the bomb-like depth gauge (which was attached here for trials only).

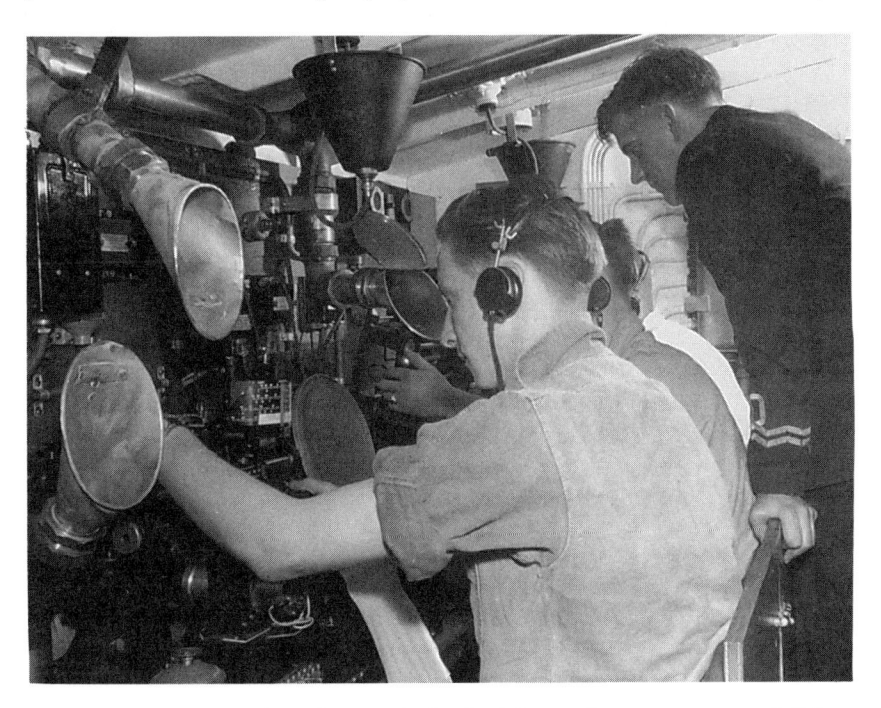

The main asdic control equipment in its final form: the type 144 set of *Coburg*, fully manned, in July 1944

The final version of the original corvette design, the 'Increased Endurance and Improved Bridge' class (like *Long Branch* seen here in 1944) came too late to help the RCN during the 1943 offensive. These corvettes did, however, form the basis of Mid-Ocean Escort Force during the last year of the war.

Castle class corvettes, such as *Hespeler*, seen here on trials in April 1944, were powerful anti-submarine ships. The single squid mounting is just behind the main gun and the type 271 radar sits high atop the lattice mast.

The Canadian-manned auxiliary aircraft carrier HMS *Nabob*, on the west coast in January 1944, shortly before sailing to the United States to embark her aircraft. Similar American carriers were successful U-boat hunters just outside the Canadian zone, while *Nabob* joined the British Home Fleet at Scapa Flow.

The River class frigate completed to Naval Staff specifications: hedgehog, twin 4-inch mounting forward and twin 20-mm secondary mountings on *Coaticook* of EG 28 in early 1945

The only designated support group in the Canadian northwest Atlantic in late 1943 was W 10. It was composed of the old, short-range destroyers *Niagara*, in the foreground, and *Columbia*, on the left, seen here at St John's on Christmas Eve. Two days later they sailed to hunt for *U543*.

One of the classic photos of the Atlantic war: *Swansea* in late December 1943 or early January 1944. Usually printed for its aesthetic quality, this shot also captures something of the 'breaking seas' and 'icy winds' that plagued the futile search for *U543* off Flemish Cap.

Reloading the twin squid launchers of *Loch Alvie* in 1944; note that she has closed upon the swirls left in the water by previous bombs.

The end of the second longest U-boat hunt of the war: a shell-torn *U744* wallowing under the guns of group C 2 with *Chilliwack*'s boat alongside, 6 March 1944

Cdr Clarence King, RCNR, captain of *Swansea*, in March 1944 after the sinking of *U845*. King was a gifted U-boat hunter and ended the war tied with Prentice: both were credited with four U-boat kills.

Vice-Admiral Nelles explaining the finer points of depth-charge attacks to his minister, December 1943. The painting captures some of the problems of using depth charges but not those of dealing with a U-boat at great depth. This unique photographic evidence suggests that Macdonald knew at least something about naval warfare.

Over the next two months as the final campaign played itself out in the Atlantic, most Canadian-escorted convoys (now in the majority) were routed clear of danger or were heavily supported. The groups sent to deal with Dönitz's last assault were powerful British ones: EG 2 composed of five superb Black Swan class sloops, and later the auxiliary carrier *Tracker*; EG 3 composed of four destroyers; EG 4 composed of the first of the new American-built Captain class frigates; EG 7 composed of sloops and frigates; and EG 8 with the auxiliary carrier *Fencer* and five destroyers. The RCN could not match these groups. The lone RCN contribution to the fall campaign in the mid-ocean was EG 5, still composed of the British frigates *Nene* and *Tweed*, and the RCN corvettes *Calgary, Edmundston, Lunenburg,* and *Snowberry*. They spent the weeks of the fall campaign supporting the movement of convoys through the mid-Atlantic, but they had little contact with the enemy.

A now familiar pattern of harassment and destruction by support groups and from the air followed the U-boats from late September into the early weeks of November. Any attempt by the U-boats to press home attacks not only failed, but also led to losses among the attackers. When a pack threatened ON 207, escorted by C 1 in October, it was supported by B 7 (acting briefly in a support role), EG 2, aircraft from HMS *Tracker*, and both RAF and RCAF Liberators. B 7 sank two U-boats around ON 207, both with hedgehogs, a weapon they had come to rely on for its accuracy.

Meanwhile, EG 5 spent the fall on the edges of the fighting. While the battle raged around ON 207, ONS 21, supported by EG 5, escaped notice by the U-boats. By early November Dönitz was forced to disperse his submarines in small groups of four to five off the Grand Banks, trusting in the convoys to come to them. Walker's EG 2, with HMS *Tracker*, cut a path through these U-boats in early November on their way to Newfoundland, and sank two.

Heavy losses and almost complete lack of success brought Dönitz's campaign in the mid-Atlantic to a halt by early November. The era of free-wheeling surface action by submarines was over. By the time EG 5 sailed through the danger area with HX 265 in mid-November, encounters with U-boats were fleeting. *Nene*'s investigation of an aircraft sighting on 16 November produced a torpedo detonation just ahead but no firm submarine contact.[41] EG 5, along with the other support groups, was left to hunt the few fugitive U-boats remaining in the mid-ocean or to chase the last vestiges of a German offensive back into the shelter of air support from French bases. The Allied victory in the Atlantic had its finale – and yet another shut-out of the RCN.

The lack of RCN U-boat kills in the open Atlantic during 1943 was matched by a much less evident failure closer to home, largely, but not entirely, owing to a lack of U-boats to sink. Apart from the failed attempt by *U262* to rescue escaped U-boat crewmen from North Point, PEI, in May (about which the RCN knew nothing at the time),[42] and minor activity south of Nova Scotia, the first nine months of 1943 passed quietly. The only known U-boat to penetrate the Gulf was *U536* at the end of September, also sent to rescue prisoners of war. Canadian authorities uncovered the plot and the RCN attempted to trap the U-boat at the rendezvous off Maisonette Point, New Brunswick, in the Bay of Chaleur. Details of the ambush were a closely guarded secret, and even the RCAF was not informed. The presence of so many warships and some bungled attempts by Canadians to communicate with the submarine from shore alerted the Germans to the trap and *U536* escaped without the RCN's ever making contact – in Canadian waters at least.[43]

The first serious attempt to hunt for a U-boat inshore during 1943 did not come until late October. By then the lack of firm U-boat presence in the Canadian zone had stultified the aggressive spirit evident earlier. Admiral Murray seemed content to allow the air force to undertake whatever offensive patrolling was called for. When *U537* was attacked by an RCAF aircraft east of St John's on 31 October, Murray had to be prodded into action by Ottawa. The result was the first 'Salmon', the Canadian code name for a 'hunt to exhaustion.'

In fairness, Murray was not the only one lacking enthusiasm for Salmon operations. The RCAF remained unsure if the concept of hunts to exhaustion were worth the effort. The RAF's Coastal Command preferred concentration around convoys where U-boats were likely to be found.[44] The RCAF's Eastern Air Command operated on the same basic premise. If such a hunt was to be launched, it was still unclear whether the British system of blanketing the area with radar-equipped aircraft was better than the American system, which used fewer aircraft but relied on the new sonobuoys to track the submerged U-boat.[45] Perhaps that indecision explains why no joint training in Salmons had yet been conducted by the RCAF and the RCN, and no operation orders had been promulgated.[46]

The object of the first Salmon, *U537*, was also in the western Atlantic on a clandestine mission, this time to land an automated weather station on the Labrador coast. This task was accomplished, and the station remained undetected for thirty-seven years.[47] *U537*'s presence, however, was betrayed by a subsequent signal ordering it to patrol off St John's. The signal was read by Allied intelligence and an 'Otter' (U-boat probability) area was promulgated. It was no accident, then, that an aircraft found and attacked the U-boat at 1227Z on 31 October seventy-five miles east of Argentia. Two hours later the first Salmon operation was ordered. Three escorts arrived on the scene seven hours after the aircraft sighting. By then the RCAF Hudson, which promptly left the scene after *U537* submerged, was long gone. The senior officer of the naval forces, aboard the minesweeper HMCS *Lockeport*, had no instructions for conducting an expanding search, and, in any event, by the time he reached the datum, the radius of possible U-boat movement was thirty five miles: too much for three ships to search. In poor weather and heavy seas and without either proper instructions or air support, the hunt was called off the next day.

A second Salmon operation was mounted following another aircraft attack on *U537* late in the day on 10 November, about 130 miles southeast of Cape Race. The area was searched briefly by an escort from a nearby convoy, but nine hours elapsed before the first of the designated hunters, *Kenora*, arrived. Eventually eleven ships were assigned, and the coordination by the captain of the British destroyer *Montgomery* was considered good. Contact with aircraft was intermittent at best, however, and was plagued from the outset by deteriorating weather. Those aircraft that did arrive cooperated well with the search, and communications appear to have been good, perhaps because *Montgomery* was carrying an RCAF liaison officer. No systematic expanding air search was attempted, however, and only the persistence of the naval escorts permitted the search to continue beyond midday on the 12th. *U537* hardly noticed. After patrolling quietly, she departed the Grand Banks for home.[48] Still no U-boat kill.

By November the lack of U-boat kills by the RCN during 1943 had taken on a new meaning. King's government was basking in the glow of the Italian capitulation, including recent RCN operations in the Mediterranean.[49] By November the First Canadian Infantry Division, now part of the famous British Eighth Army under General B.L. Montgomery, was clawing its way victoriously up the boot of Italy. The domestic political crisis was past, but the storm within NSHQ was now building to a climax. The extent to which the RCN was still 'off the pace' was evident from the text of the Anglo-American press release dealing with the fall campaign. The draft text arrived in Ottawa on 5 November for Mackenzie King's approval. It explained that some sixty U-boats had been destroyed over the previous three months. 'This brings to more than 150,' the press release noted, 'the number of U-boats destroyed during the last six months.'[50]

By this time Macdonald also had a report from his executive assistant, J.J. Connolly, who undertook a secret investigation of the handling of the RCN's expansion. On 10 November some of Connolly's evidence, in the form of an anonymous and ill-informed British critique of the RCN's handling of modernization, was unmasked to the Naval Staff. It was a scandalous document and caused quite a stir. Connolly's full case, based on evidence drawn from Nelles's own files, the mutterings of malcontents in the fleet, and some highly dubious British opinions, was placed before the Naval Staff at its regular meeting on 15 November.

Not surprisingly, the Naval Staff simply rejected the notion that they had utterly bungled the country's naval war effort. Macdonald, undeterred, now suspected an attempt to cover up the staff's previous 'failings.' On 20 November he sent Nelles a long memo listing eight 'Comments' – really indictments – on the performance of the Naval Staff. Of interest here is Macdonald's measure of operational efficiency, which – in the light of other developments in 1943 – was a very political one. He drew directly on the latest Anglo-American press release on the Atlantic war to chastise the staff for concentrating on quantity at the expense of quality. 'The price of this,' the minister wrote, 'was the fact that of the last one hundred and fifty subs destroyed not one has definitely been destroyed by a Canadian ship of War.' The fleet, Macdonald went on, 'must be competent to deal with submarines ... If the British did not realize this and insist upon the best equipment possible for ships and planes, the success which they enjoyed in the summer months would not have been possible.'[51]

The storm over the failure of the fleet to sink submarines raged on into December and became increasingly a personal feud between the CNS and the minister. Nelles intimated in his responses that Macdonald, as naval minister and member of the Naval Board, also bore some responsibility for the condition of the fleet, a charge the minister flatly denied. Macdonald's charges eventually reached the absurd. He blamed the Naval Staff for not seeking a political solution to the modernization problem in 1942 by having Mackenzie King raise it with Churchill. Had Canada not secured equality of access to modern *British* equipment – and even British shipyards – Macdonald 'would have recommended that our ships be withdrawn from the North Atlantic run [mid-ocean].'[52] The notion of Canada's abandoning a major operational commitment in 1942, at the height of the Axis tide, on a point of national pique was ridiculous and the Naval Staff knew it. But Macdonald was not interested in convoys safely escorted. He pressed Nelles over the RCN's failure to sink submarines and held steadfastly to the belief that had the British not modernized the Canadian fleet in 1942 he would have withdrawn it to Canadian coastal waters.

There is little doubt that Macdonald bore the brunt of his cabinet colleagues' complaints over the 'failure' of the fleet, and, in turn, he delivered Percy Nelles as the scapegoat. There is also evidence to suggest that Nelles's pro-British conservatism was at odds with a growing anti-British sentiment within the Naval Staff. Certainly the VCNS, Rear Admiral J.C. Jones, was known to be stridently anti-British by the late stages of the war. Jones was also an ambitious and very political man, and he had brought with him many of his retinue to NSHQ at the end of 1942 when be became VCNS. It is true that the Naval Staff, including Jones, supported Nelles in his battle with the minister in late 1943, but then Macdonald had tarred all of them with the same brush. The result was the dismissal of Nelles as CNS at the end of 1943 and his appointment to London as Senior Canadian Naval Officer. Jones took over as chief.

Ironically, the very day Macdonald's first charges of incompetence arrived in Nelles's office, the fortunes of the RCN in the war against the U-boat changed. In the first weeks of November the stragglers of Dönitz's collapsed offensive moved into the eastern Atlantic. There, as he had in the summer, Dönitz hoped to fight on more even terms, with air and possibly naval support from bases in France. As the offensive shifted eastward, so too did the support groups, and among the first to go was EG 5.

Having seen HX 265 safe home, EG 5 was diverted to support the combined convoy SL 139 / MKS 30 en route to Britain from African ports. The departure of the Gibraltar portion of the convoy on 13 November was reported by German agents in Spain, and the precise location of the combined convoys at sea was determined by aerial reconnaissance two days later. Seven U-boats were sent to attack, and the twenty-five long-range He 177 bombers of *Kampegrüppe* 40 in France were alerted. EG 5 arrived to help the close escort (EG 40) and support group EG 7 (of RN destroyers) on the 19th. By then the British frigate *Exe* had already damaged one U-boat, an RAF Wellington had sunk another, and the RN sloop *Chanticleer*'s stern had been destroyed by an acoustic torpedo (an incident attributed to a floating mine[53]).

EG 5 gained its first U-boat contact even before joining SL 139 / MKS 30, when, in the early hours of 19 November and still 115 miles ahead of the convoy, *Calgary* and *Snowberry* reported two U-boats on the surface. One escaped by outrunning the corvettes, while the other submerged to safety. No asdic contact was ever established. By 0800Z EG 5 contact was made with the convoy, and the group took station on the port bow at visibility distance.

EG 5 was no sooner on station than the first U-boat sighting, in this case by an aircraft, was reported ahead of the convoy. EG 7 moved forward to investigate and the First Division of EG 5 – the senior officer in *Nene* and the corvettes *Calgary* and *Snowberry* –

replaced them in the advanced screen. HF/DF bearings on signalling U-boats warned of more trouble directly ahead, but it was not until 2102Z that the contacts were confirmed and were close enough for EG 5 to pursue them. Thirty-five minutes after the First Division had altered course to sweep the area, *Nene* obtained a radar contact. The frigate worked up to its maximum of nineteen knots, maintaining a wide zig-zag on the contact's quarter in order to minimize the danger from acoustic torpedoes, but the target was obviously retreating. *Nene* fired starshells, which revealed a U-boat running fast on the surface and keeping her stern tubes pointed in the frigate's direction. Gunfire drove the U-boat down at 2213Z. After ordering *Calgary* and *Snowberry* to conduct a square search around the area, the frigate gained an asdic contact, which she attacked with hedgehog. Nothing came of this first attack, and it was clear that the U-boat had gone very deep – too deep for hedgehog. Ten minutes later the target was attacked with depth charges set to 550 and 500 feet. This attack, too, failed to produce results, and since the contact could not be regained, the First Division set off on another square search, this time with sides six miles long.

EG 5's initial target was probably *U648*, but the asdic search following the deep depth-charge attack actually turned up *U536* – the U-boat that had escaped the RCN's trap in the Bay of Chaleur weeks earlier. *U536*'s crew had listened to the first attack on *U648* from a depth of over 500 feet, and the sudden hammering of *Snowberry*'s asdic on their own hull took them by surprise. So too did the depth-charge attacks that followed. One of them, delivered by the corvette shortly after midnight, drove the U-boat down by the stern and damaged her ability to maintain trim. As the crew of *U536* struggled to regain control, *Snowberry* lost contact. Yet another square search was begun to relocate the submarine. Less than an hour later *Nene* regained contact, followed shortly thereafter by *Snowberry*. *U536* was still deep, so deep that the escort's fixed asdic set lost contact with the submarine while she was still a long way off. Once again the hedgehog, which had a short range, could not be used unless the time to fire was estimated by stopwatch (which greatly reduced accuracy). The depth of the U-boat also affected the accuracy of a depth-charge attack because of the dead-time between loss of contact, time taken to reach the release point, and the slow descent of the charges (this subject is discussed at length in the next chapter). *Nene* therefore marked the position of the contact by putting a flare in the water, and *Snowberry* stood by, with all engines stopped, maintaining a listening watch on the U-boat at 500 yards. With the position marked and *Snowberry* passing ranges, *Nene* came roaring in at eighteen knots to deliver her charges only 400 yards ahead of the stationary corvette. The disturbance from the explosions had barely subsided, when *U536* broke the surface off *Snowberry*'s starboard beam.

Whether this final depth-charge attack was what drove *U536* to the surface is a matter of conjecture. The Admiralty's assessment credited *Nene*'s attack as the decisive one, but survivors' testimony suggests that the problem of trim, the result of *Snowberry*'s attack an hour before, had reached a crisis and the decision to surface was taken independently of *Nene*'s final attack. Whatever the case, *U536* rose to the surface under the guns of EG 5 and the group did not let her escape. As *Snowberry*'s engines were ordered 'full ahead,' the first round from her 4-inch gun smashed into the U-boat's conning tower. Soon 2-pounder and oerlikon fire riddled *U536*'s pressure hull and swept away the crewmen who tried to man the U-boat's guns. As *Nene*, and then *Calgary*, joined in the hail of fire, *Snowberry* illuminated the scene with star shells. When it became evident that *U536* was a wallowing wreck and her crew were desperately trying to abandon ship, the order to cease fire was given. Seventeen survivors were pulled from the water as *U536* sank by the stern at 0247Z on the morning of 20 November: the first kill by an RCN support group.[54]

EG 5 pursued other contacts during the next day, but none proved successful. On the 21st the anticipated attack by twenty He 177s equipped with glider bombs took place. By then the convoy had been reinforced by HMCS *Prince Robert*, a former BC ferry and armed merchant cruiser, now refitted as an anti-aircraft cruiser. *Lunenburg*, *Edmundston*, and *Prince Robert* were narrowly missed by glider bombs, including two that fell within sixty yards of *Lunenburg*. Two merchant ships were struck, and one of them later sank; three aircraft were shot down, but remarkably few of their forty glider bombs found targets. The air attack of 21 November effectively ended the battle for SL 139 / MKS 30.

Two more battles developed around the SL/MKS convoys through late November and early December, and then Dönitz tried an operation south of Iceland. These actions were no more successful than earlier attempts. In December Dönitz finally ordered his U-boats to remain submerged as much as possible, surfacing only long enough to recharge batteries. It remained only to fight a holding action until the type XXI U-boats appeared.

The RCN's record of kills during this final crushing defeat of the U-boat fleet was thus limited to *U536* – and that was shared with a British escort. Set against the overall destruction of some sixty-two U-boats in the period September to December 1943 the RCN's was a paltry score. However, like the crisis in the spring and the bay offensive during the summer, the vast majority of the U-boat kills in late 1943 were credited to aircraft. Of the forty-six submarines sunk in the mid-ocean, the Bay of Biscay, and along the SL/MKS route from 1 September to the end of the year, twenty-four fell to land-based aircraft alone (four of them RCAF). Another nine succumbed to USN auxiliary carrier groups. Only thirteen were sunk by naval vessels alone.[55] Few of them fell to close escorts, where the bulk of the RCN remained doing vital if now much less glamorous work. Had EG 9 remained active, it is likely that the RCN's score would have been better, but not dramatically so. EG 5 performed well, but Birch was no Gretton or Walker, and corvettes were no match for sloops and destroyers.

The importance of support groups as the leading edge of the anti-submarine war was driven home to the RCN by the fall campaign in a way that had been obscured in earlier operations. During the spring and even the summer it was arguable that the issue in the Atlantic was still in some doubt – the dramatic attack on ONS 18 / ON 202 in late September provided ample evidence. The quick and evident collapse of Dönitz's fall 1943 campaign removed any lingering doubt that the nature of the war in the Atlantic was fundamentally altered.

The most tangible evidence of change was a radical reduction in escort-building programs in late 1943. Welcome as this decrease was to the fully stretched Allied war effort, for the RCN the reductions precluded the acquisition of the final – and most advanced – generation of ASW ships. That fact may well have influenced the effectiveness of the RCN in A/S operations in 1945.

When the Admiralty informed NSHQ on 30 October of its plans to reduce sharply escort construction, the RCN was planning to commission, by the end of 1945, approximately 101 frigates[56] and twenty-six corvettes, and it hoped to obtain from the RN perhaps a further thirty escort destroyers.[57] The Admiralty's decision to abandon escort building once its present contracts were completed threw Canadian plans, already in a state of flux, into something of a tizzy. The Naval Staff recommended on 19 August 1943, amid the euphoria of the earlier victory over the U-boats and the heady atmosphere of the Quebec Conference, that no further escort building be undertaken in Canada after the completion of the current programs. The thrust of ship acquisition was now directed towards a postwar fleet.[58] Under pressure from the new assistant chief of the Naval Staff, Wallace Creery, and the renewed U-boat campaign in the mid-Atlantic, however, the Naval Board reversed its decision on destroyers in early October and briefly restored priority to escort types.[59]

Canadian escort procurement plans were therefore uncertain in the fall of 1943, but the British decision in late October 1943 to abandon new escort programs struck a sympathetic chord in Ottawa, all the more so since the Admiralty, confronting manpower problems, offered more ships to the RCN, including fleet class destroyers and the manning of two auxiliary aircraft carriers. Frigates were there for the taking. The British wanted the RCN to man some of the new Captain class being built in American yards. The RCN flatly refused – the Naval Staff Minutes have a distinctly dismissive tone. However, in mid-November the RCN offered to man ten British-built frigates by February 1944, thereby freeing RN crews for the Captains.[60]

To sort out future escort needs Wallace Creery went to Britain in late November. His report confirmed the RCN's rejection of Captain class escorts. According to the Admiralty's controller, the first batch of Captains displayed 'appalling workmanship in certain instances,' were poorly balanced, stiff, and difficult to handle: a complicated 'bag of tricks.'[61] The RCN never seriously looked at the Captain option, which was a pity, since by 1945 they were among the best ships in the war against the U-boats. Quite apart from the early Admiralty reports on the U.S.-built Captain class, the professional RCN maintained its preference for British designs throughout the war, and it seems unlikely they would have warmed to this american warship. It was finally agreed that RCN frigate construction would be cut back by forty-one ships,[62] leaving a total of sixty River class frigates built for the RCN. The long-standing desire to round out the RCN's fleet of escort destroyers by acquiring at least eight more was now converted into a request for more fleet destroyers.[63] The ten frigates to be acquired from the RN were also to be – with one exception – River class. *Nene* and *Tweed*, of EG 5, and *Ettrick* would be handed over, along with six other River class frigates and one of the newer Loch class.[64]

The emphasis on acquiring British River class frigates was understandable: they were already in service and could be taken over quickly, while construction of the newer Loch class was only just under way. The only advantage the Canadian Naval Staff saw in the Lochs was that they were 'rapid construction frigates.' But there was more to the Lochs than quick building times. Their design included the new A/S weapon 'squid,' a heavy lattice mast designed to carry the new type 277 'high-definition' radar and a bridge and communications system completed to naval standards. The Lochs were, as Peter Elliott describes them, 'the ultimate in Admiralty design and North Atlantic war experience.'[65]

The squid was the latest in 'ahead-throwing' A/S weapons. It was a three-barrelled mortar, which fired very heavy hydrostatically detonated bombs well ahead of the ship. Like depth charges, the triangular squid pattern was designed to crush a submarine inside a three-dimensional explosion. Lochs carried two squid mountings, producing overlapping patterns. The squid itself was slaved to a new asdic, the type 147B, which could read both depth and range. Targeting data were passed automatically from the type 147B to the squid bombs, including depth settings, and the firing sequence was fully automated. Squid's automation, its superb asdic control, and the power of the bombs made it a far more lethal weapon than hedgehog. In 1943 the merits of squid over other systems was largely theoretical. Deep-diving U-boats could be dealt with by using 'creeping' attacks, in the manner already used by *Skeena* and *Wetaskiwin* to sink *U558* in 1942 and *Nene* and *Snowberry* to sink *U536*. Such tactics (discussed in the next chapter) were being perfected by Captain Walker's EG 2. There was nothing in the River class frigate design that precluded such tactics' being adopted: indeed, the allocation of 150 depth charges for each frigate permitted considerable expenditure.

The prospect of acquiring squid was therefore no inducement to the RCN to press for more Loch class frigates in late 1943. In fairness to the Canadian Naval Staff, *Loch Fada*, the first of her type, was not completed until December. Canadians were therefore content to settle for British River class if it was a simple matter of taking over existing ships – the most rapid form of 'construction' possible. With the promise of ten from the Admiralty and enough of the River class already being built in Canada to meet the RCN's needs, forty-one Canadian frigate contracts were cancelled in December 1943, including all the RCN's own Lochs.

The cancellation of building programs also reduced the number of Castle class corvettes the RCN operated. The Castles represented the pursuit of the basic single-screw corvette to its ultimate form and its adaptation to ocean escort duty. In terms of weapons, equipment, and bridges, the Castles were a smaller version of the Lochs. They carried a single squid along with the type 147B asdic, their lattice mainmast was designed for type 272 radar (a more modern version of the standard escort type 271 radar) or the type 277, and their bridges were to naval standards. Although comparatively slow, the Castles were superb warships with superior ASW capabilities. The twelve Castles already earmarked for the RCN in exchange for Canadian-built Algerine class minesweepers were the only ones to enter Canadian service.

The wholesale cancellation of building programs at the end of 1943 effectively stalled the Canadian anti-submarine fleet a generation behind the RN. The RCN might have equipped its 1943-4 program River class frigates with squid and heavy lattice masts while they were still being built, but such changes would have caused unacceptable delays. Creery's report noted, in any event, that the RN had no intention of retro-fitting its River class frigates with squid;[66] doing so would involve too much reworking of the entire forward portion of the ship. Moreover, the RCN remained firmly convinced of the merits of hedgehog and resisted attempts to modify the anti-submarine equipment of its River class frigates.

Policy on weapons and equipment for Canadian frigates had evolved over several years, but the fit agreed to by late 1943 for the most part saw the ships through to the end of the war. The basic search radar was a 10-cm type, officially the Canadian-developed and -produced version of the British type 271, which was known as the RX/C. Each frigate also carried HF/DF and the new A/S mortar hedgehog.[67] The main asdic was the type 144,[68] which produced a cone-shaped beam set at a shallow angle from the horizontal, and the new 'Q' device. The type 144 was the ultimate in Admiralty-designed main-search asdics in use during the war. Its principle improvements over earlier sets were in the display of information, automatic training, and its role in fire control for the hedgehog. The Q was an additional oscillator fitted beneath and slaved to the main set and permanently set to a sharper angle. It produced a thin fan-shaped beam in the vertical plain, which, along with a sharper angle of depression, allowed contact to be maintained on a target for a longer period during the run-in, particularly if the U-boat had gone deep. The Q also allowed asdic beams to penetrate more effectively the layers of inshore waters. Frigates also carried the type 761 echo sounder, designed to produce a bottom profile as an aid to navigation.

In early September 1943 the deputy director of Warfare and Training (DWT), Capt. H. McMaster (the post of DWT was not permanantly filled until December 1943), also recommended that the RCN follow the Admiralty's policy of fitting all escorts with the new type 147B asdic. His reasons were three: it gave, for the first time, an accurate reading of depth; it gave a plan of the range; and the set was being ordered in Canada in bulk (500 sets) by the British Admiralty Technical Mission. If the RCN wanted to get supplies, even from Canadian firms, it had to act quickly. As a matter of policy, McMaster

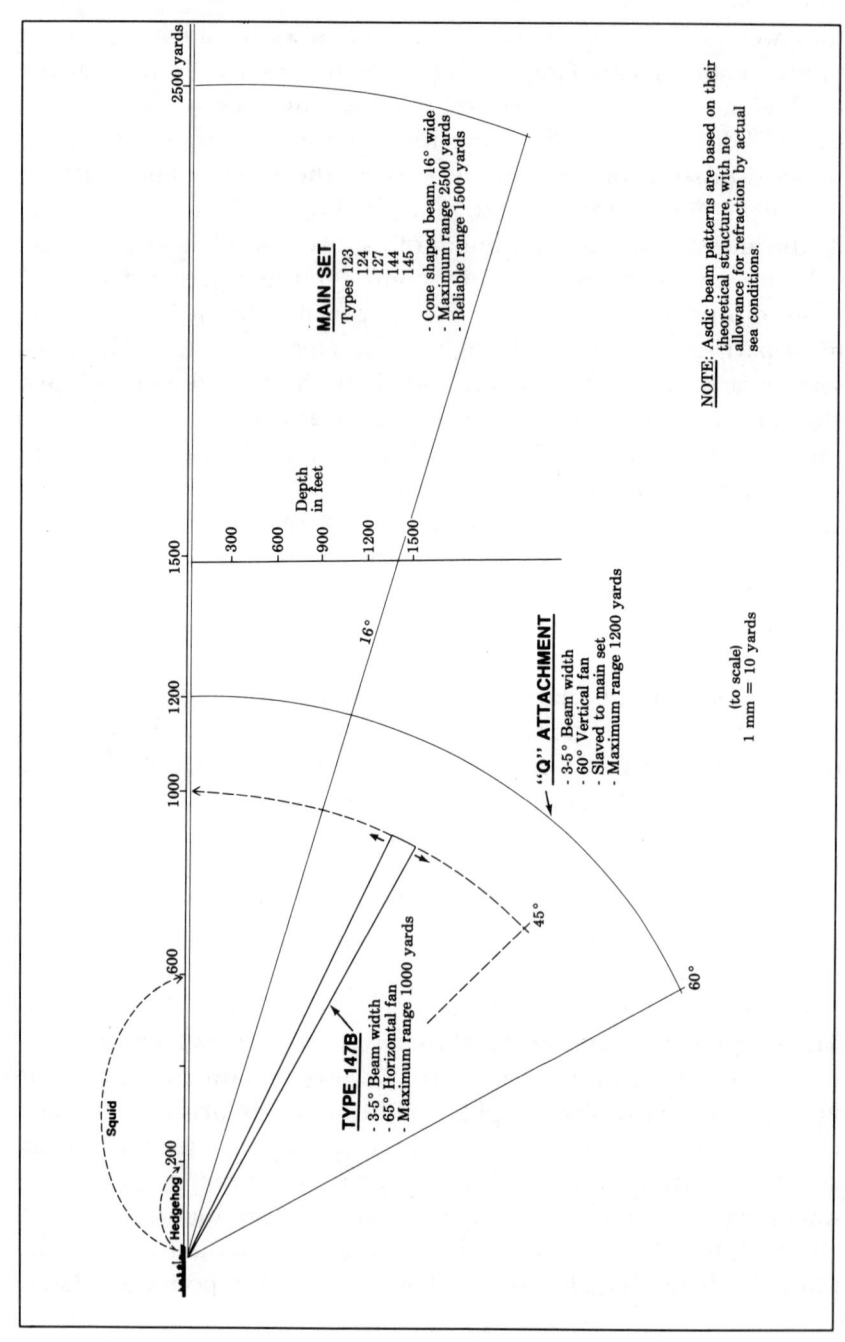

Figure 1 Theoretical asdic patterns

wanted all escorts, with the exception of Town class destroyers (soon to be retired), fitted with 147B.[69] Moreover, he wanted 150 sets ordered right away. Authorization to purchase that number of 147B asdic sets, which was all the RCN could handle 'this year,' was given by the Naval Board a week later, but it took a full six weeks actually to order the sets – a fact that the DWT brought to the attention of the Naval Staff in early November. At a time when the navy was sensitive to its poor state of equipment and to the delay that its awkward bureaucracy had on supplies, McMaster's plea that special measures be taken to eliminate delays in 'procuring equipment effecting [sic] the fighting efficiency of H.M.C. Ships found a sympathetic reception' was not surprising.[70] It did not produce the desired results, however, and it was unlikely in any event that the RCN could have received any 147B sets before the end of 1943. Moreover, the scale of the RCN's need had been underestimated, and a further ninety sets had to be ordered by January 1944.[71]

Where RCN River class frigates differed significantly from RN Rivers was in their primary and secondary armament. The British policy was to fit single 4-inch Mk XIX high-angle/low-angle guns, one forward and one aft, two 2-pounders and two oerlikons. The RCN felt that this was inadequate. The director of Naval Ordnance wanted the forward gun replaced by a twin 4-inch high-angle/low-angle mounting and the after gun replaced by a powered twin Bofors.[72] The recommendation was approved in principle by the Naval Staff at the end of May 1943, and then the RCN spent the balance of the year trying to secure supplies. Since the planned armament was more than the Admiralty authorized for the class, the British initially were less than forthcoming with the guns. In the meantime the first RCN frigates to be completed carried only a single 4-inch gun forward, and the remainder were 20-mm oerlikons. The Canadian decision to fit its frigates with the twin 4-inch gun, coupled with the battles with the Admiralty to acquire the guns, influenced the Naval Staff's subsequent policy on retrofitting the frigates with squid. Carrying the new A/S mortar would require extensive modifications to the forward portion of the ships and abandonment of the heavy, twin 4-inch gun.

Not all the necessary equipment was ready when the ships were.

Waskesiu, the first RCN frigate, was commissioned on 16 June 1943 on the west coast. She and the other early frigates got to sea without their proper asdic fit; type 128D was substituted until the 144 became available. The type 128D asdic had the same transducer under the hull as the 144 set, but the controls in the asdic hut were basic and could not get the most out of the hedgehog. *Waskesiu*, like other early RCN frigates, was armed with a single 4-inch gun forward, a 12-pounder aft, and a few single 20-mm oerlikons. When she arrived in Londonderry, priority was given to stripping *Waskesiu* of her faulty American RX/U radar, a set hastily procured when supplies of Canadian-made RX/C failed to materialize, and replacing it with type 271. Defects, trials, passage through the Panama Canal, work-ups at Bermuda, and then further repairs off and on in Halifax delayed her arrival in the mid-ocean until the end of October – nearly five months after commissioning.

By the end of the year sixteen RCN frigates were in commission, with a flood ready to follow in the spring of 1944. This schedule was better than planners had fearer two years earlier. The arrival of RCN frigates into Atlantic operations was also delayed by the decision in late 1943 to outfit the first ones as senior officers' ships and assign them to the close escort groups of MOEF. This plan conformed to the new RCN policy of appointing senior officers to groups independent of a ship command, an American concept not favoured by the RN in 1943.[73] Both the SO and his small staff would be borne in addition to the regular ship's company, and they required accommodation not included in the original frigate design. *Waskesiu* found herself the senior officer's ship of group C 5 by early November.[74]

Even as *Waskesiu* was assigned to close escort duty, plans were afoot to shift the frigates into the offensive. Admiral Murray seemed to share Nelles's conservatism, but he was no doubt aware of Connolly's 15 November report to the Naval Staff. He also closely watched the developments at sea. During the early part of the month pressure slackened on the mid-ocean convoys, while EG 2 sank two U-boats under his very nose on the Grand Banks. Murray had a chance to hear Walker's views on support group operations when the SO of EG 2 visited Halifax from 15 to 17 November to lecture on anti-submarine warfare.[75] By mid-November there was ample cause, operationally and politically, to get more involved in the offensive.

For a number of reasons then, among them perhaps EG 5's sinking of *U536* a few hours earlier, at midday on 20 November – the day Macdonald began his assault on Nelles and the Naval Staff – Murray signalled his intention to revive EG 9. 'In view of successes being achieved by Support Groups,' he argued, 'it is for consideration whether we should not aim at creating another Canadian Group to replace the late EG 9.'[76] The proposal was a modest one. The bulk of new construction ships would still be directed into the close escort groups of MOEF. But the new group would be built around new ships: the frigates *Waskesiu* and *Swansea*, one of the recently commissioned Improved Endurance corvettes, two of the Castle class 'when available', as well as two older but modernized corvettes. The plan purposely avoided any shake-up of EG 5, which was an experienced and successful group. NSHQ concurred with the re-establishment of EG 9. This would bring the number of RCN support groups back up to the level adopted as policy in August: sufficient to maintain 'one R.C.N. Support Group continuously at sea in the Mid-Ocean.'[78] Murray was reminded, none the less, of the navy's priorities: keep the best corvettes with the close escort groups and support the modernization program itself.

Apart from finding suitable ships, the biggest problem Murray faced was finding a suitable senior officer[77] and group communications, sensor, weapons, and engineering specialists. As a result of decisions taken in August to acquire cruisers and other fleet units, the professional RCN was now even more committed to its own expansion. Moreover, remarkably few RCN – as opposed to reservist – officers were either qualified or available for A/S work. The leadership of the RCN's most important close escort groups and its offensive A/S forces already relied heavily on British officers: Birch of EG 5, Burnett of C 2, Medley of C 3, Mackillop of C 4, and Bridgeman – lost with *Itchen* – of EG 9. The problem in November 1943 was, as it would be for the balance of the war, where to find Canadians who were qualified and senior enough to command groups.

Canadian intentions were signalled to the Admiralty on 2 December, the same day that Murray recommended Cdr A.F.C. Layard, RNR, to lead the new group. Layard had joined the RN as a boy, witnessed the battle of Jutland from the fighting top of *Indomitable*, and then left the navy in the interwar period. He returned to sea in 1940 and spent much of the next three years commanding destroyers. After a short stint at the Admiralty, Layard arrived in Halifax in September 1943 as senior officer of the British Town class destroyers still operating with WEF. By late fall all these tired veterans were slated for disposal and Layard soon would be posted away. Murray's chief of staff, Captain R.E.S. Bidwell, offered him command of EG 9 and he gladly accepted. A quiet man, Layard commanded the group until the end of the war.[79]

The re-establishment of EG 9 was planned for early 1944, when the new frigates and latest corvettes would be ready. In the meantime Admiral Horton anticipated the RCN's desire to get more involved in the offensive. All the sloops of EG 2 and EG 7 had developed serious cracks and were badly in need of docking and refit.[80] The renewal of the Russian convoys placed increased demands on available escorts,[81] while the general collapse of the U-boat offensive in the Atlantic offered an opportunity for further scaling down of the British effort there. Opening of the convoy cycle permitted Horton to suggest that C 2 be released for a period of support duty.[82]

Horton was likely aware of the battle raging in Ottawa over the role of the RCN throughout the 1943 offensive. The day after the proposal to assign C 2 to support duty Nelles chastised Commodore (D), Western Approaches's office for an impertinent signal, which mocked Canadian attempts to find and fit suitable 10-cm radar. 'There is no time for humour in our struggle to obtain modern equipment for our ships,' Nelles protested to the Admiralty. The animosity lasted well into 1944, and there was concern – justified – that British meddling in RCN affairs in 1943 may have caused permanent damage in the relations of the two services. Horton ordered a stop to all criticism of the RCN other than from his office. Nelles, as one of his first tasks in the United Kingdom in 1944, was instructed to mend fences.[83]

Whatever the politics of the moment, the choice C 2 was a sensible one. The core of the group, *Gatineau, Drumheller,* and *Kamloops* was of long standing, and the new frigate *St Catharines,* modified to a senior officer's ship, joined in November. *Sackville* and *Morden* had been part of the group since the battle for ONS 18 / ON 202. Perhaps more important to the choice of C 2 was presence of Cdr P.W. Burnett as senior officer.

As the last echoes of the fall 1943 campaign ebbed, the RCN had reason to hope that a year of missed opportunities was over. The enemy they sought were now, it is true, a fugitive lot, driven down and hounded incessantly by Allied airpower. They were fewer in number, too. They were also now a negligible threat to convoys, virtually immobilized by the danger of air attack, and for the time being largely limited to the area immediately west and south of Ireland. As a result, both support and close escort groups could concentrate with some leisure on pursuing contacts until the U-boat was sunk. In addition, with the Atlantic now quiet, it was time for the RN to get on with other pressing operations and to repair the damage of a year of intense activity. That preoccupation left an opening for the RCN. To use an ice hockey analogy, the second line – what British footballers would call the 'second eleven' – was now allowed to play. The results were quite surprising.

3
Triumph without Celebration

'You are not far removed from it [RN tradition] yourselves, you know. You are a part of the Empire and much of your stock is British.'
'That's so, sir,' I acknowledged. 'But ... many of us feel that we have no direct right to your traditions.'
A book lying on his desk, which I had several times seen him reading, rather demonstrated my point, I thought. It was Southey's *Life of Nelson*. 'Our tradition,' I suggested, 'is possibly being made now.'

Alan Easton, *50 North*[1]

The dismissal of Admiral Percy Nelles as chief of the Naval Staff on 1 January 1944 marked an end to the most traumatic year in RCN history. For twelve long months the navy had wrestled with the debris of unbridled growth and had done so under the critical gaze of politicians, sniping from its own men at sea, and the pernicious influence of ill-informed British officers. Events at sea accentuated Canadian shortcomings and conspired to marginalize the RCN's efforts. The result was bitterness and acrimony. For the RCN the year of the Allied victory in the battle of the Atlantic was a year best forgotten.

By 1944 attention was focused on the navy's new adventures: the operations of Tribal class destroyers, cruisers, carriers, landing craft and assault ships, and motor torpedo boats. These actions too were the fruits of 1943, of decisions taken at Quebec and late in the year. The Atlantic war was now quiet and it was time to look ahead. It was ironic, therefore, that as the excitement of 1943 diminished, RCN forces in the mid-ocean returned to their previous pattern of defending convoys *and* sinking U-boats. Moreover, the first quarter of 1944 was simply the opening act of a very successful year.

At the end of 1943 there was a sense that the north Atlantic war had stabilized. Even the Canadian cabinet and Mackenzie King lost interest. King made one final attempt in December to say something positive about the Canadian role in the great Atlantic victory of 1943. He appealed to the British for comparative figures on the Canadian and British success rates against U-boats, figures he wanted to use in his Christmas broadcast. Staff at the Admiralty refused to provide them. Captain C.D. Howard-Johnston, the director of the Anti-U-Boat Division, felt quite rightly that such a comparison would create a 'bad impression' and was 'neither appropriate nor desirable.' 'It is sufficient,' Howard-Johnston observed, 'to say that the Royal Canadian forces have bent their backs to the task, and, in the face of many difficulties arising from a very great expansion, and under conditions of appalling weather in the North Atlantic, have maintained a high standard of efficiency in the defence of trade.'[2] There was nothing in Mackenzie King's speeches at the end of 1943 addressing the role of the RCN in the recent Atlantic victories,[3] and with that his – and his government's – interest in trying to make political leverage from U-boat kills ended: just as the RCN started killing U-boats again.

By the end of 1943, therefore, the U-boat 'problem' was 'fixed' metaphorically, but it was also true geographically. The obligation to operate almost entirely submerged reduced the tactical effectiveness and range of Germany's primary submarine, the type VIIC. Since the U-boats could no longer manoeuvre into position to intercept convoys, Dönitz had to position them where they could be run over by their targets. The U-boat threat was therefore concentrated in a narrow band of ocean between 15°W and 23°W, following a wide arc west of Ireland covering all the possible approaches to the North Channel – the main funnel of overseas convoys to Britain. The Germans maintained standing patrols in the area throughout the winter. Since the U-boats still signalled their positions and made sighting reports, the Allies could easily plot the general location of the concentration through conventional DF intelligence. Support groups were now used to bring convoys through this clearly defined U-boat danger area.

This was, in fact, a subtle but significant change in the whole concept of A/S operations. Until the end of 1943 anti-submarine warfare in the open Atlantic revolved around convoy battles. Moreover, the support groups assigned to the mid-ocean were evidently seen as part and parcel of MOEF itself. Admiralty thinking at the end of 1943 specified a minimum of eight support groups as a target for the north Atlantic, and some documents speak of the support groups and the close escorts of MOEF in the same sense. Despite the Bay of Biscay Offensive, the notion of support group operations as distinct from convoy battles had not fully taken hold.

Anti-submarine warfare was evolving towards a concept of operations centred around a response to the movement of U-boat concentrations, rather than around shipping. Support to convoys was now provided for only a day – or even a few hours – at a time. The fixing of the U-boat threat also allowed the Allies to maintain standing patrols in the danger area quite divorced from specific convoy movements. For example, by early 1944 the support groups operating in the area west of Ireland came to be known as 'Task Force 31.' Its groups, with some notable exceptions, were mixed and rematched as situations warranted. One of the results of this transition in the nature of ASW is that the documentary evidence for support groups – once disconnected from convoy battles per se and before they had established an existence independent of convoy movements – is particularly sketchy. A new concept of anti-submarine warfare was therefore in development in the winter of 1943–4, one that came to dominate ASW for the last full year of the war and postwar period as well.

Submarine and A/S warfare was also entering an new era, in which the protagonists were increasingly unlikely to see one another. In December 1943, when the U-boat fleet began to operate primarily submerged, Admiral Horton advised his forces that they were seldom likely to see U-boats during daylight hours. The result would be much lower German effectiveness and greater safety for convoys. But there would also be fewer chances to sink submarines. Horton was not content to sit on his victory. He wanted the U-boat fleet hounded at every turn. Escorts now had the time to prosecute hunts at length, and Horton warned them to expect to be ordered back to a 'battle area' to mop up the U-boats after a convoy had passed. Hedgehog or the new A/S mortar squid were the weapons of choice. If the U-boat was too deep, 'creeping attacks' by two or more ships using depth charges were the solution.[4]

Horton's concern reflects the simple fact that, with sharply reduced U-boat mobility, the northeastern Atlantic remained the main theatre of anti-submarine activity for the balance of the war. In contrast operations by lone submarines in other areas – Iceland, Canada, the northeast United States, and Gibraltar – offered no clear 'concentration' upon which to fix. ASW in the western Atlantic focused on searches for specific U-boats based on various data, from HF/DF or Ultra fixes to the flaming wreckage of stricken ships. With so few targets in such a vast theatre, A/S searches in the Canadian zone during the last eighteen months of the war were difficult at best. Extremes of weather and one of the most complex underwater acoustical environments in the world (of which more later) severely hampered operations as well. For all these reasons the RCN preferred to concentrate its offensive forces in the eastern Atlantic during 1944–5. As a result, the navy's response to incursions into the Canadian northwest Atlantic remained, for the most part, in the hands of hastily assembled, ad hoc hunting groups, further reducing the likelihood of U-boat kills in the Canadian zone. Only the Americans found the northwest Atlantic a fruitful area for U-boat hunting in the last year of the war.

The significance of these factors – the change in concept, the tactics, the futility of searches in home waters, the success of forces overseas – all remained to be demonstrated in late 1943. Murray was cautioned in the end of November that the RCN's priority was still to put its best into the close escort groups of MOEF and to modernize the corvette fleet.[5] The British decision to retire many of their aged destroyers, particularly the Town class, in late 1943 put an added strain on RCN escort forces. Of the eleven short-range British destroyers in Canadian waters, nine returned to the United Kingdom in December. The remaining RCN Town class ships were then grouped into a local support group, W 10, but they too were slated for retirement. The loss of these destroyers meant that the RCN's Western Escort Force, which protected the main transatlantic convoys for fully a third of their crossing, would be reduced to a few corvettes and minesweepers. A WEF without destroyers was a radical idea in late 1943, and it appeared that new frigates, already earmarked for destroyer work in the mid-ocean groups, would now also be needed for escort duty closer to home.

After the utter and unquestionable collapse of the U-boat campaign at the end of 1943, however, it was time to wean the RCN away from its concentration on modernization and close escort. The arrival of the first of the 'Increased Endurance' corvettes and the River class frigates seemed to obviate the need for continued concentration on fleet modernization. The Naval Staff even toyed with abandoning modernization of the old corvettes at the end of 1943 but carried on because of concerns for morale. None the less, by late 1943 there was a clear sea-change in Canadian policy regarding the relative importance of close escorts and support groups. 'As the majority of successes against U-Boats are secured by Support Groups with time to carry out thorough hunts,' the 'Summary of Naval War Effort' for the last quarter of 1943 observed, 'it is felt that Canadian ships should have a larger part in them and be enabled to win more victories in the future.'[6]

The decision to abandon the navy's earlier conservatism in escort assignment may well have originated with the new chief of Naval Staff, Vice-Admiral G.C. Jones. By January, shortly after Jones assumed his new position, NSHQ was admonishing Murray to harass U-boats, keeping at them night and day. Echoing Horton's signal in December, the Naval Staff now wanted submarines sunk as a way of 'breaking the morale of the crews.'[7] It took some time for the RCN to make the switch. More importantly, it is not at all clear that Murray and his staff in Halifax were as keen as those in NSHQ to throw all the best to Horton in the eastern Atlantic. Murray had his own problems with U-boats. Moreover, he and Jones were bitter rivals,[8] and there was little love lost between Prentice and the new regime in Ottawa. These tensions surfaced over the winter of 1943-4 as the navy struggled to shape its role in the last year of the Atlantic war.

Murray, of course, wanted to deal effectively with U-boats operating within his zone, a zone that was, at once, too big to be secured effectively from all intruders and yet paradoxically too restricted to handle the new situation in the Atlantic. Allied intelligence was able to track U-boats on passage to the Canadian zone, and Murray's aircraft were at liberty to pursue them to the eastern limits of their range. But the control of naval forces transferred to the Americans just outside Murray's small zone, or to Horton at 47°W. In practice Murray ignored that limitation, but operations by his naval forces outside the Canadian northwest Atlantic poached on other turf. And of course the RCN had no control over British or American groups operating on the fringes of the Canadian zone. The changing nature of the war had, by the end of 1943, made a nonsense of the command boundaries established at Washington nine months before. The mid-ocean was no longer the key theatre; rather, it was the coastal zones, where submerged U-boats lay, extending to the mid-Atlantic that constituted the new locus of action. Horton had command of his crucial zone. But the approaches to the Canadian zone were carved up between the British and Americans. In fact, the Canadian and U.S. approaches overlapped east of Nova Scotia, and the USN operated its own hunter-killer groups against westbound submarines. Under these circumstances it was difficult for Murray to argue that he needed offensive forces of any size.

Murray's only solution was to ignore the limits of his command. He did so in the lengthy search for *U543* on the Grand Banks at the end of 1943. *U543* was the first of three U-boats to patrol in the vicinity of Flemish Cap, a 50- by 100-mile expanse of shallow water 400 miles east of St John's (see endpapers). The U-boat was fixed by HF/DF stations just before Christmas, and on Boxing Day a 'hunt to exhaustion' was ordered. Between 26 December and 6 January, when the hunt was finally abandoned, a series of search zones was promulgated which, in total, equalled an area the size of the Bay of Biscay.[9] To find the lone intruder the RCAF flew twenty-one Liberator and seven Canso sorties. Despite intelligence that the U-boat would remain submerged during the day, only three patrols occurred at night. With only 76 per cent of the search area swept during the day and virtually none at night, it was not surprising that no RCAF aircraft gained contact.[10]

The navy was scathing in its criticism of the RCAF's contribution to this 'hunt to exhaustion,' but its own effort left much to be desired. The only support group available, W 10 with the aged Town class destroyers *Columbia* and *Niagara*, was ordered to the search area on 26 December, and two British frigates, *Goodson* and *Balfour*, were diverted from escort duties. The next day Admiral Murray sent help in the form of an ad hoc group from Halifax composed of frigates still undergoing work-ups: the core of the new EG 9, *Matane*, *Stormont*, *Swansea*, and *Montreal*, and the destroyer *St Laurent*. Over the next two weeks ships came and went, particularly the destroyers of W 10, which operated at the limit of their range. *U543* was well aware of the presence of searching forces; she could detect their radar transmissions and remained submerged during the daylight hours.[11] She had to surface at night to recharge batteries, however, and that need made her vulnerable. It was during one of those occasions, late in the evening of 2 January, that *Swansea* made a radar contact which she classified as a definite U-boat. That classification was confirmed about a half an hour later when an explosion was heard nearby, believed at the time and confirmed since to have been a torpedo from *U543*. Despite the certainty of the contact, twelve hours of underwater searches in the area proved fruitless.[12] In all likelihood *U543* slipped beneath a sharp temperature change layer (thermocline) in the water, which the hunting ships' asdic could not penetrate. Alan Easton, *Matane*'s captain, recalled, 'We searched with, against and across the breaking seas for three weeks in the teeth of icy winds which cut our faces and made our eyes bloodshot. We only succeeded in stopping her broadcasts.'[13] *U543* escaped further detection and on 6 January the search for her was called off.

The hunt for *U543* demonstrated that Murray, too, needed a

support group with sufficient endurance to tackle U-boats well offshore. Unfortunately, there seemed to be no official requirement for such a group, nor was there much likelihood of his acquiring one. Frigates were entering service, but slowly. This delay was true of both those built in Canada and those the RCN hoped to acquire from the British. By the end of 1943 sixteen RCN River class frigates were in commission, but only three were yet on operations: *Dunver* in C 5, *St Catharines* in C 2, and *Waskesiu* in EG 6. Some delays were occasioned by the decision to modify new ships to accommodate senior officers and some by teething troubles, and some of the sixteen frigates in commission at the end of 1943 were rushed into service to get them clear of winter ice. The real deluge of new construction had to await the opening of the St Lawrence River in the spring. The net result was that RCN River class frigates took a long time between commissioning and assignment to operations over the winter of 1943–4: four months was average and some like *Wentworth* took six.[14] There are also indications that Captain (D), Halifax, was not prepared to let the new ships go off to sea ill equipped or poorly worked up – or too quickly. Setting the stage for the battle between Halifax and NSHQ over training and allocation of the new frigates in the winter of 1943–4 therefore requires a slight digression.

Concern for the dwindling strength of WEF in part governed RCN considerations of frigate assignments and the speed with which they were committed to operational groups. So long as the ships were working up at Halifax, Murray had a semblance of a strike force and Prentice seemed to show the same enthusiasm for training his ships at sea against the enemy as he had earlier in the war. Once the frigates were committed to operations, Murray was likely to lose them to Western Approaches Command, which already controlled the best of the Canadian A/S fleet (including MOEF ships).

Prentice retained responsibility for work-ups of new construction throughout 1943. His staff was small, he had operational groups to look after, Halifax was crowded, and winter weather off the Canadian coast was often brutal. From early 1943 the Naval Staff wanted to move the work-up of new construction out of Halifax. It preferred to have Canadian ships go through work-up at the British establishment in Bermuda, which was set up to handle American-built ships, and obtained permission to do so. Prentice objected. Training of RCN ships he saw as his responsibility. Admiral Murray supported him, although for other reasons. 'From the operational point of view,' Prentice wrote Murray in March 1943, 'the training group (at Halifax) forms the only reserve of Canadian ships available between St John's and New York.'[15] And so it would for another year.

NSHQ won the battle in early 1943 and the frigates were ordered to work up at Bermuda. Lt Cdr M.A. Medland, the training officer of the Directorate of Warfare and Training in Ottawa, also wanted to see a separate 'sea training command' established on the east coast entirely divorced from Captain (D), Halifax. He was supported by Captain A.R. Pressy, then the interim head of the DWT. The problem was thrown back in the Naval Staff's hands in August 1943, when the Admiralty rescinded its offer of British facilities for Canadian use, effective 1 October. The reasons for this change of heart are unclear. It is interesting to note that the withdrawal of the offer to let RCN ships work up at Bermuda was made about the same time that Commodore (D), Western Approaches Command, told Captain Strange that something quite unorthodox was needed to shake up NSHQ. It was also about the same time that the British rescinded their offer to refit a number of Canadian corvettes in U.K. yards. It stretches credulity to see this retraction as part of some vast conspiracy, but these developments are consistent with a widespread belief within the RN that the Canadians had got into a complete muddle and ought to sort it out themselves.

Only *Waskesiu* and the new minesweeper *Sault Ste Marie* worked up under the RN in Bermuda in 1943. The issue refused to die and the situation on the east coast was exacerbated by severe winter weather at the end of the year off Nova Scotia which impeded all attempts at local training. In late December the RCN sought permission to 'base' ships under training at Bermuda. Their only demand on the local RN establishment would be occasional fuel, emergency repairs, and anchorage for four to five ships. The local British admiral suggested that the Canadians use Jamaica, but the Admiralty relented. By February four new ships and one of the training submarines assigned to the RCN were exercising off Bermuda.

The case for a separate 'sea training command' was renewed in January by the new Director of Warfare and Training, Captain K.F. Adams. With a flood of new construction expected in the spring, something clearly had to be done, and the ineffectiveness of the 'hunts to exhaustion' conducted by ships still under training was not lost on the staff in Ottawa. Moreover, the RCN now had its eye on the RN work-up establishment at Bermuda, which was scheduled to close once the wave of American-built escorts passed in the spring of 1944.[16] The Naval Staff's victory came in February, when Murray finally agreed to the establishment of a sea training command separate from Captain (D), Halifax, and that Bermuda was the logical place. Two obstacles – Prentice and the British – remained.

Whether the RCN's frigates would have entered operational service faster in the winter of 1943-4 had Captain (D), Halifax, not insisted on retaining control over ships under work-up and employing them on local hunting tasks remains a mystery. The importance of frigates to the RCN and the delays in getting its own into commission in numbers were probably at the root of the Canadian suggestion, made in November, that the RCN man ten British-built frigates by February 1944. These ships would have filled the most pressing gaps in the fleet over the winter of 1943-4. NSHQ appears to have believed that they could be taken over and sailed directly into battle. The ships were slow to arrive, however, and the first to do so were badly in need of work. *Ettrick*, commissioned into the RCN at Halifax on 29 January 1944, required a refit that lasted until April, and she did not return to operations until early June. *Tweed* was to be next. Her loss upset plans and resulted in *Nene* not being handed over until April. *Meon* was commissioned into the RCN on 7 February and was not assigned to operations until May. Not until October 1944 – nearly a year after the RCN made the offer – were all ten frigates operational with the RCN.[17]

While the staffs in Halifax and Ottawa haggled over responsibilities and escort allocations, the support group already on operations in the eastern Atlantic got on with sinking U-boats. EG 5, now redesignated EG 6 to avoid confusion with C 5 of the MOEF, ended 1943 supporting a series of convoys passing through the danger area southwest of Ireland. Apart from the rescue of seventy-four of blockade runner *Alsterufer*'s crew from their boats,[18] the first days of January 1944 passed quietly, too. The corvettes were forced to return to Londonderry to refuel and to land their prisoners. The frigates refuelled at sea, and then on 6 January *Nene*, *Tweed*, and *Waskesiu* set off on a combination of blockade-runner patrol and pursuit of HF/DF fixes on transmitting U-boats. It was while pursuing one of the latter that *Tweed* was sunk by a single GNAT fired by *U305* on 7 January. The attack came without warning and the frigate sank quickly. While *Nene* rescued survivors, *Waskesiu* searched for the attacker, whose periscope was briefly glimpsed. George Devonshire, a seaman torpedoman on *Waskesiu*, recalled that one of the counter-attacks was delivered close – perhaps too close – to some of *Tweed*'s survivors floating in their life jackets.[19] The search for *U305* was unsuccessful. *Nene* then took *Tweed*'s survivors back to Londonderry while *Waskesiu* joined the corvettes, now augmented by *Camrose*, who were back at sea supporting OS 64. That convoy was combined with KMS 38, escorted by the RN's EG 4 under Cdr E.H. Chavasse, for passage through the danger area.

As the combined convoy OS 64 / KMS 38 sailed south on the evening of 8 January, Chavasse's Captain class frigate *Bayntun* obtained a radar contact about nine miles ahead, which she drove down and attacked with depth charges. *Camrose* and *Edmundston* were ordered to assist and arrived to find *Bayntun* with her asdic, radio-telephone, and one engine out of service from the shock of her own depth charges. While *Edmundston* screened *Bayntun*, *Camrose* gained contact with the U-boat, assumed that the submarine was deep, and attacked with a ten-charge pattern. Contact was then lost. Chavasse ordered Lt Cdr L.R. Pavillard, RCNR, of *Camrose* to take charge of the hunt and pursue it for twelve hours. *Bayntun*, having made repairs, later rejoined. While *Edmundston* and *Snowberry* conducted square searches around the area, *Camrose* and *Bayntun* attacked a series on underwater contacts. All of the charges were set for great depths, although later assessments concluded that the U-boat was not particularly deep. None the less, the attacks were accurate in plan, and it was agreed that the sixth attack, made by *Camrose*, probably mortally damaged the U-boat, producing both oil and wreckage. Pavillard was chastised later for not using his hedgehog, but the error in overestimating the U-boat's depth may have contributed to his reliance on depth charges. The result in the end was the same: *U757* was destroyed, *Tweed* was avenged, and EG 6 had its second kill.[20] EG 6 stayed at sea until 21 January, when it returned to Londonderry for a much deserved rest. It had been a busy month. *Waskesiu* logged thirty-two consecutive days at sea, refuelling from convoys but in the end running 'uncomfortably short of food stores.'[21]

By the middle of January EG 6 was finally joined in the support role by C 2, which sailed on its first such operation in support of ONS 27 on 15 January. C 2's mixed group followed the same pattern as that of EG 6. The destroyers *Gatineau* and *Icarus* (RN) were often detached for special tasks, while the corvettes, *Chilliwack, Fennel,* and *Morden,* along with the senior officer in the frigate *St Catharines,* stayed close to the convoys themselves. And again, operations were concentrated in the area west and south of Ireland, where the patrolling U-boats lurked. C 2's first patrol, which lasted a week, passed largely without incident and was characterized by severe weather, which sharply reduced its effectiveness and was hard on the men in the corvettes.[22]

The problem of corvette endurance and habitability did not go unnoticed. Even the newer 'increased endurance' corvettes, with the same endurance as frigates and fitted with modern sensors and weaponry, suffered from heavy weather and lack of plotting facilities. As Peter Gretton observed following his own experience in a support role in November 1943, corvettes 'were handicapped by insufficient speed, lack of plotting facilities and their size and movement in rough weather imposed great strain on personnel after long periods at sea.'[23] The Canadian experience was no different, and it reduced the effectiveness of RCN support groups in the winter of 1943-4. For example, in the sixteen-day period between 9 and 25 February, when EG 6 supported five convoys and did an A/S sweep, its corvettes were detached twice – on the 16th and 24th – to Northern Ireland to refuel. In many ways, EG 6 and C

2 were support groups in name only. In practice, they augmented close escort groups and allowed the best to spend time pursuing contacts, much as had happened with ONS 18 / ON 202 in September 1943. By way of contrast, the sloops of Walker's EG 2 were at sea at the same time (actually from 29 January until 24 February) for twenty-seven days without interruption.

The superior quality of Walker's sloops was evident to the Canadians, and the director of Warfare and Training requested at the end of December 1943 that the RCN acquire some of them. They could not be built in Canada, however, and, in any event, escort-acquisition policy was already set.[24] The ultimate solution, and one towards which the RCN moved in early 1944, was all-frigate support groups. Canadian frigates under operational conditions had an eighteen-day endurance at twelve knots, and much A/S searching required slower speeds.[25] Meanwhile, the two components of RCN support groups typically found different uses. Indeed, at one point in early February the corvettes of both C 2 and EG 6 were temporarily combined into 'EG 16,' while the large ships of the two groups operated briefly as 'EG 18.' None the less, by mid-January the RCN finally had two support groups in operation once again, representing half of the support-group strength then committed to the main convoys. It remained simply to get EG 9 into service and the RCN would be up to speed on its support-group plans of the previous November.

Sending EG 9 off to Horton was certainly the RCN's intent at the end of January 1944. After a final exercise off Newfoundland the group was to sail eastward, in support of SC 153 and HX 279, and then operate in the area of highest U-boat density.[26] Horton was anxious to have command of the group, and he prodded Murray on its state at end of January. The RN was planning to put a couple of escort carriers into the U-boat concentration west of Ireland and needed additional support groups to screen the carriers. Horton had only four: EG 2, EG 6, and two groups released from MOEF, B 1 and C 2.[27] He wanted EG 9 to escort the carriers.[28]

EG 9, composed of the frigates *Swansea* and *Matane* and the modified corvettes *Frontenac*, *Owen Sound*, *North Bay*, and *Atholl*, became operational under Cdr A.F.C. Layard, RNR, in *Matane* on 7 Febru-

Triumph without Celebration 109

ary. It was a good combination, all things considered, and showed a commitment on the part of the RCN to put its best new ships into offensive operations. The modified corvettes were the only Canadian ones yet in service. By the time EG 9 was ready, however, Murray was reluctant to send it overseas. The Canadians were watching the movement of another submarine onto the Grand Banks. Air patrols along its line of advance were flown in late January, and its anticipated area of operations, again off Flemish Cap, was too far afield for the aged destroyers of W 10 to maintain any presence. The frigates and increased endurance corvettes of EG 9 were another matter. On 8 February Murray informed Horton that 'owing to the submarine threat to convoys west of WESTOMP, propose to retain EG 9 in this area for hunting and supporting convoys until situation clarifies.' Horton pointed out, politely, that the U-boats were concentrated on his side of the Atlantic and that the 'probabilities of good killings are much better on this side.' For the time being neither Murray nor NSHQ could be persuaded.[29]

EG 9 was therefore retained to search for *U845*, believed to be off Flemish Cap early in the second week of February. On 9 February *U845* torpedoed the merchant ship *Kelmscott* off St John's, much farther west than the Canadians had anticipated.[30] The hunt to exhaustion ordered, in which EG 9 participated, ended three days later without success. On the 14th *U845* was relocated – this time off Flemish Cap – by an RCAF Liberator, which attacked the U-boat and gave it a good shaking. By that time a second U-boat, *U539*, had joined *U845* in the area, but both were too far seaward in the worst of winter weather to mount an effective hunt. The Canadians contented themselves with suppression of the U-boats, which was perhaps enough: *U539* left empty handed and *U845* scored no more hits.[31] Ironically, EG 9 would have another chance at *U845*, but not in the western Atlantic.

If the failure of the hunting off the Grand Banks was not enough to convince the RCN that the place to kill U-boats was in the eastern Atlantic, events in early February drove it home. At the end of January the RN began its carrier offensive against the U-boat concentration west of Ireland, *Activity* and *Nairana* sailing with EG 2 and *Fencer* and *Striker* with EG 16. The carrier operations were plagued by severe weather and terrible flying conditions: only *Fencer* recorded a kill.[32] For the surface escorts operating in support the situation was entirely different. Over a two-week period – a single patrol – Walker's sloops in EG 2 accounted for six U-boats. Most of these submarines went deep, but EG 2's skill at creeping attacks and barrages destroyed them, each in turn. One of the U-boats, *U264*, carried a new 'schnorkel' breathing device, the first to do so on operations. Walker's accomplishment was not without cost. *Woodpecker* had her stern blown off and was latter lost while in tow, but EG 2's patrol was exceptional – 'The most outstanding success of the Anti-U-boat war.'[33]

Walker's remarkable patrol in February 1944 may have owed something to preferential treatment by Western Approaches staff. Bill Willson, at sea over the winter of 1943–4 in command of *Kootenay*, recalled that Walker was given the frequency settings of U-boat radio traffic for the period of his epic patrol. According to Willson, this information allowed Walker's group to anticipate U-boats signals, 'and would have been as good as an extra radar or sonar.' Despite his protests to Western Approaches staff, Willson was never given the same edge and the RCN 'poked around convoys with information coming ages late.'[34] If Willson is right, it also meant that EG 2's searches were very well directed. It may be that the destruction of *U264*, the first schnorkel-equipped U-boat, was not simply blind luck.

Walker's exploits in the winter of 1943–4 also represented the perfection of anti-submarine warfare based on depth charges and fixed asdic transducers. Anti-submarine warfare up to the end of 1943 was largely predicated on the submarine's dependence on surface manoeuvrability, on the comparatively shallow depths to which submarines usually dived, and on their slow (three- to six-knot) submerged speed. At the start of the war, for example, depth charges had a maximum setting of 150 feet. By 1942 settings of 350 feet were possible, and by 1943 depth charges could be set to fire at 550 feet.[35] So long as the Germans pressed home their attacks on the surface and the escort's primary task was to ensure safe and timely arrival of the convoy, ASW tended to be both hasty and limited to a thin veneer on the surface of the sea; the technology, tactics, and doctrine of both sides aligned to make it so. The designed maximum diving depth of the type VIIC U-boat, the workhorse of the Atlantic war, was 100 metres, roughly 333 feet, although in an emergency they were capable of 500 feet or better.[36] For most of the war few U-boats had time to go that deep before the first depth charges exploded around them. If these first charges were well placed, the U-boat either sank or – more often – suffered enough damage to require an emergency surfacing, typically into a hail of gunfire followed by a ramming. If the escort's first charges missed, there was little time for it to develop a deliberate hunt before being called back to the convoy.

Prolonged anti-submarine hunts prior to mid-1943 were rare, and few escorts had the opportunity to track a U-boat at depths (in 1942 up to 600 feet) beyond the limits of existing A/S equipment. The Germans, too, understood how increased depth complicated the A/S problem, and some U-boat captains risked crushing depths to escape their hunters. In late 1941 the type VIIC U-boat design was modified with heavier plates in the pressure hull. The new type VIIC/41 had a standard diving depth of 400 feet, a test depth of 600 feet, and a destruction depth of as much as 1,000 feet.[37] Plans for a type VIIC/42 pushed these limits even further, to 666, 1,000, and 1,666 feet, respectively. The proposed type VIIC/42 was overtaken by the decision to build the type XXI, and it was never built. By August 1943, however, the first of the type VIIC/41 U-boats began to appear, and even the older types were pushing the limits of their pressure hulls as a way of escaping searching vessels.

The ability of U-boats to dive very deep did not surprise the Allies. But by the fall of 1943 escort commanders complained that while they could often track U-boats at great depths, current depth charges could not reach them. Cdr E.H. Chavasse, senior officer of EG 4, recalled being reprimanded by Western Approaches Command in late 1943 for excessive expenditure of depth charges with no U-boat to show for it. 'With some acidity,' Chavasse wrote after the war, 'I pointed out to my superiors that U-boats, when attacked, had recently developed the habit (and ability) of diving to the previously unheard of depth of 800 feet ... and were jolly nearly safe as houses.'[38] That was not strictly true. New pistols for depth charges were just coming into service: the Mk IX^X had settings to 700 feet and the Mk IX^{XXX} could get down to 875 feet. In addition, depth charges fitted with the new, 70 per cent more powerful, explosive 'minol' were now available, although they were initially thought to be too powerful for use by slower escorts like corvettes.[39]

But new pistols with deeper settings were not a solution to the problem. Depth charges had to be laid astern of the attacking vessel, and the charges themselves took a long time to sink. Submariners could easily anticipate the depth-charge attack because of the increase in the attacking vessel's speed, which was intended to keep it clear of its own charges. The run-in to the firing position and the long descent time of the charges gave the U-boat ample time for evasive action. North Atlantic veterans had long known that the best way to overcome this problem was to employ two ships. One held the U-boat in asdic contact, preferably astern of the target so the submarine could not hear the sound, while the other was directed over the target at a slow speed (hence the term creeping attack). The order to fire was normally given by the contact keeper, who maintained a plot of both the attacker and the target. In essence, the contact keeper used a second ship as a form of 'ahead-throwing weapon.' *Skeena* and *Wetaskiwin* used this method to sink *U558* in July 1942, and EG 2 had used it in the fall of 1943 to sink *U226* and *U842* on the Grand Banks. The Admiralty promulgated the tactic in August 1943, but the success of early February 1944 brought creeping and barrage (more than one attacking ship) attacks to prominence.[40]

Even as EG 2 boiled the ocean with its depth charges, the wave of the future was upon them. A new support group, EG 10, operating in the vicinity of EG 2 and supporting the same convoys, sank two U-boats on 18 and 19 February using the new type 147B asdic and the Q attachment on the type 144 set. The type 147B, it will be recalled, produced a thin, fan-shaped beam in the horizontal plane, and the transducer was articulated so that the angle of the setting coupled with the very thin beam could be translated into an accurate depth. The Q attachment produced a narrow fan-shaped beam in the vertical plane and was slaved to the main set but aligned at a sharper angle in order to maintain contact on a U-boat

Triumph without Celebration 113

until the last possible moment of the run-in to fire depth charges. In the main asdic-Q system the depth of the contact had to be estimated from the range at which the fixed beam lost the target: a simple question of geometry. As the main set swept over the target during the approach, the asdic operator maintained contact by switching to his Q attachment. This combination of skill, experience and educated guesswork was not necessary with the type 147B set. It provided accurate measurements of depth of the target right up to the last moment.

The British frigate *Spey* of EG 10 had all of this asdic equipment when she encountered *U406* and *U386* while supporting convoy ON 224 in mid-February. *U406* was found on asdic at 1,700 yards and using 'Q' and the main set, *Spey* held her as close as 300 yards: indication of a shallow target. No creeping or barrage attacks were necessary. Only one ten-charge pattern was dropped, which inflicted mortal damage on the U-boat. *Spey* obtained her second U-boat contact on asdic less than a day later at a range of 1,800 yards. The frigate began her attack at 750 yards with depth charges set shallow, to 50 and 140 feet. Moments later *Spey*'s type 147B revealed (still 500 yards from the target) that the U-boat was actually 350 feet deep. Depth-charge parties were immediately ordered to reset the charges to 250 and 385 feet, and they barely had time to do so before the first left the ship. As *Spey* turned to renew contact, *U386* broke surface 800 yards away and tried to escape. Once again the U-boat was riddled with gunfire. *Spey*'s captain resisted the temptation to ram, contenting himself with shattering the U-boat's final resolve with a shallow pattern of depth charges. Sixteen submariners were recovered.[41]

The difference between the methods of EG 2 and those of EG 10 were apparent and were commented on by the Admiralty. Walker's methods required considerable time and many depth charges. His kills averaged four hours and 106 depth charges each. 'The danger of the creeping attack,' the *Monthly Anti-Submarine Report* for February 1944 observed, 'lies in the fact that it tends to develop into a long, drawn-out struggle during which asdic conditions or weather may deteriorate, thereby causing the hunting vessels to lose contact and giving the U-Boat the opportunity to escape.' Creeping attacks also required 'exceedingly close cooperation between ships and almost unlimited time and depth charges,' and it was necessary to force the enemy to go deep. In contrast, *Spey* destroyed her two U-boats in minutes, blowing them to the surface with single ten-charge patterns and then riddling them with gunfire. 'It is obviously tactically advantageous,' the *Report* went on, 'to kill the quarry as quickly as possible and the latest developments in asdic and anti-submarine weapons are designed to assist in the accomplishment of this object.' While the RCN could not hope to compete with Walker's success, Canadians understood his methods and used them successfully.

The RCN's skill at deep-ocean ASW was demonstrated in late February and early March during an exceptional three-week period. It began when EG 6 sailed from Londonderry on the 7th for another patrol southwest of Ireland. The first few days were routine, with the usual false contacts and refuellings at sea. A week of rough weather followed, which forced the corvettes back to Northern Ireland for fuel on the 16th. As they left, *Nene* and *Waskesiu* and the destroyer component of C 2 (*Gatineau, Icarus,* and now *Chaudière* – the last of the escort destroyers promised by the RN a year before) went to help ONS 29 – the convoy that *Spey* was supporting when she got her two kills. The corvettes rejoined briefly on the 18th, but poor weather forced them to refuel at Moville again on the 22nd, by which time *Nene* and *Waskesiu* had joined SC 153, escorted by C 5, as it moved through the danger area. They were joined by all of C 2, EG 3, and EG 1. Eventually twenty-seven escorts were assigned to defence of SC 153, an exceptionally strong force.[42]

It was while on a extended screen around SC 153 in the early hours of 24 February that *Waskesiu* got an asdic contact on *U257*, a type VIIC U-boat on her way home after seven uneventful weeks at sea. *Waskesiu* attacked immediately with hedgehog. Since contact could not be maintained closer than 400 yards, however, it was considered too deep for hedgehog. Working alone, in the dark, *Waskesiu*'s commanding officer, Lt Cdr J.P. Fraser, RCNVR, now conducted a variation of the creeping attack. He marked the location of the U-boat with a flare and then ran in quickly to drop a ten-charge pattern. A second attack followed with charges set to 350 and 550 feet.

The third attack, again delivered largely by eye, was thought to be well placed, but contact was then lost. When Birch arrived in *Nene*, both frigates obtained contact, but following another attack by *Waskesiu* neither ship could regain it. A search re-established contact about two hours later, when it was decided that the U-boat was very deep and that it was best to hold it until morning when a proper creeping attack could be arranged. Birch, however, tired of waiting. When his asdic operator reclassified his contact as 'non-sub,' *Nene* set off for the convoy. Fraser was given permission to fire one more pattern before he too was to give up the search.

Birch picked a poor time to leave, but the sound of his departure may have triggered the final act of the drama. Thirty minutes after *Nene* left, *Waskesiu*'s asdic operator reported that the U-boat was no longer deep: for whatever reason the captain of *U257* was gradually bringing his boat to the surface. Fraser had depth charges set for 350 to 550 feet and launched his final attack. As the frigate opened the range to renew contact, *U257* could be heard blowing her ballast tanks. She surfaced 1,800 yards ahead of *Waskesiu* into an area already illuminated by star shell from the frigate, and to a welcome of 20-mm and 4-inch gunfire. *U257*'s crewmen were cut down as they attempted to man the guns, and not a shot was fired in return. *Nene* rejoined in time to fire a few rounds and to illuminate the sinking sub with her searchlight: nineteen survivors were recovered. It was the first kill by a Canadian frigate.[43]

Fraser's action drew praise from all quarters. Commodore (D), Western Approaches, commented that *Waskesiu* had done very good work, a sentiment shared by the staff of the Anti-U-Boat Division of the Admiralty. Horton commented to the Admiralty that 'H.M.C.S. *Waskesiu*, Lieutenant Commander J.P. Fraser, RCNVR, showed commendable determination in maintaining contact with a target which could not be confirmed by H.M.S. *Nene*... and is to be congratulated on surfacing *U-257*, subsequently destroyed.'[44] Hard words were saved for Birch, who one officer charged with conducting a 'homeward bound' style of hunt, ready to give up the chase. Perhaps for that reason *Waskesiu* was rewarded for her perseverance with sole credit for the kill.[45]

If the RCN needed evidence that the place to destroy U-boats was in the eastern Atlantic, the events of February provided it. The day after *Waskesiu*'s destruction of *U275*, and in the wake of Walker's exceptional cruise, the Naval Staff agreed to transfer EG 9 to Horton's command: they would not be disappointed.[46] In the meantime it was C 2's turn to sink a U-boat, in one of the classic hunts of the north Atlantic war.

On the morning of 5 March C 2 was deployed in an advanced screen five miles ahead of HX 280, escorted by B 7 with Gretton now in the frigate *Chelmer*. U-boats were known to be nearby and *Icarus* had already obtained a contact with her HF/DF set. So it was no surprise when *Gatineau*, on her way to Londonderry for repairs, obtained an asdic contact at about 1000Z. The U-boat presented no immediate threat to the convoy. *U744*, a type VIIC commanded by Lt Blischke, had been submerged for the previous six hours. Blischke was drawn towards HX 280, unsure of what it was, by hydrophone noises. While *Gatineau* positioned herself for an attack, Burnett arrived in *St Catharines* to take charge of the hunt. Having got the measure of the enemy, Burnett lined up *St Catharines* for a hedgehog attack only to discover that *U744* was too deep. *Chilliwack*, which arrived a few minutes later, was allowed to attack with hedgehog despite the great depth of the target; it produced no results. It was at this point that C 2 realized it was up against no ordinary U-boat commander. Blischke moved *U744* inside the larger ships' turning circles and hid his U-boat in the tangle of wakes and disturbed water following each attack.

With *U744* too deep to attack with hedgehog, Burnett switched to creeping attacks delivered over the next few hours by *Gatineau*, *Chilliwack*, and *Icarus*. These assaults were followed by another pointless hedgehog attack by the British destroyer. The threat to HX 280 from other U-boats forced Burnett to detach *Chaudière* and *Icarus* to the convoy in mid-afternoon, while *Gatineau* set off for Londonderry. Only *St Catharines*, *Chilliwack*, and *Fennel*, were left to continue the hunt. To assist them, Gretton dispatched the new Castle class corvette *Kenilworth Castle*, complete with type 147B and squid, the first use of squid on a deep U-boat. *Kenilworth Castle* arrived shortly before 1500Z and delivered three squid attacks, the second and third punctuated by a series of creeping attacks. The third squid attack was plagued by asdic failures and by *U744*'s now familiar zig-zag manoeuvres at 600 feet. All seemed to no avail at the time. Prisoner-of-war evidence later revealed that the shock of *Kenilworth Castle*'s third attack shook the U-boat and she had trim problems thereafter. As darkness closed in, *Fennel* and *St Catharines* made the last creeping attack of the day. Without a destroyer on hand to direct night attacks with the aid of a range finder, Burnett decided to simply hold the contact and wait until morning.

Attacks were made on *U744* just after midnight, 'to dispel any ideas he might have of counter-attacking with acoustic torpedoes,' and when the weather deteriorated during the night, there was some concern that contact would be lost. Burnett held on to *U744* from *St Catharines* and organized the corvettes around the contact in a box to ensure that it would not escape. *Icarus* and *Chaudière* rejoined at 0500Z on the 6th, but Burnett waited until full daylight (three hours later) before the pounding of *U744* began again. From 0751Z until after 1000Z C 2 tried hedgehog and creeping attacks, with no apparent success. In fact, by 1000Z *U744* was seriously damaged. The diesel engine base plates and cylinder block were broken, cooling water poured from the engines, the pressure hull leaked, and the air purifier was damaged. Moreover, *U744* had been submerged for over twenty-four hours and her crew was weakening from the stale air. Blischke made one last bid to escape in the disturbances of the final attacks, and contact was briefly lost until *Icarus* regained it at 1145Z. Frustrated after over a day of trying everything in his arsenal and now affected by a rising sea, Burnett instructed C 2 simply to maintain contact and wait. Exhaustion of the U-boat seemed the only solution.

The next series of attacks, scheduled for 1600Z, was intended merely to keep *U744* in check. *Fennel* was given a warning order at 1520Z and was preparing her depth charges when *U744* suddenly surfaced directly ahead of *Chilliwack*. The corvette, having heard the blowing of ballast, was ready and smothered *U744* in gunfire. The first round from the 4-inch gun blew one set of anti-aircraft guns off the U-boat and others pierced the conning tower. So too did rounds from the 2-pounder pom-pom, while incendiary rounds from the 20-mm guns caused several small fires. *U744*'s crew was too exhausted to offer any resistance. When this fact became evident, firing ceased; the hunt ended thirty-two hours after it began, the second longest of the war.

As *U744* lay wallowing under the guns of C 2, something of a race began between several ships to get boats alongside, a race eventually won by *Chilliwack*, but not without incident. The heaving seas capsized the whalers from *Chilliwack* and *St Catharines*, leaving the motor boat and whaler from *Chaudière* and *Fennel*'s whaler the task of rescuing friend and foe alike. In the end all the RCN personnel and thirty-nine Germans were recovered. *U744*, a tempting prize but heavily damaged and settling slowly, was torpedoed by *Icarus* at 1830Z on 6 March.[47]

The destruction of *U744* was a signal accomplishment. Burnett, of course, had more than a little to do with it. As Horton noted, 'The hunt and destruction of *U-744* is a classic example of Anti-U-Boat warfare in which the operations of the opponents were conducted by experts of their profession.'[48] The length of the hunt and C 2's inability to destroy the U-boat outright were attributed to the excellence of the U-boat captain. 'The commanding officer of *U-744* was evidentially an expert,' the director of the Anti-U-Boat Division of the Admiralty wrote, 'and by means of violent alterations of course at a depth of about 500 feet managed to avoid lethal damage from 24 attacks which were made on him.'[49] Burnett might have done better, one staff officer observed, had he kept *Kenilworth Castle* at hand to provide depth readings from its type 147B asdic. As it was, the failure of the squid attacks was attributed to poor training and particularly to the sound of a U-boat at extreme depth, which was entirely new and unexpected. These shortcomings were understood, and the asdic operators of *Kenilworth Castle* apparently retained 'an undiminished faith in their asdic gear and the squid.'[50]

Burnett later recalled that his slow and deliberate development of the hunt for *U744* was based on his realization that the asdic officers of C 2 were 'inexperienced.' This was no indictment of the RCN; Burnett held the same opinion of the asdic officers of his British group, EG 10, later in the war. Rather, he simply wanted to ensure that *U744* was firmly held before attacking and to ensure that a perimeter was maintained around the hunt to prevent escape. These goals he achieved superbly, and the ships of C 2 performed well under often difficult asdic conditions. The destruction of *U744* in the second longest A/S hunt of the war was the kind of dramatic victory the RCN had long needed.[51]

Happily for the Canadians, there were more successes to come. The next U-boat kill occurred around convoy SC 154, escorted by C 1. EG 9, which was en route to Western Approaches Command in support of nearby HX 280, was ordered to assist SC 154 on 2 March. Bad weather delayed EG 9's arrival, and several of the group's corvettes detached to the Azores for fuel. By the 10th the frigates *Swansea* and *Matane* and the corvette *Owen Sound* had joined the convoy, as had *St Catharines* from C 2. U-boats were known to be in the vicinity. Their HF signals were plotted by the escorts, but the contact with *U845* was entirely fortuitous. On the morning of 10 March the destroyer *St Laurent*, commanded by Lt Cdr G.H. Stephen, RCNR, was well astern of the convoy helping the crew of the *San Francisco* to fight the fire that had caused her to straggle astern of SC 154. U-boat transmissions were plotted between the lagging ship and the main body of the convoy, and so *St Laurent*, which had left *San Francisco* in the morning to rejoin SC 154 and was now in company with *Owen Sound*, set a course to investigate one of these transmissions at 1500Z about thirty miles astern of the convoy. Meanwhile *U845*, the large type IXC commanded by Lt Cdr Weber that had eluded Canadian searches on Flemish Cap in February, was attempting to close SC 154 from astern as well, largely on her electric motors. With the batteries nearly exhausted, Weber surfaced to restore power. Fifteen minutes later he was spotted by a lookout on *St Laurent*. To escape, *U845* went deep, at times to as much as 700 feet, where the pressure was so great that submarine bubble targets – a cylinder ejected from the submarine that released a cloud of bubbles simulating a U-boat echo – could not be released.

Once again a macabre dance of death ensued. Asdic conditions were good and the escorts were able to maintain contact. After three depth-charge attacks by *Owen Sound* and *St Laurent*, a hedge-

120 The U-Boat Hunters

hog attack was tried. It failed because of the extreme depth of the U-boat. Once that fact was determined, a series of creeping attacks by *St Laurent* and the British destroyer *Forester* were tried, following which contact was temporarily lost. Twenty minutes later, in gathering darkness, *St Laurent* regained contact. Stephen now decided to wait until the moon rose to provide illumination and until *Swansea* joined, before proceeding with more attacks. As the three escorts were preparing to renew the action, *U845* suddenly surfaced 400 yards away, manned her guns, and set off at high speed. Weber hoped that clouds would obscure the moon and that his pursuers, which he believed were corvettes, would be too slow to catch him. He was wrong on both counts, but he nearly escaped all the same.

U845 raced away at full speed – contemporary accounts say twenty-one knots, about three knots faster than the U-boat's rated speed[52] – and, while steering wildly evasive courses, maintained a steady return fire with her heavy anti-aircraft armament. The chase lasted an hour. Initially only *St Laurent* responded effectively, streaming her CAT gear and setting off in pursuit. *Forester*, which had arrived to replace *Owen Sound* (sent to look after *San Francisco*), took time to work up to speed. Meanwhile, *St Laurent* fired away at *U845*, at times correcting her fire with radar-generated ranges. Eventually the superior gunfire from the escorts began to tell. *U845* slowed and *St Laurent* surged past, firing every gun that could bear and laying a shallow-depth charge pattern in her path. By the time Stephen turned his destroyer around, *Swansea* and *Forester* were picking up survivors. It had been a furious night battle, *St Laurent* alone expending 119 rounds of 4-inch, 1,440 rounds of 20-mm, and 1,400 rounds of small-arms ammunition. *U845*'s conning tower was half shot away, and Weber lay dead on the deck. As one of *Swansea*'s officers observed, 'No one in the conning tower during the battle could have escaped.'[53] *U845* sank at 2348Z on 10 March: forty-five of the U-boat's fifty-four crewmen were saved.

Although in the end C 1, with some help from EG 9, finally sank *U845*, she too was safe from her attackers at 700 feet. If the U-boat's battery had been fully charged, the hunt might well have developed along the lines of that for *U744* – victory through sheer stamina. Comments from senior officers in Britain were limited largely to the escorts' apparent disregard for the possibility of an acoustic torpedo attack from *U845*, particularly during the surface action. That one was not fired was attributed to the accuracy of the ships' gunnery. Stephen was commended for his 'tenacious' maintenance of contact. Moreover, since this was the first time in a long while that close escorts had scored a kill, 'the careful handling of the hunt and the patient approach before attacks on a deep U-Boat' were considered 'most satisfactory.'[54] Here, at least, was grudging recognition that the RCN's close escorts in MOEF were quite capable of sinking submarines, which that had, in fact, been the case for some time.

Three days later the RCN shared in another kill, this time alongside the USN. Early in the morning of 13 March aircraft from the USS *Bogue* found and attacked *U575* a short distance from ON 227, which was escorted by C 3. *U575* should not have come to grief so easily, particularly at the hands of aircraft. She was the second schnorkel-equipped U-boat to undertake operations, and her new breathing tube ought to have kept her safe from aircraft. But she habitually ran on the surface at night for four hours, without incident, during her passage of the Bay of Biscay. She had also transmitted a lengthy – and anxiously awaited – report on her schnorkel under operational conditions. Like Walker's location of *U264*, the first operational U-boat with schnorkel, it was probably no accident that *Bogue*'s group found *U575* on 12 March.[55]

It was while she was attempting to run on the surface on the night of 12 March that radar-equipped aircraft from *Bogue*, assisted by DF fixes, located *U575*. The carrier's aircraft used sonobuoys to maintain contact with *U575* until the destroyer escort *Haverfield* arrived on the 13th. The whole operation demonstrated the very high level of skill obtained by American hunter-killer groups. As the American search developed, the frigate *Prince Rupert* was detached from C 3 to assist. She arrived to find aircraft from *Bogue* and an RAF Flying Fortress from the Azores circling while the USS *Haverfield* attacked. *Prince Rupert* waited her turn and then attacked with depth charges and hedgehog. The two ships alternated for a while, until *Prince Rupert* finally suggested that she direct *Haverfield* in a creeping attack. Communications problems between two navies not accustomed to operating together foiled the attempt. When the USS *Hobson* arrived, a variation on a creeping attack was conducted, *Haverfield* and *Prince Rupert* holding contact on either side at 1,200 yards, passing range information to *Hobson*, whose depth charges finally forced *U575* to the surface. Few U-boats ever faced such withering firepower. While the warships pounded *U575* with every gun they could muster, an Avenger aircraft from *Bogue* hit her with rockets and 20-mm fire. Survivors were picked up by all three ships.[56] The loss of *U575*, and that of *U264* earlier, left the Germans anxious about the viability of schnorkel under operational conditions until the very eve of the landings in France.[57]

Prince Rupert's participation in the destruction of *U575* capped an unprecedented period of success for the RCN in the offensive against the U-boat. Of the seven submarines sunk by escort vessels alone between 24 February and 10 March, three fell to the RCN and two each to ships of the RN and USN. This Canadian success was particularly noteworthy, since U-boat operations – like those of Allied escort forces – were winding down in preparation for the invasion of France. In fact, following the destruction of *U575*, only one more kill was recorded in the north Atlantic for the balance of March, shared by aircraft from *Vindex* and EG 2 on the 15th. Whatever their previous shortcomings, in the first three months of 1944 the RCN's escorts demonstrated their ability to sink U-boats when given the chance.

Creeping attacks and barrages offered only a partial solution to the problem of increased diving depths of U-boats. *U358*, sunk by EG 1 on 1 March after the longest hunt of the war – thirty-eight hours – ended only when the U-boat became exhausted. The same was true of the hunt for *U744* by C 2. U-boats at a depth of 600–700 feet were, as Chavasse described them, 'safe as houses.' The British considered the type 147B and Q attachment to the type 144 asdic sets the most immediate solution. Both could be used to direct hedgehog attacks, but the weapon of choice was now squid. Without such equipment escorts were to resort to creeping attacks.[58]

Senior RCN officers were generally well informed about these developments at sea. At the end of February, for example, the staff authorized the fitting of one-metre-long rangefinders, giving priority to frigates and corvettes in support groups. Walker made excellent use of rangefinders to direct creeping attacks, and it was impossible to conduct such attacks at night without them.[59] Other equipment shortfalls were not so easy to resolve, in part because the decision to abandon construction of the final generation of wartime A/S ships left the RCN stuck with technology that was rapidly being eclipsed by events at sea.

The RCN had a chance to see first hand the new combination of squid mortar and type 147B in early 1944, when *Hadleigh Castle* called at Halifax. No specific reaction to *Hadleigh Castle*'s visit has surfaced, but the War Diary for C-in-C, CNA, for January observed that trials of the squid and type 147B in St Margaret's Bay (near Halifax) 'were satisfactory.' In early February the comparative effectiveness of A/S weapons was discussed in a report by the RCN's own operational research scientists. The Canadian report claimed that depth charges had a 'blind time' of 'over 65 seconds for 300 ft. depth.' That time included approximately half a minute to cover the distance between the loss of contact and the firing of the first charge and thirty seconds for the charges actually to sink. The sinking rate for ordinary Mk VII charges was ten feet per second, or 16 feet per second for the 'heavy' charges (those with extra weights attached).[60] With nearly a full minute for the charges to reach 600 feet coupled with the thirty seconds taken by the run-in, U-boats at extreme depth had ample opportunity to take evasive action. The only real solution to attacking deep U-boats with depth charges was a barrage, which covered an area so large that no evasive action would work. This was ASW by brute force – and with no guarantee of success.

Hedgehog, the RCN's weapon of choice, fared little better in the evaluation by the Canadian scientists. The hedgehog bomb had a fairly fast sinking rate of twenty-two feet per second (allowing for 7.5 seconds in the air, that meant about thirty-five seconds blind time on a target at 600 feet). Only about 30 per cent of attack situations favoured the use of hedgehog, however, and its theoretical lethality of 60 per cent on targets at 300 feet was never realized. In fact, hedgehog had obtained only 13 per cent lethality by the end of 1943. The RCN report accepted the British argument that the problem of hedgehog effectiveness was due largely to poor training and operation, but there were other considerations as well. The limited range of the weapon, only some 200 yards ahead of the ship, and the fixed beam of the type 144 asdic meant that hedgehog was accurate only down to about 350 feet. The repeated hedgehog attacks on deep U-boats in early 1944 had to be made using a stopwatch and firing the weapon manually – which accounts for their universal failure. The accuracy and usefulness of hedgehog against deep targets could be improved by using the Q device or a type 147B asdic. Even so, the bombs were contact fused and a miss was as good as a mile. Moreover, the 32-lb torpex warhead was not proof against the reinforced pressure hulls of the type VIIC/41 U-boats and could not be counted on to breach the hull should it strike other parts of the U-boat (such as fuel tanks or deck plates).[61]

The Canadian OR report of early February pointed towards squid as the only real solution to deep-diving U-boats. Squid was slaved to the depth determination asdic type 147B and featured fully automated fire control. The asdic passed depth-setting information to the squid bombs, set the fuses and fired the weapon. These features and the fast descent time of the bombs (the RCN report claimed thirty-eight feet per second; the RN figure was forty-four) pointed to a high lethality, estimated to be 60 per cent at depths up to 650 feet. The British estimated that the blind time of the squid on a target at 600 feet was approximately twenty-five seconds. Equally important, squid bombs were filled with 200 lbs of lethal minol II. They produced a large and powerful explosion, which accentuated the extreme pressures already faced by deep U-boats.

The RN was sold on squid in early 1944; the RCN wavered. The best its own operational scientists would say in February was that squid 'would appear to be a very promising weapon' – hardly a ringing endorsement. The report was probably the basis for the Naval Staff's discussion of squid on 7 February. The director of Warfare and Training, Captain K.F. Adams, defended hedgehog, which he believed would 'become more valuable as an A/S weapon as experience is gained.' He also objected to the fitting of squid,

Triumph without Celebration 125

because such a change on the RCN's frigates involved the removal of their twin 4-inch gun forward, a modification the RCN had pushed for in the face of British resistance. Adams recommended that a decision on squid be held in abeyance but that the necessary drawings be obtained. The Naval Staff concurred. The lengthy refit needed for an unproven weapon was not yet considered justified.[62]

While the staff vacillated over the issue of A/S weapons, it was clear on the merits of the latest asdic. On 12 April the RCN's final wartime asdic policy was promulgated to senior officers: 144Q and 147B for all escorts larger than corvettes and 145Q and 147B for corvettes.[63] A few days later NSHQ asked the senior Canadian naval officer (London), now Vice-Admiral Nelles, to request that the Admiralty place HMC Ships on an equal priority with the RN for the modernization of asdics, a priority that, Nelles replied, was already in effect.[64] The RCN also took steps to ensure that ratings trained in 144/145Q and 147B went to the appropriate ships and that no untrained asdic ratings were substituted.[65]

The balance of 1944 was a rush to obtain supplies of 147B and Q and get them to sea as quickly as possible. By the spring of 1944 supplies of type 147B asdic were short and Canadian production was behind schedule.[66] The British provided some assistance. Based on EG 10's quick kills in February, the RN adopted a policy of having at least one 147B-equipped ship with each group. Perhaps for that reason the last three of the ten British frigates handed over to the RCN for manning in 1944 were Loch class. Horton advised the RCN in May to assign them to EG 6 and EG 9.[67] By the end of the summer *Loch Alvie* was in EG 9, and *Loch Achanalt* and *Loch Morlich* were in EG 6.

Apart from asdics and the ongoing discussion over squid versus hedgehog, over the winter of 1943-4 the RCN was also vexed by radar problems. Shortages of 10-cm radar in 1942 encouraged the RCN to work with the National Research Council and Research Enterprises Limited to develop and manufacture a Canadian set known as the RX/C. The set was supposed to be an improvement over the British type 271 radar, roughly the same range but easier to operate. It was adopted as the main surface search radar for the new frigates. Design and production problems seriously delayed the RX/C, and once in service it proved a disaster. The RX/C was a delicate instrument, and the RCN had not adequately prepared its radar branch for proper operation or maintenance of sets at sea. In the spring of 1944 the RCN issued a warning to its radar personnel not to 'turn knobs unless you know the result you will get by so doing.' Turning knobs indiscriminately with the hope that doing so would restore the set to operations would, the warning observed, 'in general, have one result: complete failure and possible damage to the RX/C.' It is remarkable that with such a tender piece of equipment no maintenance manual was issued until the spring of 1944 – about the time it was withdrawn from service.[68]

The RX/C was a good set, when it worked, but it was not suitable for operational conditions. In the end thirty-two RCN frigates, including most of those in the first building program, got to sea with the RX/C as their main radar.[69] Those operating with support groups based in the United Kingdom very quickly switched to the RN's type 271Q, leaving a concentration of RX/C-equipped ships in the less important close escort roles. Many frigates assigned to support groups throughout 1944 operated with RX/C, however, including groups based in Canadian waters. The frigate *La Hulloise* kept her RX/C set operational by assigning one man, Lloyd Dawson, the sole and singular responsibility for continuous maintenance. Dawson recalls that *La Hulloise*'s captain excused him all other duty. As a result, the RX/C set worked well. It worked even better following a rebuild by Dawson in early 1945, after a heavy sea had smashed the antenna and the entry of water into the set blew it to pieces. He rebuilt most of the set during a layover a Scapa Flow using British components. Dawson's persistence would eventually pay off, but few RX/C-fitted ships enjoyed the benefit of such skilled and dedicated technicians. It is hard to contest David Zimmerman's conclusion that 'The debacle of the Canadian naval centimetric program left the RCN escorts ill-equipped for the ASW role.'[70] Perhaps the failed searches closer to home during 1944 provide a clue to the impact of RX/C on operations.

Those same searches also highlighted the ongoing problem of proper training and the use of ships working up for operational purposes. These problems boiled to something of a crisis by late

Triumph without Celebration 127

March, following yet another failed Salmon operation. Once again the Canadians were aware of the U-boat's approach. By late winter, however, the Germans had drastically reduced their routine HF radio traffic, and that fact made it difficult to plot *U802*'s movement westward. Searches were based on decrypts of her tasking signals and on educated guesswork. By 22 March *U802* was much further inshore than intelligence estimated, and her position was betrayed only by the sinking of the small steamer *Watuka* off Halifax.

This time Murray did not have even a semblance of a support group to send out. EG 9 was overseas and the last of the RCN's Town class destroyers was retired from operations: W 10 no longer existed. In the absence of a local support group, a new W 10 was formed for this operation. It consisted of the corvettes *Battleford, The Pas, Alberni, Kenogami,* and *Shediac,* the Bangors *Truro* and *Red Deer,* and, a few days later, the new frigate *Wentworth.* The group was reasonably well equipped, since most of the old corvettes had reliable type 271 radar, but it was ad hoc. W 10 was joined two days later by Captain (D), Halifax, himself, in the new Algerine escort *Wallaceburg* and by the corvette *Dunvegan,* at which time Prentice took over the search. On the 24 March, in response to a sighting by *Swift Current,* yet another ad hoc group of corvettes and Bangors, under the assistant training officer at Halifax, was sent to sea. Several contacts were made on the U-boat, including a night-time challenge by searching aircraft that forced *U802* to crash dive. But the U-boat easily avoided her searchers and by early April was on her way home.

The failure to exact any retribution from an assailant in the very approaches to the RCN's principal base drew the ire of officers both at Halifax and in Ottawa. Not for the last time Cdr P.B. Ashe, who as Commander of the Port (COP) was responsible for the Halifax approaches, described the attempt as 'uniformly unsatisfactory.' Apart from the problems of searching with ad hoc formations, coordination of naval and air patrols remained appalling. 'Considering that Dartmouth airport is between 5–10 miles of the ship in question [conducting the search],' wrote the staff officer (Operations) at Halifax, Cdr P.M. Bliss, 'and *no* aircraft has ever arrived in under 1/2 hour, [I] agree with COP that the RCAF are not going to be much use.' The staff officer in charge of signals put his finger on the major problem of communications. 'This whole thing has been a recital of muddle and inefficiency,' he wrote. 'I wish I knew how to rectify short of being in both the ship and the a/c at the same time with a sledge hammer.'[71] One way to rectify the situation was to issue operational orders for Salmons and start training airmen and sailors in their conduct. In early May of 1944 the first 'general operational order' for Salmons finally appeared. It was not until the late summer, however, that an RCAF section was posted to the Tactical Unit at Halifax and joint exercises began.[72]

The only reasonable explanation for the appalling delay in laying the groundwork for joint RCN-RCAF action against U-boat incursions into the Canadian zone is the overall low level of German activity in the area between late 1942 and mid-1944. Clearly, it was not impossible to sink U-boats in the western Atlantic. The USN carrier *Coatan*'s group sank *U856* on 7 April, 250 miles southeast of Sable Island, a U-boat that the RCAF had pursued for several days.[73] The continuing success of USN carrier groups, particularly the sinking of *U856* right under the RCN's nose, prompted Canadian operational research scientists to press once again for the acquisition of an escort carrier in the spring of 1944.[74]

Most of the Canadians' problems in the western Atlantic endured for the balance of the war. In part they were due to the complex ocean environment of the inshore waters of the Canadian northwest Atlantic (of which more later). But in early 1944 the area was still without its own, permanent, long-standing and well-trained support group. The tendency to use ships under training to fill the gap was no solution. Murray officially relied on diverting escorts away from nearby convoys as a means of providing hunting groups,[75] but the 'Training Report' for his command for the month of March suggests otherwise. 'The use of unworked-up ships for operational purposes,' the report observed, 'has considerably delayed the progress and will almost certainly lead to congestion at the work-up base [St Margaret's Bay] in April.'[76] The search for *U802*, for example, drew escorts largely from WEF groups W 8 and W 4, but the frigates *Port Colborne* and *New Glasgow*, the corvette *Matapedia*, the Bangor *Ingonish*, and the Atlantic coast training ship *Clayoquot* all were drawn in.[77]

Quite apart from the shortfalls of the existing training system, the expected spring freshette of new construction from the lakes and St Lawrence River brought some urgency to the matter by February. Over the next six months the RCN planned to commission over forty new escorts, twenty-six of them frigates.[78] By March the Naval Staff was committed to moving its work-up training to Bermuda, while plans were afoot to forestall any opposition that might arise from Halifax. Citing the recent failures of searches off Flemish Cap, which suggested a poor level of training among ships under work-ups, the ACNS planned to relieve Prentice as Captain (D), Halifax, and replace him with someone more in sympathy with the new arrangements. On 10 April the Naval Staff agreed to the establishment of an RCN training base at Bermuda. After nearly five years of war the RCN finally had a fair-weather establishment devoted entirely to the proper preparation of its new construction for war.[79]

The question of a permanent support group for the Canadian northwest Atlantic was also resolved in the spring of 1944. The availability of new frigates and the decision to concentrate them in support groups provided an opening for Murray to act. So, too, did another U-boat incursion into the Canadian zone in late April and early May. *U548* arrived on the Grand Banks around 23 April, when it was spotted by an RCAF aircraft supporting a nearby convoy. The same dreary pattern of ad hoc hunting groups and bungled searches followed, punctuated by the loss of the frigate *Valleyfield* of C 1 as she steamed through a search area south of St John's on 6 May. A single acoustic torpedo, fired from an almost perfect angle, ripped her apart and sent her to the bottom so fast that it was some time before the other ships knew what had happened. Many of those not killed outright perished in the bitterly cold water: only thirty-eight of *Valleyfield*'s 163 men survived.[80] Searches for *U548* off Newfoundland were finally abandoned on 10 May, and although she was by then operating off Halifax, no further contact was made. In fact, it was only after DFing her departing signals on 30 May and 1 June that further searches, again unsuccessful, were mounted.[81] Another intruder had come and gone, this time inflicting heavy casualties on the RCN. There was little surprise in the failure of yet another hunt to exhaustion in Canadian waters, particularly on the Grand Banks, where flying conditions were so poor. But the bedrock of the problem was the RCN itself. As Commodore Reid, Flag Officer, Newfoundland, observed at the end of March, 'Recent unsuccessful hunts off Halifax and Newfoundland, where a U-Boat was known to be present, by motley assortments of ships in various states of efficiency and training, lends emphasis to the fact that none but a highly trained, thoroughly co-ordinated and ably led team can hope to destroy U-Boats at this stage of the campaign.'[82] The dismal search for *U548* served to drive home the point: Murray needed his own permanent support group. On 14 May he signalled his intention to put the next available frigates into a designated support group for the northwest Atlantic.

Murray's case for his own support group was strengthen in the spring of 1944 by the looming influx of new frigates, and by the major rearrangement of escort forces in preparation for D-Day (discussed in the next chapter). As part of that reorganization, C 2 returned to close escort duty in March, while the weight of frigate allocation shifted to support groups. During the spring the corvettes of EG 6 and EG 9 were replaced by new frigates, as these two groups were left to carry the RCN's colours in the offensive role through April and May.

April turned out to be another successful month. EG 9, now composed of the frigates *Matane, Swansea, Stormont,* and the modified corvette *Owen Sound* sailed on 9 April to patrol west of Ireland. When HF/DF confirmed the presence of a U-boat in their area on the 11th, EG 9 was joined by the carrier *Biter* and the sloops of EG 7. At 1218Z on the 14th, and while just two miles ahead of the carrier, *Swansea* obtained an asdic contact. In view of the threat to *Biter,* later confirmed by U-boat survivors, *Swansea* attacked promptly. *U448* immediately went deep, to an estimated 820 feet, and *Swansea*'s next attempt to attack using hedgehog and then depth charges failed to produce results. Shortly before 1300Z HMS *Pelican* joined and contact was regained. Since it was clear that *U448* was very deep, two creeping attacks were organized, the British sloop directing. The first, at 1414Z, was unsuccessful. The second at 1615, which was followed within a minute by a ten-charge pattern from *Pelican*, seemed to be equally unsuccessful. But in fact *Swansea*'s second attack did extensive damage to *U448*. The seventh charge was a direct hit, wrecking the U-boat's diesels, batteries were damaged, depth gauges were broken, and a 6-inch hole was made in the after part of the pressure hull. Leaking badly, *U448* surfaced at about 1625Z into heavy fire from both ships. Several crewmen were killed before firing stopped: forty-two survivors were rescued. The entire action had lasted four hours and had expended fifty-six depth charges.[83] Both *Swansea* and *Pelican* were commended for the patience shown in the creeping attacks. The fact that they 'waited until the U-Boat was on a steady course made the accuracy of the attack much greater.'[84]

EG 9 enjoyed another success in its hunting role during April, but that kill would remain unknown for over forty years. On the night of 22 April the group parted from EG 7 and was on independent patrol when, in a tossing sea and gathering darkness, *Matane* obtained an underwater contact. Confirmation was obtained a few minutes later when the swirl of a U-boat was plainly seen, as was its periscope. Fearing that the U-boat was a threat to the rest of the group, *Matane* made a hasty attack, which was considered at the time to be inaccurate. By the time *Swansea* arrived, *Matane*, low on fuel and buffeted by heavy seas, was having difficulty maintaining contact. *Swansea* fared better, but she could deliver only one deep pattern before she, too, lost contact. It was assumed that the U-boat had escaped. The Admiralty believed that there was 'insufficient evidence of a U-Boat.' Commodore (D), Western Approaches, trusted his sailors' judgment but concluded, 'The U-Boat must be considered to have escaped undamaged.'[85] But *U311* never returned to port. Her fate was listed as 'unknown' until 1986, when further research attributed the kill to *Swansea*. Horton had promised better hunting in the eastern Atlantic. EG 9 shared in the destruction of *U845* in March and *U448* in April as proof of his claim: credit for *U311* would have been welcome news.

While EG 9 ended its highly successful month of April supporting the carrier *Vindex* west of Ireland, EG 6, now composed of the frigates *Waskesiu*, *Grou*, *Outremont*, and *Cape Breton*, set off for Russia, along with a sizeable RN force, to bring back convoy RA 59. U-boats still presented a major threat to the Russian convoys, and RA 59 was sighted by German air patrols at midnight on 28–29 April and a group of eleven U-boats was sent to intercept. Two nights later one ship was sunk, following which EG 9 had a number of contacts. One of them, pursued by *Cape Breton*, was on a U-boat whose conning tower was spotted amid ice on the edge of the permanent pack. Asdic conditions were deplorable, and the only kills were made by aircraft from *Fencer*, which sank three U-boats found on the surface.[86]

May was a quiet month in the Atlantic war, on all fronts, as both the Allies and the Germans made their final preparations for the great showdown in the English Channel. The eyes of the world, including those of Canadian politicians, were focused on the looming invasion of France. In that fixation an important chapter in RCN history went largely unnoticed. The RCN's 'Summary of Naval War Effort' for the first quarter of 1944 described the period as 'one of the most satisfactory experienced by the RCN.' Between the middle of November 1943, when EG 5 sank *U536*, and 22 April 1944, when *Swansea* destroyed *U311*, the RCN sank or shared largely in the sinking of seven U-boats and played a significant role in another (*U575*). Comparisons are always invidious, but in view of the events of 1943, some are warranted here.

Between the sinking of *U536* and that of *U311* approximately thirty-one U-boats were sunk by warships in what the British official historian called the 'North Atlantic.' Nineteen of those sinkings were claimed by RN ships, six by the USN, and six by the RCN (the kills of *U757* by *Camrose* and *Bayntun* and *U448* by *Swansea* and *Pelican* are included here as split evenly, but *Prince Rupert*'s share of *U575* is not credited in the RCN tally). All of these numbers warrant some qualification. Most of the kills by U.S. ships came as part of hunter-killer operations with auxiliary carriers, an edge the RCN did not enjoy. Similarly, a third of the RN score was accounted for in a single cruise by an exceptional group – EG 2 – whose training, equipment, leadership, and (apparently) access to special intelligence the RCN could not hope to match. The RCN's proportion of

Triumph without Celebration 133

U-boats destroyed looks even better during the high period between 24 February – when *Waskesiu* sank *U257* – and the end of April. Of the fourteen U-boat kills by surface vessels alone during that period, four were primarily RCN: one less than the RN, and one better than the USN. Thus, through late February, March, and April one-third of the U-boats sunk in the areas of principal Canadian interest were sunk by Canadian warships. This was the kind of news both the RCN and Mackenzie King had desperately needed a year earlier.

In the spring of 1944 it was possible to believe that Allied A/S forces, especially ships, had the measure of their opponent. Just what the men at sea thought of those new breathing tubes glimpsed quickly on battered U-boats in the winter of 1944 remains a mystery. Few, if any, understood that the tubes portended a much harder campaign ahead. None anticipated just how difficult and different the next phase of the anti-U-boat campaign would be. After all, they had nearly five long years of wartime anti-submarine warfare under their belts and might assume they knew their business. The director of the Admiralty's Anti-U-boat Division's observation at the end of March 1944 that 'in view of the forthcoming operations it would be prudent to study the operations of U-boats in coastal waters,' came late and suggested that even specialists did not really understand what lay in store.[87]

4
The Summer Inshore

> Methought I saw a thousand fearful wracks;
> A thousand men that fishes gnaw'd upon; ...
> All scatter'd in the bottom of the sea.
>
> Shakespeare, *Richard III*, act I, scene iv

The central act of the war in the west in the spring of 1944 was the long-awaited Allied invasion of France. The success or failure of that operation would determine the course of the war for the next two years, and perhaps even its outcome. Since a landing on the scale envisaged by the Allies could not easily be remounted, a repulse would leave the Germans free to concentrate in the east. At the very least they might regain the initiative and prolong the war for many years; at worst the Germans might secure a negotiated settlement and an end to the war on terms favourable to themselves.

The importance of the Normandy landings, as the west's best chance to win the war, was not lost on either side in the early months of 1944. Most understood well enough that the ultimate decision rested on what happened ashore, between the two armies. But that decision, particularly in the early days, would be influenced by the extent to which the Allies were free to establish and reinforce their bridgehead; that success, in the end, depended upon the unimpeded use of the sea. As winter gave way to spring, all eyes turned to the narrow waters separating Britain from the continent. And it was there, from 6 June to the end of August, that the anti-submarine action was concentrated. On the whole, the A/S campaign associated with the landings in France failed to live up to expectations. For the RCN, however, it was a summer of accomplishment. The Canadians put their best into the campaign and were rewarded with a performance second to none.

By the spring of 1944 the Allied invasion of France was an open secret, but there was little the Germans could do to forestall the initial landings. Attempts during March 1944 to maintain surface patrols in the western channel were met by heavy Allied attacks, which drove them in and prevented further deployments.[1] Only small craft, especially submarines, could hope to operate directly against the invasion fleet. The U-boats' ability to move and survive in the channel depended largely on the fitting of the new schnorkel breathing tubes.

The idea of fitting a breathing tube to submarines to allow them to operate their diesel engines while submerged was not new. The German submarine engineer Professor Hellmuth Walter had designed one in 1933, and the Dutch had fitted several of their submarines before the war. Under the prevailing tactical and strategic situation in the first four years of the war, however, the German U-boat fleet required freedom of movement on the surface. The schnorkel, as the Germans dubbed it, was not adopted for general use until after the heavy losses of 1943. Fitting began in the winter of 1944. The loss of *U264*, the first U-boat to take the tube on operations, dampened enthusiasm for the new equipment somewhat.[2] But fitting went ahead, and schnorkels proved increasingly effective as crews gained experience in their use.

The schnorkel adopted for existing U-boats consisted of two tubes. The tall intake tube rose twenty-six feet from the deck, and was capped by a float value to prevent water from entering. Later the schnorkel head was often covered with rubber in an attempt to reduce radar signatures, and it also carried late-war radar detectors. The exhaust tube, fitted just behind the intake, was shorter and expelled gases just below the surface. In most applications the two tubes were combined within a single streamlined casing. The schnorkel was a major addition to the older U-boats and had to be

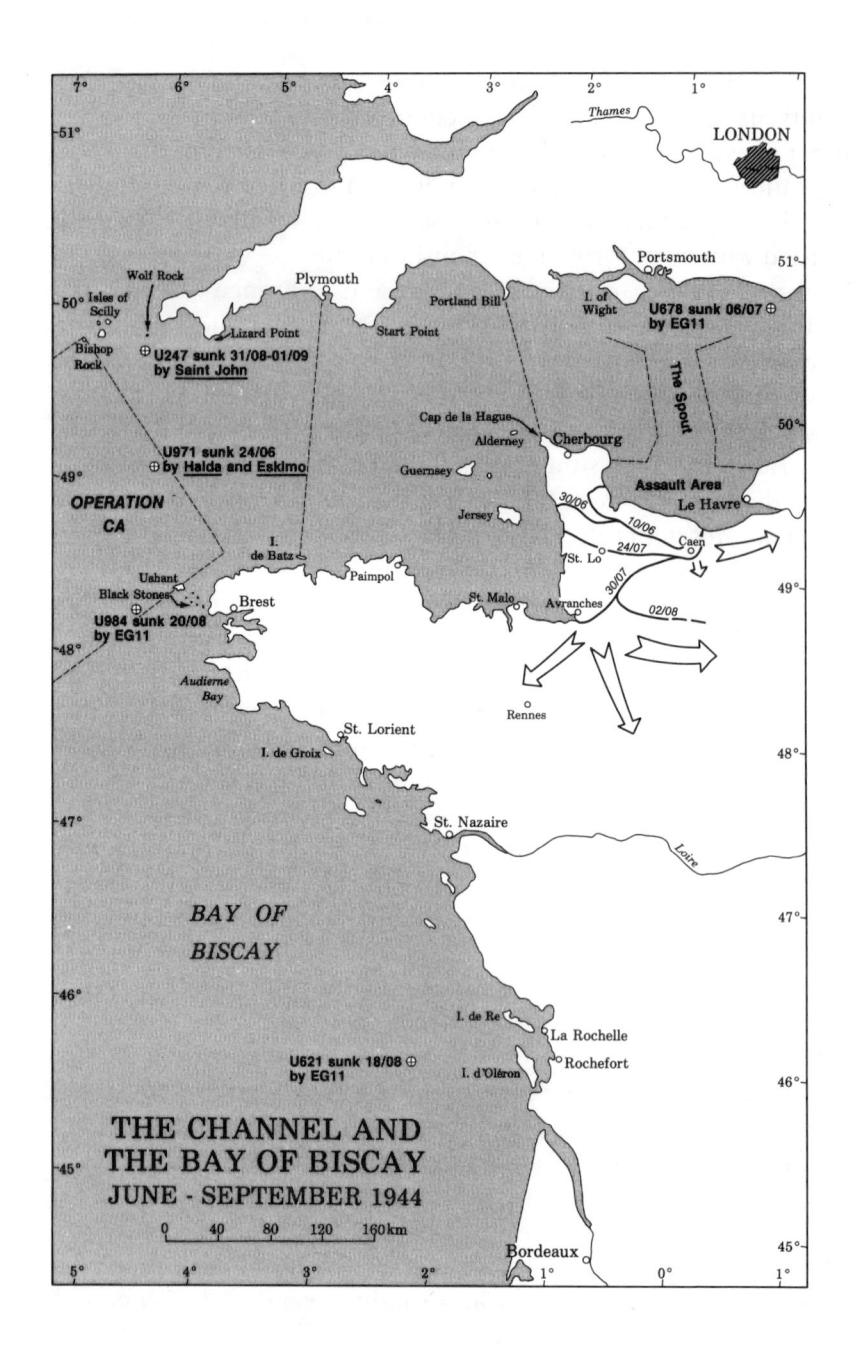

THE CHANNEL AND THE BAY OF BISCAY
JUNE - SEPTEMBER 1944

fitted externally, hinged at the base of the conning tower and stowed flat along the forward portion of the hull. In the new types XXI and XXIII U-boats the schnorkel was built into the conning tower, from which it was raised in the same way as the periscopes were. The retrofitted schnorkels were therefore an awkward arrangement. There were major initial problems, and even in moderate seas the device could not be used. Maintenance of perfect trim was crucial. Dipping the schnorkel caused the diesels to draw air from the U-boat itself, resulting in rapid and painful changes in air pressure within the submarine.[3] It was not until late 1944 that confidence and expertise in the use of schnorkel were translated into increased operational proficiency.

None the less, Dönitz put enormous emphasis on schnorkels in the preparations to meet the Allied invasion. A patrol of eight U-boats sent into the western channel in late May proved that schnorkel-equipped submarines could operate there without serious interference from air and surface forces. Unfortunately, Allied bombing of both production and, especially, the French rail and road systems in preparation for D-Day seriously delayed deliveries of schnorkel equipment. When the invasion finally happened, only nine U-boats based in French ports were fitted with schnorkel and seven more were en route from Norway.[4]

Dönitz observed in his *Memoirs* that 'German light naval forces found themselves confronted with a vastly superior enemy'[5] on 6 June 1944. That superiority, in itself, was part of the Allied plan; for in their darker moments the Allies believed that the Germans would launch a desperate assault on the invasion forces, inflicting enormous destruction. Cdr H.J. Fawcett, of the Admiralty's Anti-U-Boat Division, estimated at the end of March that 170 U-boats were available for use in the channel. If, Fawcett speculated, all of them hit with just 20 per cent of their torpedoes, the Allies might lose thirty-five ships per day – 240 per week. This possibility, he admitted, was a worst-case scenario. But 170 submarines could none the less do considerable damage. Even Coastal Command, flying at four times the rate employed in the bay offensive, could not promise to stop all of them, especially if the U-boats used coastal waters and schnorkel. Moreover, if the Germans employed radically new submarines, like the Walter boat (the 'W-boat') driven by hydrogen peroxide and not dependant on surface manoeuvrability, aircraft could not expect either to sink U-boats or to stop them from sinking ships. By the end of April Fawcett believed the Germans had a total of about 400 U-boats and perhaps fifty W-boats within range, and might sink or damage 800 ships over forty days. The Germans, he observed, would accept the loss of 150 submarines in the process.[6]

Fawcett was not alone in predicting a Wagnerian climax to the U-boat war. The more sanguine estimate produced by the naval intelligence staffs based on Ultra decrypts were alarming enough. Operational Intelligence believed in early March 1944 that ninety U-boats would be present in the invasion area by early June and a further sixty-five dispatched within three to five days (roughly Fawcett's estimate of 170). Within a few days of the landings, OIC estimated, approximately seventy-five U-boats could be in the channel.[7] In the final analysis, Allied plans were based on a direct and immediate threat from some 120 submarines.[8] Moreover, like Fawcett, intelligence officers expected the Germans to suffer appalling losses to check the invasion.

Allied naval planners were particularly anxious through early 1944 with the potential threat from the radically new W-boat. A small submarine, of approximately 250 tons, it was driven by an engine that was air independent and was apparently designed for extremely high speeds (forty to fifty knots) just below the surface. Intelligence on new German submarines was muddled, however, and the W-boat never got past the prototype stage. What misled Allied intelligence was the evidence of large-scale production of small submarine hulls. It was not until late May that firm information indicated that the small submarines under construction were the conventionally powered type XXIII: the coastal version of the type XXI.[9]

In anticipation of a *Götterdämmerung* in the channel, the Allies drew every available escort into the D-Day operation. With fifteen escorts or A/S vessels for every 100 miles of channel, there was a high likehood of destroying all U-boats sent in to attack the invasion fleets. In a worst case scenario, it was expected that the ASW success rate for the broad ocean, presently running at 20 per cent of U-boats deployed, could be achieved. The U-boats, after all, could not dive deep to get away.[10] That being said, operational scientists warned at the end of April that the Germans had much going for them and should be able to saturate defences around the D-Day convoys. Schnorkels would get them in and GNATs provided an effective weapon. It would be best, the scientists went on, to hit the U-boats hard and fast at the outset with a 'knock-out blow.' 'This would seem a strong reason,' their report of 24 April advised, 'for having some of our best and most experienced anti-submarine forces available for immediate deployment in an entirely offensive role.'[11] In the end they mustered 286 destroyers, sloops, frigates, corvettes, and trawlers to act as A/S support for the invasion forces (itself a fleet of 4,000 vessels).[12] Among those 'best and most experienced' anti-submarine forces in the front line was the cream of the RCN.

To find these ships major changes were made in deployments world wide, including those for transatlantic convoys. In March NSHQ and the Admiralty agreed to the withdrawal of all the remaining British groups from MOEF. The remaining five C groups would run the convoys with an operational strength of one frigate and four to five corvettes. The strength of WEF was to remain at seven groups, although in practice only four were needed for the oceanic convoys. The corvettes still in EG 6 and EG 9 would be replaced by frigates. More importantly, the RCN's River class destroyers would also be withdrawn from MOEF and formed into two new support groups, EGs 11 and 12, for service in support of Operation Overlord.[13]

To facilitate these changes the frequency of transatlantic convoy sailings was reduced and the size of individual convoys increased. Starting in April, all eastbound convoys originated at New York as simply 'HX' fast, medium, or slow (i.e. 'HXM' for medium-speed eastbound convoys).[14] Westbound convoys, the ON series, were sailed in fast, medium, and slow versions as well. Few transatlantic convoys sailed with less than 100 ships over the summer months. The largest of the entire war, HXS 300, crossed in late July with 167 ships, carrying over a million tons of materials and covering thirty

140   The U-Boat Hunters

square miles of ocean.[15] Sole responsibility for complete transatlantic close escort of the Allies' most important convoys was for the RCN a signal accomplishment: Canadian 'hegemony' on the north Atlantic run. It says a great deal, however, about even German preoccupation with operations in the eastern Atlantic that such massive convoys moved with impunity under the slenderest of escort.

At the tactical level it seems fair to say that there was a certain smugness among Allied A/S practitioners about the pending inshore campaign. Most of them had, by 1944, long years of wartime A/S experience. It was understood that shallow water, submerged features, and coastlines would limit the U-boats' mobility. The *Monthly Anti-Submarine Report* of February 1944 contained a short passage under 'miscellaneous information' on what to expect from U-boats inshore. In general it advised that German submariners, used to operating in deep water, would be timid inshore and unlikely to remain there, particularly when being hunted. Rather, the U-boats would probably 'crawl away at slow speed, deep,' up wind or along the most direct course to seaward. Hunts to exhaustion were 'the thing most feared by any submariner,' the *Report* concluded, and ironically German submariners were likely to move instinctively into deeper waters, where conditions for such hunts were favourable. The possibility of U-boats' resting on the bottom was discounted, except for special circumstances. There is nothing in the *Report* that would indicate a need to modify the practice of the previous four years.[16]

The problem of sorting out a U-boat from the litter on shallow seabeds was not new. Confidential Admiralty Fleet Orders (CAFO) were issued periodically throughout the war on the 'Use of Wreck Charts by A/S Vessels.' These orders warned A/S specialists to be wary of U-boats lying on the bottom to avoid contact and of the difficulty of classifying such contacts. To reduce the problem, wreck charts were issued to 'certain of H.M. ships for use in connection with A/S work,' showing the positions of both wrecks and other non-sub contacts inshore. The order in force by the spring of 1944, CAFO 1544 issued in July 1943, was unchanged from the earlier orders except for methods of reporting new contacts. It reminded escort officers that, 'The possibility of a U-boat lying on the bottom

The Summer Inshore 141

to avoid detection must not be forgotten.'[17] The RCN also issued wreck charts. They echoed the Admiralty instructions and advised A/S forces not to abandon an inshore contact until 'at least one pattern of depth charges, set to fire on the bottom, has been dropped on the position.'[18]

Which opinion, that of the CAFOs or that of the *Monthly Anti-Submarine Report*, held sway among the escort commanders as they gathered for the D-Day operations remains a mystery. The distribution of wreck charts was not universal among support groups. The destroyers in EGs 11 and 12, committed to patrols within the English Channel itself, had them from the outset, but the frigates of EGs 6 and 9 managed without for the first weeks. Dan Hannington, the navigating officer of *New Waterford*, recalled that he still did not have wreck charts when his frigate moved into the channel proper at the end of June.[19] Accurate wreck charts were central to effective inshore ASW, particularly in an area like the English Channel with its heavy traffic and history of conflict. By 1944 it was estimated that 700 wrecks littered the bottom from Dover to Land's End, producing an average of eight non-sub features every ten square miles, or one every mile and a half. In theory, A/S ships in the channel should have obtained an asdic contact every half hour. In practice contacts were fewer than that. Over one thirty-eight-day period the Canadian frigate *Grou* classified seventy non-sub contacts, about one every twelve hours, or roughly one in every twenty possible. Familiarity with local conditions and delays caused by the need actually to classify new contacts probably accounted for the failure to approach the theoretical limits of non-sub contacts.[20] But the simple fact remained: no one in early June 1944 anticipated how time consuming and exhausting sorting out the U-boats from the background would be.

To facilitate use of the wreck charts and to aid navigation in the channel, some ships were also fitted with new radio navigation aids. All the aids were variations of the same principle: the transmission of radio signals from fixed stations from which it was possible to determine a position. This equipment was initially developed to help bombers find their targets, but it was quickly adapted for use at sea. As a rule, low-frequency signals, which tended to follow the curvature of the earth, were much less accurate than high-frequency signals, which were line of sight. Both types had maritime applications. To assist in general navigation a low-frequency system know as LORAN was under development. For tactical purposes very accurate fixes were needed, which involved high-frequency, but comparatively short-range, signals. The D-Day minesweepers used a very accurate HF system known as DECCA (known initially as QM), accurate to within 100 feet at a 100-mile range. However, there were only three DECCA stations and its range was limited.[21] The system adopted for tactical purposes by the supporting forces was a variation on the GEE system, a high-frequency aid developed for bombers. Known as 'QH' by the RN, GEE lacked both the range and the accuracy of DECCA, but within U.K. waters it could place a ship within 400 yards of a desired position.[22] GEE made it possible to relocate previous bottom contacts with enough precision to discriminate between a known non-sub and something that apparently had moved off since the last contact. In much of the channel escorts could also rely on shore-based high-frequency radar stations, which, over the summer, were used to vector individual ships and groups onto suspicious targets with considerable precision.[23]

Veterans have little recollection of briefings on inshore ASW prior to June 1944, and nothing yet unearthed in the documents sheds much light on the matter. By all accounts routine refresher training on ASW was undertaken by the frigate groups prior to D-Day, but nothing special was laid on. The same was true of the destroyer groups, EGs 11 and 12. *Gatineau's* first lieutenant, E.M. Chadwick, dismissed any notion of special A/S training with the observation that by the spring of 1944 the men were considered skilled enough by virtue of their north Atlantic experience. Instead, the support groups concentrated on the novel forms of threat expected – aircraft, E-boats (the term applied to the German equivalent of motor torpedo boats), R-boats (essentially motor gun boats), and of course U-boats, a term that usually included the high-speed Walter boats. Some training was undertaken on submarines in April, but as the days wore on, the groups concentrated on the unfamiliar. *Chaudière*'s commanding officer, C.P. Nixon, recalled that surface gunnery was the main thrust of preparation. Bill Willson, the captain of *Kootenay*, noted in his diary that the only 'really alive exercise' was a simulated attack by RN Martlet fighters, in which they 'hurtled around us like hornets roused, firing live ammunition all around us.'[24] Willson summed up his pre-D-Day training by observing, 'Until now our training has been very mixed-up with everything from subs to W-boats to night encounters with other destroyers. No one could figure out what we were being prepared for, but it now appears that we are to expect anything and everything and may be sent anywhere.'[25] According to Chadwick, everyone expected asdic conditions to be poor inshore and they simply had to get on with it.[26] Nor were the Germans expected to have it all their way. As an RCN historical officer observed later, the 'confined spaces, strong tides, shallow water and wrecks – appeared to be in *our* [emphasis added] favour.'[27]

It was expected that the Germans would deploy their submarines well to seaward in waiting positions, then surge up the channel in waves to attack the D-Day shipping. The Allied plan called for a layered defence to check the inrush of U-boats (see map p. 136). To the west, an outer rectangular patrol area 280 miles long and 200 miles wide was anchored at one corner on Bishop's Rock in the Scilly Isles, with its short side running southeast and the long axis of the box covering Brest and the entrance to the channel. For this operation, known as 'CA,' C-in-C, Western Approaches, disposed six groups, EGs 1, 2, 5, 6, 9, and 15, supported in turn by two aircraft carriers outfitted for fighter aircover, lest the Luftwaffe attempt to attack the escorts, and one carrier equipped for reconnaissance. The remaining groups, EGs 3, 11, 12, and 14, were assigned to C-in-C, Plymouth, and provided a middle barrier, operating between lines drawn from Plymouth to the Ile de Bas in the west and Portland – Cap de la Hogue in the east. Support groups were to move between the two lines conforming to the movement of waves of U-boats as they entered and left the channel. These middle patrols were in support of a massive aerial operation in the western Channel known as 'Cork.' Its intent was to make it impossible for non-schnorkel-equipped U-boats to reach the assault area. Twenty-one squadrons of radar-equipped aircraft were to fly patrol lines thirty miles apart from the Scilly Isles to Alderney, a pattern specifically designed to prevent submarines from surfacing to recharge batteries. Surface forces were to respond to aircraft contacts and to pursue HF/DF bearings. In the end, the Allies anticipated that only schnorkel-equipped U-boats would be able to reach the assault area.[28] The cross-channel invasion shipping lanes themselves between the Isle of Wight and the Baie du Seine were covered by eastern and western 'walls,' the final inner barrier.[29] The 'distant cover' of the A/S plan was not strictly part of Operation Neptune, the naval side of the invasion. Rather it was part of a general covering operation provided by Western Approaches, Plymouth, and the Home Fleet, coordinated by the Admiralty.[30] It is significant, however, that the four RCN groups represented 40 per cent of naval forces committed to offensive A/S operations in support of the Normandy campaign: a fair proportion of the 'best and most experienced' and due recognition of the accomplishments of the previous winter. EGs 11 and 12 represented the best ASW force in terms of men and equipment that the RCN could muster. The River class destroyers were powerful A/S ships, well equipped, built to naval standards (and so were more capable of sustaining damage), and manned by a higher percentage of long-service professional officers and men. It was perhaps no accident, then, that they would be among the best – perhaps *the* best – of the support groups in the channel during the summer of 1944. Certainly EG 11 would take second place to none, and its senior officer, J.D. Prentice, got the opportunity he had longed for. By the spring of 1944, after nearly five years of continuous service, Prentice was tired and in poor health. He had been plagued by chronic intestinal problems since 1940, and over the winter of 1943-4 he complained of dizzy spells. By the spring of 1944 Prentice ought to have been sent on extended leave. It was also believed, at NSHQ, that he must be removed as Captain (D), Halifax, to facilitate the implementation of the new sea-training schemes. The solution, to the latter problem at least, was to send him back to sea as senior officer of EG 11.

Although tired and worn, the choice of Prentice for EG 11 was unquestionably sound. 'He has the fighting instinct of a terrier,' Admiral Murray wrote in April 1944, 'and a peculiar mixture of leadership and slave driving which has vastly improved the efficiency of the RCN escort forces during his term as Captain (D).'[31] He was, in fact, the closest thing the RCN had to the RN's ace, Johnny Walker, and Prentice was senior – and pugnacious – enough not to get pushed around by RN officers. In late April Prentice was posted to *Ottawa* in command of EG 11. *Ottawa*'s captain, Cdr H.F. Pullen, RCN, was a seasoned north Atlantic veteran, previously senior officer of C 5 and already in line to command EG 11. Had Pullen stayed on to command EG 11 over the summer of 1944, he would have been the only professional RCN officer to command a Canadian support group. As a full RCN commander, however, Pullen was inappropriate as the captain of the senior officer's ship. In any event Pullen knew Prentice well enough to know who the real captain of *Ottawa* would be. The plum offered to Pullen was the post of 'commander' (executive officer), of the new RCN cruiser HMCS *Uganda*. Pullen took his leave in mid-May when Prentice arrived to see EG 11 through its final training.[32] In addition to *Ottawa*, EG 11 contained the destroyers *Gatineau*, *Kootenay*, *Chaudière*, and *St Laurent*.

The RCN's other River class destroyers went to EG 12: *Qu'Appelle*, *Restigouche*, *Saskatchewan*, and *Skeena*. *Assiniboine* was also assigned to EG 12, but she was in refit until August. The RCN chose as senior officer Cdr A.M. MacKillop, RN, who had commanded one of the RCN's MOEF groups since 1943, and gave him command of *Qu'Appelle*. Like Prentice, MacKillop was older than most of his colleagues, in his forties. Easton, who served under MacKillop as captain of *Saskatchewan*, described him as a small, 'somewhat weather beaten' man, with a 'cherubic face ... and a lower lip which stuck out a little, providing a feature of determination to his otherwise innocent appearance.'[33]

EG 11 and EG 12 assembled in the United Kingdom during April. There they were joined by the two frigate groups in May. EG 6 under Birch was now composed of *Waskesiu*, *Cape Breton*, *Grou*, *Nene* (now RCN), *New Waterford*, and *Teme*. With the exception of *Nene*, this core of EG 6 remained unchanged until April 1945. The same stability – and size of group – is evident in the history of EG 9 from the spring of 1944 onwards. Under Layard in *Matane*, EG 9 consisted of *Swansea*, *Stormont*, *Port Colborne*, *Saint John*, and *Meon*. As with EG 6, this corps remained intact to the end of the war with the exception of *Meon*, which was replaced by *Nene* in September. The only significant addition to both EG 6 and EG 9, the Loch class frigates, did not take place until late summer.

The ships of the four RCN support groups were reasonably well equipped. All carried some form of 10-cm surface warning radar, and all the frigates except *Waskesiu* and all the destroyers except *Ottawa*, *Kootenay*, and *Assiniboine* were fitted with some version of the type 144 asdic. *Chaudière*, *Meon*, *Restigouche*, and *Qu'Appelle* were also fitted with Q attachments, and *Chaudière* – alone – was equipped with a type 147B asdic set.[34] All of the ships were armed with both hedgehogs and minol-filled depth charges. *Kootenay* also carried a warning device that alerted her to the gun-laying radar of German coastal batteries, which came in handy. All ships improved their equipment over the course of the summer.

By the end of May the training and preparation was complete. While the long columns of men, guns, and tanks snaked their way down to embarkation ports throughout southern England, the ships of the A/S support forces lay quietly in northern ports. The destroyers of EG 11 and EG 12 were the first to leave Londonderry for the channel, when they sailed for Plymouth on 3 June. After a brief layover, the destroyers departed on the 5th to their patrol areas. On that day EG 6, EG 1, and EG 5 sailed for waiting positions south of Ireland, along with the auxiliary carriers *Tracker* and *Pursuer* escorted by EG 15. They were to be joined by the other groups, but only EG 2 and half of EG 9 arrived before the landings began. The rest of EG 9 was delayed when the carrier *Emperor*, which they were escorting, developed defects; they did not arrive on station until 7 June.[35] In the end they all had to wait an extra day, since bad weather delayed the assault.

The landings on the 6th, despite some tense moments, were enormously successful and a complete operational and tactical surprise. For the A/S barrier forces the excitement in the Baie du Seine on 6 June was far removed from their experience. Groups operating in the channel proper saw convoys passing eastward, but for the support groups the greatest amphibious assault in history was marked by quiet patrolling, even for the destroyers in the middle

The Summer Inshore 147

barrier, and no aircraft threatened the outer patrols. Both 6 and 7 June were, to quote Easton, 'quiet and peaceful, like a summer excursion.'[36] Patrolling aircraft, their undersides brightly marked in 'invasion stripes,' passed overhead, covering every corner of the channel at twenty-minute intervals. The Germans were conspicuous by their absence. Not only had they been caught by surprise, they were overwhelmed by the sheer size of the Allied armada.

Continuous patrolling by surface and, especially, air forces made German movement largely impossible. On the afternoon of 6 June patrolling forces were given their first warnings that U-boats had departed French bases. In fact, thirty-seven had sailed. Half of them were deployed in the Bay of Biscay in an utterly futile gesture designed to guard against further Allied landings. The Biscay deployments served only to put vulnerable submarines at risk, and a heavy toll was taken by aircraft. By 9 June five had been sunk – two in the space of twenty-two minutes by Flying Officer K.O. Moore, RCAF, with 224 Squadron RAF – and a further nine were damaged by aircraft. Attempts to move non-schnorkel-equipped U-boats up the channel were equally pointless, and by the 10th the Germans gave up trying.[37] Airmen had delivered the first, and decisive, check to the U-boat offensive against the landings.

It was quite a different story for the naval A/S forces during the first weeks. Support groups obtained contacts, some very early on, but conditions inshore worked in the Germans' favour. EG 12 was the first Canadian support group to make a confirmed contact on a U-boat, but despite a thirty-hour hunt it could not make a kill. The destroyer groups (EGs 11, 12, and the British 14) were sent almost immediately to the French coast to prevent U-boats moving along the shore line and into the channel. Late on 7 June, as EG 12 patrolled thirty miles northwest of the island of Ushant, *Restigouche* reported an underwater contact. At about the same time a deep underwater explosion was heard, which was attributed to a torpedo. MacKillop ordered the group to reduce speed and stream CAT gear as a precaution. While *Qu'Appelle*, *Skeena*, and *Restigouche* slowed to a safe seven knots to conduct a deliberate search, *Saskatchewan* boxed in the contact with a two-mile-square search around the area. Asdic conditions were poor. *Saskatchewan*'s asdic officer, after outlining the difficulty evident over the past two days, complained to Easton that the 'outlook isn't particularly bright.' The rest of the group appeared to be at a loss over what to do next.[38]

Over the next hour or so the presence of the U-boat was confirmed by several torpedo explosions and near misses, but no firm contact was gained until the periscope of *U984* was sighted from *Restigouche* at 2126Z. Gunfire was brought to bear but the destroyer could not turn quickly enough to ram and was travelling too slowly to drop depth charges. Twenty minutes later it was *Qu'Appelle*'s turn to sight a periscope, which she fired on and attacked with depth charges. These moves were followed immediately by a huge explosion which vaporized *Saskatchewan*'s CAT gear; the torpedo, the third fired at EG 12 in less than two hours, was intended for *Qu'Appelle*.[39] The group attacked further contacts while *Saskatchewan* streamed new CAT gear, but these attacks simply produced the same result as all previous ones – dead fish. MacKillop informed C-in-C, Plymouth, of his definite contact and his plans to pursue the U-boat with a retiring search plan (a search that shifted position in the direction the U-boat was expected to go) to the north, since he estimated that the submarine was intent on entering the channel. For the rest of the night the retiring box searches produced only more non-sub contacts. At 0600Z on 8 June the searches were abandoned and the group swept the area at fifteen knots in line abreast. At 0930Z a deep underwater explosion was heard near *Qu'Appelle*, followed quickly by an explosion between *Skeena* and *Restigouche* that threw up a large column of water. Not surprisingly, MacKillop believed that he had relocated *U984*: in fact EG 12 had stumbled upon *U953*. Almost at once *Skeena* obtained a firm contact, the best the group had had so far. The target showed movement and a pronounced doppler effect was heard. A disturbance in the water, believed to be the U-boat blowing ballast tanks, was also seen, and *Skeena* attacked with hedgehog just as a torpedo was seen passing her bows. Several of the hedgehog bombs exploded only one second after entering the water, and oil immediately appeared, but the detonations were attributed to contact with the U-boat's wake. *Skeena* followed up this attack with depth charges fired by eye after the *U953*'s periscope had submerged.

Now back in contact, the group slowed to seven knots and that reduction helped to save *Qu'Appelle* and *Saskatchewan* at 0947Z from a torpedo that detonated between them. As series of depth-charge and hedgehog attacks on a great many contacts produced nothing, MacKillop ordered a halt to attacks and instructed the group to wait for the U-boat to surface. EG 12 reverted to patrolling the area with CAT gear streamed. As MacKillop wrote later in his report, 'An all day search in terrible asdic conditions produced nothing more than a few tons of fish.'[40]

While EG 12 patrolled, C-in-C, Plymouth, ordered EG 3 to go to its assistance. Before the group arrived, an aircraft reported a U-boat on the surface eight miles east of EG 12 at 1700Z. MacKillop set off at twenty-five knots and arrived to find EG 14 already searching. The two groups combined efforts only to be drawn away two hours later when a U-boat was spotted on the surface. It was forced down by gunfire from the destroyers, but no asdic contact was ever gained and the two groups moved off westward. Puffs of black smoke were soon seen through the rain about four miles away, and it seemed possible that the U-boat was getting away with the aid of its schnorkel. *Qu'Appelle* and *Saskatchewan* pursued this contact, and *Saskatchewan* obtained a good asdic echo which she attacked to no effect. The hunt was finally abandoned late on 8 June, about thirty hours after it began. 'The net result of this patrol,' MacKillop wrote, 'was the slight damage of a U-Boat, the expenditure by the enemy of seven torpedoes to no avail and the strengthening of our confidence in catgear.'[41]

The decision to abandon the hunt off Ushant was confirmed by C-in-C, Plymouth, who warned that a force of German destroyers, including two Narviks (large and very powerful ships), an ex-Dutch destroyer, and one small Elbing class destroyer, were expected to leave Brest headed for the assault area that night. The destroyers of EG 12 and EG 14 were no match for these warships, and an interception by the 10th Destroyer Flotilla (DF), of which the RCN Tribal class *Haida* and *Huron* were a part, was already planned. The support groups cleared the area and the heavyweights had their battle; one of the Narviks and the ex-Dutch ship were destroyed, and the British destroyer *Tartar* was damaged.[42] The destruction of the Germans' lone major surface force in the western channel opened the French coast to more aggressive patrolling by A/S forces. EG 12 was back off Ushant by the morning of the 9 June, pursuing U-boats but with no better luck than before.[43]

Meanwhile the two frigate groups, EGs 6 and 9, operated further to seaward and closer to the English coast. Apart from the alarms caused by numerous non-submarine contacts, there was little initial excitement in their patrols. Fighters from the auxiliary carriers provided cover, but in the absence of any German threat the carriers themselves became a hazard. Escorts responded nervously to the many contacts inshore, and their anxiety was heightened by the need to protect the carriers. In the early hours of 10 June *Teme* of EG 6 was nearly cut in half by *Tracker*. *Teme* responded to a contact inside the screen just as the carrier altered her own course. *Tracker*'s bow hit *Teme* squarely on the portside, just behind the bridge, slicing through twenty-four feet of the frigate's thirty-six-foot beam. The two ships lay jammed for some time, the gentle swell turning *Tracker*'s bow into a great saw, slowly grinding its way through *Teme*'s hull. Miraculously, only four crewmen were killed and three injured, and the frigate remained afloat after *Tracker*'s bows had pulled free. All but essential crew were removed and *Teme* was towed in by *Outremont*. The collision seemed to underline the pointlessness of the carriers' role and on the 11th they were withdrawn. The support groups were thus freed for more aggressive action, and by the 18th all of the ten D-Day support groups were passed to C-in-C, Plymouth, to command. Plymouth also took charge of operation CA, which they ran until it was abandoned a few days later.[44]

It was logical that Plymouth should take a direct hand in control of channel operations by mid-June. On 15 June the first of the Norwegian-based schnorkel boats to arrive in the channel, *U767*, sank the frigate *Mourne* off Land's End, and on the same day *U764* crippled the frigate *Blackwood* in mid-channel; she later sank under tow, only to be 'rediscovered' later in the month by EG 9. Moreover, losses began to occur in the assault area as the Germans followed the rocky and difficult French coastline. Proof lay in EG 14's sinking of *U767* off Paimpol, under fire from coastal batteries, on the 18th: the first by an A/S group. By late June then, as EG 6 rested and restored itself in Plymouth, EG 9 was off the Ile de Bas, where by 23 June it was briefly in contact with *Haida* and *Eskimo* of the 10th DF.

These two Tribal class destroyers were drawn away westward on 23 June to help EG 2, which was in contact with a U-boat about half way between Land's End and Ushant. *Haida* and *Eskimo* arrived to find an aircraft dropping depth charges. They took over the hunt, pending the arrival of EG 15. When the latter heard that two fleet class destroyers were attacking an immobile target that seemed to be on the bottom, they left *Haida* and *Eskimo* to pursue their own folly. Two hours of attacking produced no results. *Eskimo* then classified the contact by passing over it with its echo sounder. The trace produced the unmistakable silhouette of a U-boat, showing the hull and – more importantly – the conning tower. The attack was renewed with vigour.

Haida and *Eskimo* had stumbled upon *U971*, a type VIIC U-boat without schnorkel, commanded by Oberleutenant zur See Walter Zeplien. *U971* was on her first war patrol. She had been attacked repeatedly by aircraft during her passage from Norway and had reached the channel with all but two torpedo tubes and most of her anti-aircraft armament damaged. Zeplien was en route for Cherbourg in the early hours of 24 June when he first encountered warships, which he could not attack because of his damaged torpedo tubes. Realizing that to be of any use *U971* needed repairs, Zeplien altered course for Brest. Running along on the surface, *U971* was attacked twice by aircraft but was unable to respond because of her damaged guns. When the U-boat's batteries also failed, Zeplien decided to lie on the bottom in hopes of getting into Brest on the surface the next night.

U971 was attacked during 23 June by a number of unknown forces, but it was not until mid-afternoon, when she surfaced briefly and was attacked by waiting aircraft, that *Haida* and *Eskimo* arrived. A series of attacks followed, one of which, by *Eskimo* at 1825 hours, produced leaks. With his diesels damaged, his batteries exhausted, his weapons unusable, and just enough air left to bring

U971 to the surface one last time, Zeplien finally gave up. He decided to bring the U-boat up, abandon ship, and scuttle the submarine. The crew was assembled, the plan explained, beer was issued, which they drank standing knee deep in rising water, and then the U-boat's ballast tanks were blown. *U971* surfaced into a hail of gunfire and as the crew poured out the U-boat was scuttled. Despite the fire from *Haida* and *Eskimo*, all but one of *U971*'s crew of fifty-three were saved.[45] After nearly three weeks of tireless effort by the A/S specialists during which they had managed to sink only one U-boat, the A/S novices of the 10th Flotilla sank one on their first try.

EG 11's first days in the channel were devoted to quiet patrolling, first off the English coast and then, by 7–9 June, in the same area where EG 12 pursued its U-boats. Once the German destroyer threat was defeated on 9 June, Prentice wasted no time getting in along the French coast. On that day EG 11 steamed around the Channel Islands, in broad daylight under the guns of the German garrison, screened from easy view by limited visibility: a bold, but carefully calculated show of brass – typical Prentice. The next day EG 11 received a change of tasking well suited to Prentice's aggressive nature. EG 11 was designated the 'A/S Killer Group' – a title that suggests Prentice had a hand in its selection – for C-in-C, Portsmouth, and became the first of the support groups to move into the central channel, an area know as the 'Spout' between the Isle of Wight and the assault beaches.

The movement of EG 11 into the central channel anticipated the arrival of the first U-boats in the area, which did not occur until the 14th. With the U-boats came EGs 1, 4, and 5, as these groups followed the 'wave' of attackers up the channel. The British groups accounted for two U-boats off Portland Bill before the end of the month, and so that only three U-boats actually reached the main assault convoy routes by the end of June.[46] This remarkable accomplishment on the part of Allied naval and air forces left EG 11 with little A/S action through the first weeks of the Normandy campaign. Moreover, Prentice soon found that he had to temper his previous enthusiasm for speed and aggression in pursuit of submarine contacts. 'After two days of operations in the shallow and wreck infested waters of the Channel,' Prentice wrote after the war, 'we revised our tactics completely from those which I had advocated for corvettes acting as close escorts in the deep waters of the Atlantic.' Prentice now substituted a 'slow hunt,' the first ship to gain contact holding the target until at least one other was in firm contact as well. Deliberate and unhurried attacks followed.[47] The tactics were similar to those employed during the previous winter by support groups on deep-diving U-boats. The variation inshore depended a great deal on proper classification of the contact, however, and EG 11 became noted for its skill and persistence in this process.

When EG 11 first arrived in the Spout, there were no U-boats, but there was considerable action all the same. On the night of 17 June *Gatineau* was attacked by a glider bomb, and on the night of 22–23 June *Chaudière* narrowly avoided a torpedo from a Ju 88. Three nights later *Gatineau* and *Chaudière* received another surprise in the form of three German motor torpedo boats, which were making their way from Alderney to Dieppe.[48] Such small, nimble, and high-speed craft were difficult targets at the best of times, and their torpedoes were lethal. In a surprise night action shrouded in smoke the two escort destroyers had little opportunity to hit them. The torpedo boats were illuminated by star-shell and, although they moved quickly out of oerlikon range, they were engaged with some effect by the destroyers' 4.7-inch guns. *Gatineau* avoided torpedoes fired her way at the last moment when they were heard by an alert asdic operator. One torpedo boat was damaged; all three escaped.[49]

On the whole, June was a frustrating month for the support groups. They had expected the fight of their lives, and, like the expected windfall of U-boat kills, it had failed to materialize. The plethora of non-submarine contacts and alarms frayed nerves and exhausted personnel. The RCN's wartime naval historian spoke for all when he wrote of EG 9's experience in the channel in June 1944: 'The first three weeks of Operation Neptune had given EG 9 little reward for its effort save for possibly, in common with all such groups, the satisfaction derived from an unprecedented expenditure of depth charges and hedgehog and of having assisted materially in causing a heavy mortality rate among the fish in the English Channel.'[50] Conditions that were supposed to work to the advantage of the escort groups seemed to work for the Germans. But on the whole, A/S forces were enormously successful. Of the twenty-five U-boats ordered up the channel in June, five abandoned the attempt, seven were sunk, and three were damaged sufficiently to send them back to base. The Germans accomplished nothing, and the kill rate of U-boats, at nearly 30 per cent, was much higher than predicted. By the end of June only four had reached the Spout and six were still trying to do so. The successful prevention of any serious U-boat threat was attributed by all concerned to radar-equipped aircraft. Their success was sharply reduced by the advent of the schnorkel and its progressive fitting to the U-boat fleet. When the schnorkel was added to the new high-speed U-boats known to be under development, the prognosis for Allied A/S measures was not good. As Stephen Roskill, the RN's official historian observed in summarizing the events of June 1944, 'unless victory could be gained before the two developments had come into general use, we might find ourselves struggling against an enemy who was once again possessed of the inestimable benefit of the initiative.'[51]

The Germans did not possess the initiative in the channel over the summer of 1944. They fought the Allies to something of a draw, however, when (to steal more lines from the passage of Shakespeare quoted at the outset of the chapter) they 'woo'd the slimy bottom of the deep, / And mock'd the dead bones that lay scatter'd by.' For contrary to Allied expectations, the Germans were quite prepared to settle on the bottom and mimic a wreck. Evidence of this tactic came early, and by 16 June the Admiralty promulgated a warning to that effect. The 'likely' course of action for a U-boat being hunted was to bottom and lie 'with the tide' (bow resting on the bottom and the stern pointing down the direction of the tide stream) no more than 2–4,000 yards from its would-be target.[52] This trick is what *U971* played on *Haida* and *Eskimo* two days after the Admiralty's signal.

A bottomed target, particularly close inshore, presented a difficult problem. In deep-ocean ASW the hunter had only to deal with problems of sound propagation in the sea: the extent to which temperature distorted the beam or prevented its penetration of certain layers. Inshore these problems existed as well and were complicated by differences in salinity in areas where there was fresh-water run-off. But the seabed itself and the presence of a tide added measurably to the problem of classifying an underwater contact. As all the escort groups discovered in the channel during June, schools of fish, rock ledges and pinnacles, and U-boats returned echoes. The bottom could hinder or help a search depending upon its type: smooth sandy bottoms absorbed sound, and therefore echoes returned in those areas were likely to be something man-made. Boulder-strewn rocky bottoms bounced echoes in all directions, creating so much reverberation it was impossible to distinguish a U-boat from the background noise. The set of the tide and its speed were key factors. The asdic operator of a searching warship drifting unwittingly on the tide could easily ascribe motion to motionless objects on the bottom. Moreover, a very fast tide surging around a submerged object produced cavitation noises, the sound of the collapsing of minute air bubbles formed by the passage of an object through a body of water. Tidal action turned a rock into a U-boat in motion. As Commodore (D), Western Approaches, observed, 'to classify correctly in the Channel is almost all the battle.'[53] The practitioners were more emphatic about the importance of classification: in inshore ASW it was (and remains) everything.[54]

The difficulty of classification of bottomed contacts inshore led A/S specialists to hope that the Germans would stay in motion, since getting the submarine moving in the water mass greatly simplified the problem. The Germans also were aware of this fact, and their adoption of bottoming and schnorkel inshore created a completely new tactical problem. Schnorkel was virtually impossible to locate with existing radar, and it was estimated that schnorkel reduced the effectiveness of air searches by as much as 90 per cent. Aircraft had forced the U-boats down; it was now up to warships to locate and sink them. But without help from aircraft and given the Germans' great reluctance now to transmit on HF radio (by which they could be DFd), it was hard to narrow the search area. HF/DF in the channel was never a success, since the receiving stations were too far away, and the accuracy of the fixes was seldom better than thirty miles. Without hot 'fixes' provided by sightings from aircraft, naval vessels had greater difficulty in localizing their targets. Once a U-boat was localized, all the problems of classification of the contacts and dealing with a cagey opponent remained. Troublesome asdic conditions, a tide to move silently on, and schnorkel allowed a skilled submariner a great deal of latitude in the last year of the war. The extent of this problem remained to be fully demonstrated by the end of June 1944, but the trends were already apparent.

The arriving U-boats in the Spout, the area between the Isle of Wight and the assault beaches, were met by a 'picked force' of groups operated by C-in-C, Portsmouth, which included EG 11.[55] With the thoroughness one would expect from Prentice, his group made absolutely sure of the identity of each underwater contact. On 27 June off Cherbourg EG 11 'classified' a contact by towing a depth charge on a length of cable fitted with a grappling hook. The idea, initially that of Prentice's staff officer Lieutenant Bob Timbrell, became a standard feature of EG 11's operations.[56] Once the grappling hook had snagged whatever lay on the bottom, the charge was fired electronically from the ship. In its first use the towed charge produced evidence of a merchant ship. There were other ways of determining the nature of a bottom contact. *Kootenay*'s commanding officer, W.H. Willson, preferred to lay depth charges alongside the contact with the aid of an echo sounder. The practice was to creep slowly along, place the charges set to their maximum depth, and then go full speed ahead while the depth charges 'cooked off' on the bottom. As Willson recalled, it was an effective 'but damned dangerous' procedure. More than one ship was seriously damaged by the concussive effect of her own charges in shallow water. Depth charges placed by echo sounder became the most common – and officially sanctioned – method of opening a wreck, or a bottomed U-boat.

But Prentice preferred his own system, and he got a chance to use it in early July. By 3 July EG 11 was off New Haven helping HMS *Forester* pursue a contact that Prentice was convinced was a U-boat.

Hedgehog bombs striking the water ahead of *Ottawa* in the English Channel, July 1944, while personnel on the bridge check their watches

Depth-charge explosions in shallow water, as evidenced by the dark sediments in these plumes astern of *Ottawa*, December 1943, posed a serious threat to anti-submarine ships operating inshore.

Chaudière and another unidentified ship of EG 11 in the English Channel in early September 1944

The courage of their conviction: a second swastika goes up on the funnel of *Chaudière* in late August 1944. The first was for *U621*, but the Admiralty did not award the kill of *U984* to EG 11 until after the war.

The frigate *Saint John*, seen here working up at St Margaret's Bay in April 1944, was a highly successful U-boat hunter. Note her hedgehog mounting well forward, the single 4-inch guns forward and aft, and the HF/DF aerial on her masthead.

A close view of what an acoustic homing torpedo was capable of: *Chebogue* in Port Talbot, Wales, in October 1944

It helped to be young: Lt Cdr Bill Willson, RCN, and the officers of *Kootenay* in the English Channel, August 1944. L to R, S/Lt P.C. Berry, RCNVR, Lt R.A. Creery, RCN, Lt A.H. McDonald, RCN, and Willson

Sleek type XXI U-boats at Lisahally, Northern Ireland, 1945. *U2582*, on the left, has her schnorkel mast raised. The submarine in the middle is *U2511*, the only type XXI to undertake a war cruise.

Magog, shattered, adrift, and bleeding oil, shortly after being hit by an acoustic homing torpedo from *U1223* in the St Lawrence River, 14 October 1944

The schnorkel head of the type XXI U-boat, *U2548*. The float valve that controlled access to the intake tube (on the left) shows clearly, as does the basket-like radar detector mounted on the top. The checkered surface is the rubber 'Albericht' coating, intended to absorb radar signals.

Lt Cdr Craig Campbell, RCNVR, late captain of *Clayoquot,* confers with Rear Admiral L.W. Murray and, on the right, Captain (D), Halifax, W.L. Puxley, RN, following his rescue on Christmas Eve 1944.

The Naval Staff in the last winter of the war: L to R, Captain D.L. Raymond, RN, director of Warfare and Training; Captain H.S. Rayner, RCN, director of Plans; Captain H.G. DeWolf, ACNS; Vice-Admiral G.C. Jones, RCN, CNS; Cdr K.C. Cooper, RCNVR, secretary; Captain E.S. Brand, RN, director of Trade; Captain D.K. Laidlaw, RN, director of Operations; and Captain S. Worth, RCN, director of Signals

An unlikely place to find an air force photographer! But a good shot of the dummy schnorkel mast fitted to the British training submarine *Unseen*, based at Digby, NS, in early 1945

Survivors of *U877* on the quarterdeck of *St Thomas*, December 1944. The dearth of depth charges carried by squid-equipped ships, in this case a Castle class corvette, is evident in this shot.

Lt D. Miller, RCNVR, surveys some of the damage in *Strathadam*'s communications mess following her hedgehog accident in March 1945.

A hard target for radar or the human eye: the dummy schnorkel and search periscope of the British training submarine *Unseen* off Digby in early 1945

The Bangor class minesweeper *Esquimalt*, the last RCN ship lost to enemy action, in St Margaret's Bay in April 1944: one year and a few miles away from her ultimate end

The RN Captain class escorts *Redmill*, her stern shortened by an acoustic torpedo, and *Rupert*, west of Ireland on 27 April 1945. The small drum on the top of their masts contained the American 3-cm SL radar, a key weapon in the battle to find schnorkel-equipped U-boats.

U889 surrenders to the RCN at Shelburne, NS, May 1945: note the schnorkel arrangements and the mast itself stowed forward.

As a junior officer, Ted Simmons led the boarding party onto *U501* in 1941, until recently the first known RCN U-boat kill. As captain of his own corvette, *Port Arthur*, he sank the Italian submarine *Tritone* in 1943, and his group, EG 26, claimed the last RCN U-boat kill of the war, *U1003*, in March 1945.

The enemy too was young. Lt Werner Meuller, twenty-two, and Lt Ernst Glenk, twenty-one, both of *U190*, were three-year veterans of the war at sea by the time they posed for this shot in May 1945.

The conning tower of *U190* at Sydney, NS, 1 October 1945. Her schnorkel mast, with its rubber coating and radar receiver, is raised, as are her two periscopes (search on the left, small attack one to the right). The rectangular frame raised to the left of the periscopes is the antenna for the Hohentwiel radar.

EG 9 in a British port, probably Londonderry, May 1945. The differences between River class and Loch class frigates are shown well in this photo: *Loch Alvie* (K428), *Monnow* to her right, and *Nene* far right. *Waskesiu* of EG 6 lies along the wharf, while the stern of Castle class corvette *Coppercliff* is in the foreground.

A remnant of the German arctic fleet en route to Loch Eriboll in May 1945 under the guns of *Loch Alvie*

Lt J.J. Coates, RCNVR, poses with Signalman W. Parrish and Petty Officer E. Massey (in the non-issue parka) on one of their prizes in Loch Eriboll, May 1945.

Unfortunately by the end of the day that contact and six others were confirmed to be wrecks. The group remained along the English coast for the next two days and so was in position to respond to the British corvette *Statice*'s call for help late on 5 July. *Ottawa* and *Kootenay* arrived at 0715Z the next morning; by 0938 *Ottawa* gained contact, and *Kootenay* joined two minutes later. *Kootenay* at once began to maintain a plot of the action, recording the movements of all the ships, the target, and the tide as a way of maintaining a continuous picture of events. Good plotting was crucial to effective ASW, and *Kootenay* was commended during her training in April for the superior quality of her plots.[57] In *Kootenay*'s case much depended on the skill and stamina of Sub. Lt Peter C. Barry, RCNVR, who remained at his plot – except for a brief spell of three hours – for most of the next two days.

It was initially assumed that the contact was on the bottom, in about 225 feet of water. The set of the tide, running at eighty degrees at roughly two knots, ruined the initial attacks, and Prentice paused to assess the situation more carefully. The tide quickly removed the local disturbances from the previous attacks, and it seemed that the contact was moving easterly at two knots, the same speed and direction as the tide. At 1017Z *Ottawa* attacked with hedgehog, which produced an explosion at about 100 feet and then some light oil. The target then changed direction, cutting across the tide stream. *Kootenay* attacked again with depth charges set to 225 feet at 1029Z on what was, by now, a much clearer contact. Passes over the contact, however, using echo traces by both ships revealed an object about sixty feet deep, now stationary relative to the bottom but stemming the tidal stream – which made it appear to have a forward movement of two knots. *Statice* now followed at 1123Z with a hedgehog attack, which produced 'one definite hit,' further oil, and some wood. Then the U-boat appeared to bottom, although apparent motion and sounds produced by tidal currents frequently suggested otherwise. Once these elements were eliminated, it was clear that the target never moved again.

A depth-charge attack by *Kootenay* at 1159Z produced further wreckage, including equipment from a U-boat. This evidence was normally sufficient to confirm a U-boat kill, but as Timbrell recalled later, 'The ground rules for getting credit for sinking a U-boat were very simple: ... if you didn't get a body, you couldn't claim a destruction.'[58] No human remains were recovered from *Kootenay*'s attack at 1159, 'although an object resembling a lung was seen floating past.'[59] Prentice now resolved to hold and hammer the contact 'until destruction was absolutely assured.' A series of depth-charge, hedgehog, and towed-charge attacks followed, intended to break up the U-boat. None produced clear evidence. Prentice ordered *Kootenay* and *Statice* to hold the contact throughout the night. Meanwhile, *Ottawa* and the RN frigate *Rowley* pursued a contact nearby. At 0649Z on the morning of 7 July *Kootenay* and *Statice* attacked again. The evidence they obtained, including more oil and a locker door of German origin, was no more convincing than that from nearly a day before, and it proved to be the last. At 1000Z Prentice left for Portsmouth bearing the evidence of a kill and *Ottawa, Kootenay,* and *Statice* were credited with *U678*.[60]

In the absence of survivor evidence it was impossible to know just which attack was responsible for *U678*'s loss. Commodore (D), Western Approaches, believed that the hedgehog attacks by both *Ottawa* and *Statice* obtained hits and considered the depth-charge attack by *Kootenay* at 1159Z accurate. 'The U-Boat must have been killed by 1200,' Simpson wrote, 'although the Senior Officer in *Ottawa* is considered correct in suspecting a ruse from the cascade of floating books.' It was, none the less, 'inconceivable that a submarine should live to withstand two hedgehog hits and the explosion of a 300 lb. minol depth charge in contact with its hull,' and so Simpson considered that Prentice was a little over-zealous in his pursuit of confirmation after noon on 6 July. '[T]he necessity therefore for the pulverising of the dead hull,' Simpson concluded, 'is not clear.' Other British officers thought the same, and Horton believed – as did many others, including Willson of *Kootenay* – that charges laid alongside the bottomed target by echo sounder were more effective than Prentice's towed charge for 'opening tins.'[61]

Prentice was commended for his thoroughness in classification and for the tactical handling of his group, which Simpson considered 'absolutely correct.' Prentice was pleased, too. EG 11, he wrote

The Summer Inshore 159

in mid-July, was 'beginning to work as a Group' and had become 'tolerably efficient at hunting A/S contacts as a team in the Channel.' At the same time he warned of trouble ahead with the mechanical reliability of his ships. Most were overdue for refits and were kept running only through the extraordinary efforts of the engine-room departments. The men, too, were worked extremely hard by these inshore operations. Many of the specialists worked in only two watches – four hours on, four hours off – for days on end. Neither the ships nor the men could endure such a pace.[62]

Prentice's persistence in classifying contacts would pay off twice again, six weeks later, but for the moment the sinking of *U678* was the signal accomplishment of EG 11. It was, perhaps, a measure of the group's skill that it was selected in July to test and perfect the new procedure, using echo-sounding traces, for classifying bottomed contacts.[63] This application, in fact, had been one of the foremost lessons of the destruction of *U678*.[64] By the end of July groups were busy using echo sounders to classify and buoys to mark wrecks. By early September there were so many wreck buoys in certain places in the channel that they had become a navigational hazard.

The destruction of *U678* was not the only victory for RCN groups in the channel in early July. Even as *Statice* lay in wait for EG 11 on the night of 5 July, EG 12 was slipping out of Plymouth on a quite different mission. The Germans escorted their submarines in the approaches to Brest, and C-in-C, Plymouth, wanted to stop them by destroying the escorts. When it was clear that the remaining German destroyers had withdrawn to Bordeaux, EGs 12 and 14 were ordered to carry out 'Operation Dredger.' The plan called for the two groups to enter the approaches to Brest during the night, one to attack and destroy the surface escorts before they rendezvoused with any U-boats, and then the support group would move in to tackle the U-boats themselves. EG 12 formed the 'striking force' while EG 14 remained in the offing ready to hunt U-boats.

By 2340Z on 5 July EG 12 was in position well south of Ushant. Mackillop planned to make his approach from the direction of Brest itself, which meant a sweep in along the rugged shores of Brittany. A QH set on *Qu'Appelle* and local navigation lights greatly reduced the fear of grounding as the destroyers – *Qu'Appelle, Saskatchewan, Skeena,* and *Restigouche* – swept-in on an eastward course at twenty knots in single line ahead. Radar contacts were obtained by *Qu'Appelle* at 0113Z to the northwest and they were confirmed by two other ships. Speed was increased to thirty knots, the course gradually altering northward and then around to nearly due west. As the distance closed, speed was dropped to sixteen knots to reduce the white bow wave which might alert the enemy. Finally at 0137Z, while they were steering 290 degrees, the enemy was sighted just off the starboard bow at 3,000-yard range. MacKillop ordered speed increased again to thirty knots, in part because all of the old destroyers had fewer vibrations to disrupt their gunnery at that speed and in part because he did not want to give the Germans any extra time to get the range with their small-calibre weapons. At that point the scene was illuminated by rockets fired from the charging destroyers and the 'Battle of Pierres Noires' began.

What the rockets revealed were three or four (accounts differ) 'trawlers' in line ahead with two U-boats in station on either quarter of the middle trawler. They were heading westward out to sea along a series of rock ledges known as the Pierres Noires, or Black Stones, which arced from the mainland to Ushant. This discovery upset the plan slightly, since MacKillop had expected to meet the escorts on their way out without U-boats in tow or, at best, to find U-boats being escorted into Brest. EG 12 had the small flotilla trapped against the Pierres Noires, however, and it set about dealing with the escort. At 0144Z *Qu'Appelle* opened fire, as did each of the destroyers as their guns came to bear, and each destroyer also fired individual torpedoes at specific targets – to no apparent effect. Hits from gunfire were observed on this first pass, and, while the U-boats sought the cover of Brest at high speed, the trawlers returned fire with rather heavier weapons than MacKillop had anticipated, 3-inch and 4-inch as well as 20-mm guns. From *Saskatchewan*'s bridge it appeared that no one on *Qu'Appelle* could survive the hail of fire pouring into her from the leading German ship. Easton watched in amazement as tracer, 'like high pressure water from half a dozen hoses' pumped 'streams of liquid fire' into

MacKillop's ship. He then realized that his ship was next, and that *Saskatchewan* was already receiving much the same treatment from her opposite number in the German line of battle. In fact, the heavier armament of the destroyers had a telling effect from the outset. By the time the first run was complete, one trawler was stopped and smoking heavily, another was on fire, and a third was adrift and smoking. The second pass poured more fire into the disabled German vessels as MacKillop closed the range, and the whole battle drifted on the setting tide towards the Black Stones. The leading German ship was soon sinking. The second, now burning from stem to stern, continued to fire back until the last possible moment. EG 12 then closed with the third German ship to point-blank range, which was more daring than wise, since the trawler still had a lot of fight left. While the destroyers pounded her with 4.7-inch shells and oerlikon fire, the trawler responded with accurate 3-inch or 4-inch and 20-mm gunfire. MacKillop's *Qu'Appelle* caught the worst of it, her bridge being struck and MacKillop himself seriously wounded. *Saskatchewan* was next in line, and the doomed trawler retained enough fight to blow away the destroyer's 271 radar antenna and inflict casualties. Soon the weight of the destroyer's armament silenced all opposition, and the burning trawlers were finished off with impunity. By the time EG 14 arrived, the Battle of Pierres Noires was over, and the U-boats had escaped. With MacKillop wounded, command of the group passed to Lt Cdr P.F.X. Russell, RCN, who ordered EG 12 back to Plymouth, where it took four days to repair the battle damage.

On the whole there was little criticism of MacKillop's handling of the action. The staff officer (Tactics) in the Directorate of Warfare and Training in Ottawa offered him a mild rebuke for allowing the U-boats to escape, observing, 'Inasmuch as the object of the Striking Force was the destruction of the enemy surface escorts encountered, this operation was a complete success.' EG 12 followed its orders, sinking the three heavily armed trawlers in a well-executed night engagement. By MacKillop's own admission, EG 12's gunnery was not good, but this situation was to be expected in ships without director control (removed to fit type 271 radar). *Saskatchewan* alone expended 274 rounds of 4.7-inch ammunition, plus several thousand rounds from her secondary armament. In contrast, 'The enemy's close range fire was much too good,' MacKillop wrote, 'and there was too much of it to close initially to point blank range.'[65] The action had an exhilarating effect on the personnel of EG 12. 'It was all so different from the U-boat warfare of the Atlantic,' Easton recalled, 'I felt that I would prefer this kind of thing anytime to the long-drawn-out, anxious nights of the submarine war.'[66] The Battle of Pierres Noires and the sinking of *U678* established a clear pattern that the two destroyer groups followed through July and August, one the surface 'specialists,' the other the A/S leaders.

EG 11 patrolled for only a few more days following its destruction of *U678* before enjoying a two-week layover in Londonderry from 12–28 July. EG 12, now under Birch, who moved from EG 6 to take command, was back at sea on 16 July off Brittany, chasing U-boats and classifying contacts. On the 24th, while patrolling with EG 14 south of Plymouth, it was attacked by aircraft in poor visibility. Their aircover was drawn off by a lone Ju 88, and then at least seven aircraft equipped with glider bombs attacked. *Skeena* and *Restigouche* sustained near misses, and *Qu'Appelle* managed to shoot down one bomb that was clearly tagged for her. By the end of the month the group was alongside at Londonderry for a well-earned forty-hour layover.

By comparison, the operations of the frigate groups in July were almost uneventful. EG 6 sailed from Londonderry on 27 June to patrol off Land's End in support of convoys passing the area. By early July they were off Ushant alongside EG 14 before returning to Plymouth, where Cdr Birch left the group to take over EG 12. In his place the commanding officer of *New Waterford*, A/Cdr W.E.S. Briggs, RCNR, became the senior officer of EG 6, a post he held until the end of the war. In a navy short of qualified talent for senior commands, Briggs was an excellent choice. Before the war Briggs had been a radio announcer for the CBC and was a man of impressive presence and pith. He had made something of a name for himself in 1939 during the royal visit to Canada. While doing a broadcast from the Hotel Nova Scotia in Halifax, Briggs had had to fill forty minutes of live air time because of the late arrival of the king and queen. He did so by giving a graphically detailed description of the paintings 'lining the hotel ballroom': a remarkable feat, since there was not a painting in the place.[67]

But Briggs was also professionally qualified to lead a group by 1944. In 1943 he was noted by the training staff of Western Approaches for having 'made a deep study of anti-submarine warfare in the Atlantic both from the practical and theoretical stand points.' Briggs's opinions were 'sound and well reasoned.'[68] By January 1944 both Prentice and Murray had tagged him as a potential group senior officer. In July 1944 it remained for him to earn his spurs as the SO of EG 6, the first Canadian reservist to take command of a support group. Unfortunately, the chances for Briggs to distinguish himself did not come easily. Although EG 6 moved into the Spout in mid-July to relieve EG 11, it patrolled for the rest of the month uneventfully.

EG 9 fared little better. It went back to sea on 28 June, and the next day *Port Colbourne* illustrated the perils of improper classification when she attacked the wreck of HMS *Blackwood* off Portland. The frigate's hedgehog bombs detonated *Blackwood*'s magazine or countermined her depth charges, resulting in a huge explosion which damaged *Port Colbourne*'s asdic, radar, gyro, and engines. She was able to make port on her own power. Later that day EG 9 shifted to the Channel Islands, which remained in German hands until the end of the war, and drew uncomfortably accurate fire from the coast batteries on Guernsey. The group remained to the west of the Cherbourg peninsula for several days, chasing a schnorkel, drawing fire from Alderney after an error in navigation, and being attacked by aircraft. On 6 July they moved to the English coast, then to the west of Ushant and back to Plymouth, from which a special convoy was escorted to Northern Ireland followed by a four-day layover from 11–15 July.

The British assessment of EG 9's operations up to mid-July gave them a passing grade – but just. The director of the Anti-U-Boat Division of the Admiralty considered the group's performance a little 'amateurish,' and he planned to raise the matter with the staff at Western Approaches. Whether he did so or not remains a mystery, but his staff did note that EG 9's standards would improve with the fitting of radio navigational aids in the near future.[69] In fact, *Matane* was withdrawn briefly from operations on 4 July to fit QH.

By 16 July EG 9 was back at sea off Ushant and it remained in the Brest area for the next four days. On the 20th, in generally good weather but with high broken cloud, an explosion was heard astern of *Meon*. The natural assumption was that an acoustic torpedo had detonated in her wake. A few minutes later, however, aircraft were seen high in the clouds, well out of range of even the main armament. In the absence of an air-warning radar set in the frigates, the aircraft had achieved complete surprise but fortunately had missed with their first glider bomb. The second bomb fared better. The release was watched by the men of EG 9, but they could only wait for it to come within range before trying to knock it down. This they failed to do, and before *Matane*'s order for full ahead both engines took effect the bomb struck the frigate a 'glancing blow just before "Y" gun', plunged into the water alongside, and exploded.

Four more glider bombs were aimed at the ships, but miraculously none struck the immobile *Matane*. Her bomb had carried away some topside fittings and equipment, and two ratings who were standing in the bomb's path on the after deckhouse were never seen again. Much of the blast was absorbed by the water. The shock nonetheless ruptured *Matane*'s side and flooded the engine room, bursting steam pipes, scalding everyone on duty, and killing one rating. Crewmen on deck, many of whom would have been killed by a detonation in the air, escaped largely unhurt. Bulkheads held, power was restored, meals got to the crew, and *Meon* took the crippled frigate in tow. *Matane*'s newly installed QH3 navigational aid was thrown to the deck of the charthouse by the explosion. However, 'when replaced and switched on, [it] was found to be in perfect working order.'[70] *Matane* made port and eventually returned to duty, and EG 9 went back to work immediately, for a few days under Lt Cdr C.A. King of *Swansea*. The rest of their month was one of patrolling, largely without incident, off Cornwall.

In July the burden of anti-submarine work inshore shifted to warships, but the return on effort – although better than that for June – remained disappointingly small. Of the nine U-boats destroyed

The Summer Inshore 165

off the French coast and in the channel, six fell to surface vessels. The only air successes, two U-boats destroyed, occurred in the Bay of Biscay. One U-boat was sunk by a mine in the approaches to Brest. All of the six sunk by surface vessels were credited to different ships or groups: one each to EG 1, EG 2, EG 3, EG 11, 10th DF, and the British escorts *Wanderer* and *Tavy*. The sinking of *U333* on the very last day of July by *Loch Killin* and *Starling* of EG 2 well to the west of Land's End was the first kill by a squid-equipped ship.

If there was pattern to the July U-boat kills it was geographical. Whereas action in June was concentrated either around Ushant or between Cap de la Hague and Start Point, in July activity was in the area from Start Point – Cap de la Hague eastward to the Spout.[71] The shift was not surprising. The assault area was the principal operational zone for U-boats, and the schnorkel-equipped submarines revealed their presence by attempting to attack shipping.

August brought yet another shift in pattern, one occasioned by changing fortunes ashore. For nearly two months the Normandy campaign ground along in a battle of attrition within the narrow confines of a small beachhead. Cherbourg fell at the end of June, but on the whole the Allied armies remained locked in by fierce German resistance until late July. Finally, on 25 July the Americans broke the cordon at St Lo. By the 31st they were at Avranches, and in the early days of August, as the British and Canadians fixed the weight of the German army around Caen with a series of brutal assaults, the Americans were driving through western France as fast as wheeled vehicles could carry them. The German army's collapse in Normandy spelled the end for the U-boat bases in France. Hasty preparations were made to get every submarine to sea as quickly as possible, while those still on patrol were ordered to bases in Norway.

The hasty evacuation of bases from Bordeaux to Brest presented the Allies with an excellent opportunity to destroy what remained of the German navy in France – especially the U-boat fleet. By early August patrols by striking forces and A/S groups were extended into the Bay of Biscay. EG 11 began Canadian anti-submarine involvement in these operations by screening the carrier *Striker* from 30 July to 3 August and then by operating off the great U-boat base of St Lorient alongside EG 2. It was there, on 6 August, that *Loch Killin* of EG 2 began the new Bay of Biscay campaign by sinking *U736* with her squid. When EG 11 returned to the area on the 8th after refuelling at Plymouth, Prentice pushed its patrols well inshore off the Ile de Groix, which lay just off St Lorient. This rashness brought upon the group a night-time glider-bomb attack. One bomb passed directly over *Kootenay*'s masthead and crashed into the sea close by, shaking the whole ship violently. Then, as the group headed south, eight miles from the Ile de Groix the shore batteries fired a couple of parting salvoes, two rounds landing close enough to *St Laurent* to displace a gun mounting and force her back to Plymouth for repairs.

For the next few days EG 11 pursued targets inshore, trying to intercept coastal traffic. It was tense and nervous work. The fear of mines, long-range gunfire, submarines, fast attack craft, and air attack was componded by navigational hazards. Reports and memories of the period indicate the men of EG 11 were tired and edgy, and some were less than anxious to push their luck. That tension was evident on 10 August in the ill-conceived bombardment of the wharfs of the small fishing village of Concarneau in the belief that a minesweeper and a ship of 1,000 tons lay alongside. *Ottawa* and *Chaudière* entered the harbour and, while making considerable speed and turning tight circles, blasted away at a range of 8,000 yards. The only apparent result was a great deal of damage to the waterfront and a terrorized local population.[72] Later on 10 August EG 11 pursued and attacked a contact reported as a submarine by a patrolling aircraft. Prentice considered the contact a good one and the U-boat destroyed, but the Admiralty disagreed, noting 'insufficient evidence of damage.'[73] Three days later, while *Kootenay* and *Chaudière* were refuelling in Plymouth, *Ottawa* and *St Laurent* were close enough to witness the last acts in a battle between *U270* and an aircraft of 461 Squadron. Responding to flak burst on the horizon, the two ships arrived in time to watch the U-boat sink and to rescue seventy-two survivors, who included German army and air force personnel.

Meanwhile EG 12 returned to sea on 7 August, minus *Saskatchewan*, which was to leave for a refit in Canada, but having added *Assiniboine.* After a brief patrol off Land's End the group moved to the Ushant area, and there on the 11th they were ordered to conduct a repeat of Operation Dredger. This time they were joined by the small British Hunt class destroyer *Albrighton,* which had superior fire-control equipment and heavier armament but was capable of only twenty-three knots. With *Qu'Appelle* leading, *Albrighton* behind and *Assiniboine, Skeena,* and *Restigouche* following along, EG 12 set off on its second surface action.

At 0210Z on the morning of 12 August what appeared to be three contacts were obtained on radar at 13,000 yards. Birch positioned his ships so that the targets, which were just a mile off the coast of Ushant, were silhouetted by the moon. During the first run one trawler was hit, and in the confusion the Germans were seen to be firing at each other. During the second run two other ships were struck, and they headed for shore while the lead ship – a guncoaster – attempted to draw EG 12 away. By the third run two trawlers were beached on Ushant and were on fire, and a nearby farmhouse was also ablaze. In the darkness and smoke it took two more runs to find and finish off the larger ship. The only serious damage to EG 12 resulted from a collision. On the fifth and last run *Skeena* came careening out of the smoke to find *Qu'Appelle* directly ahead of her. Orders for all astern both engines were too late to prevent *Skeena* from crashing into *Qu'Appelle*'s starboard quarter. *Qu'Appelle* had repaired her steering by 0600Z and the rest of her damage was not serious. *Skeena*'s bow was wrecked, and her speed had to be cut to six knots so as not to put undue pressure on the forward bulkheads. Although both ships returned to Plymouth safely, the collision took them out of action for the balance of the summer.[74] Their loss reduced EG 12 to only *Assiniboine* and *Restigouche.* On the 12th *Assiniboine* was seriously damaged by a premature depthcharge explosion, and *Restigouche* now found herself on that same day serving with EG 11.

By mid-August EG 11 was maintaining a presence in the Bay of Biscay through a rotation of ships to Plymouth for fuel, ammunition, and stores. On the 18th *Ottawa, Chaudière, Kootenay,* and *Restigouche* were about seventy miles off the Ile D'Oleron (the approaches to Rochfort), when at 0953 hours *Kootenay* reported another in a seemingly endless series of contacts. It was, none the less, a good contact, and while *Chaudière* and *Restigouche* boxed in the area with square searches, Prentice and Willson went to work on the target. As *Ottawa* lined up for a hedgehog attack at 1012Z, the target showed movement and appeared to be taking avoiding action, but it was not enough and came too late. One hedgehog bomb exploded about 160 feet off the bottom, followed by the appearance of both bubbles and oil on the surface. Echo sounder traces shortly thereafter showed the target at rest on the seabed. It never moved again, and Prentice considered that he had 'disabled' the U-boat with this attack.

Attempts to deal with the target on the bottom got off to a shaky start. *Kootenay* tried to lay a single charge over the contact using echo sounder, but the charge – set for 500 feet – detonated at thirty feet, shaking the destroyer. Willson then fired his Mk X depth charge, a ton of high explosive carried in a torpedo-shaped canister and ejected from the destroyer's torpedo mounting. Much to everyone's dismay, the charge failed to go off. Dave Groos, *Restigouche*'s captain, was well aware of this failure when he was then asked to attack the contact with depth charges. Attempts by *Chaudière* around noon to obtain a better fix on the target with her type 147B failed, owing to defects in the set, and she withdrew to continue her box search.

Prentice did not need reminding that air bubbles and fuel oil were insufficient evidence of a kill; so the process of tearing open the bottomed submarine now began. It was not an easy task. The tide produced a false sense of motion in the target, and the oil oozing from it returned traces on the echo sounder and produced false echoes (and therefore erroneous depths) on the type 147B sets of both *Chaudière* and *Restigouche*. A series of attacks with both depth charges and hedgehog produced further oil and some wood, until an attack at 1529Z which produced 'A fountain of bubbles.' Prentice now considered the U-boat destroyed, but he lacked firm proof. As the group moved off, *Chaudière* was left to watch over the contact. She stayed with it all night, and when, after losing contact on the morning of 19 August, she left to rejoin the group C-in-C, Plymouth, ordered her back. *Chaudière* dutifully maintained contact throughout the day. When the wind drove the rising oil into a well-defined slick, she got a firm enough position on the wreck to attack it with hedgehod at 1607Z and 1609Z. These final attacks produced a letter in German, dated 11 August, more wood, oil, and a chocolate wrapper with 'Berlin' written on it. One final attack produced a separate, rending explosion from the wreck but no additional evidence. The Admiralty assessed the attack as 'probably sunk,' and the kill was confirmed after the war. EG 11 had destroyed *U621*.[75]

Since all ships were now short of fuel and *Restigouche* was already detached to Plymouth with defects, the three ships of EG 11 headed for Londonderry on 20 August. While en route, *Ottawa* obtained a good asdic contact twelve miles due west of Ushant at 1935Z. *Chaudière* at once conducted a box search around the area while *Ottawa* and *Kootenay* prosecuted the contact. The U-boat was lost briefly and then regained by *Kootenay*, who classified it as definitely submarine. *Chaudière*, because of her 147B and Q sets, was called in to fix the contact. She reported a strong and distinct echo and gained a firm contact with her type 147B, from which she obtained targeting data for a hedgehog attack. Contact, with a definite 'metallic ring' was then regained, and *Chaudière*'s very last hedgehog bombs were fired using her Q attachment. When that attack failed to produce the desired results, *Chaudière* resorted to depth charges delivered with the aid of Q, an attack that Nixon estimated 'must have burst a few more rivets.' But *Chaudière* could not wait around for results, owing to a 'dire shortage of fuel.' By 2300Z it was too dark to see, all the ships were low on fuel, and *Kootenay* had sprung another leak in her main feed pump. The hunt was abandoned.

The Admiralty considered this attack initially as a U-boat 'probably damaged,' an assessment later altered to 'insufficient evidence of the presence of a U-Boat.'[76] The Admiralty's assessments were usually sound, but in this case they were far off the mark. As Nixon commented many years later, 'We did not believe the Admiralty's assessment ... we were the people on the spot and by this time after several bouts with known U-boats (some got away) we knew a real one when we heard it.'[77] When *Chaudière* arrived in Plymouth, her crew promptly painted two swastikas on her funnel. EG 11 had sunk *U984*, the U-boat that had eluded EG 12 on 8 June, and was the second kill by EG 11 in as many days.[78]

Although at the time EG 11 was credited with only one U-boat during its Biscay patrols, senior British officers were impressed with the group's efforts. 'This is a very fine patrol,' a staff officer at the Anti-U-Boat Division of the Admiralty observed, 'and the report is of considerable interest.' EG 11 was estimated to have been in contact with perhaps five U-boats between 10 and 20 August, sinking one certainly and damaging two others.[79] Commodore (D), Western Approaches, Commodore G.W.G. Simpson, credited EG 11's success to Prentice's 'tactical directions.'[80]

The destroyer groups lay in port through much of the end of August 1944, crippled by age and war weary. Their place in the Bay of Biscay was taken, in part, by EG 9. That group spent the first three weeks of the month screening convoys along the Cornish coast, where the concentration of wrecks made classification of contacts difficult. The value of new navigational aids for plotting contacts was noted by Layard in his report for the period. 'The value of this navigational aid for pin-pointing the position of wrecks, contacts, submarines, etc., in tidal waters out of sight of land,' Layard observed, 'cannot be overstated.'[81]

On the 19 August EG 9 was ordered to the Bay of Biscay, where it patrolled off La Pallice. By the 29th it was off the Ile de Re, close enough for *Stormont* to be straddled by fire from the coast batteries. The balance of the Bay of Biscay patrol was rather routine: numerous wrecks were attacked and classified, some Spanish trawlers were investigated, and the body of an American P-47 pilot was recovered and buried at sea. It was, as C-in-C, Plymouth, observed, a disappointing patrol 'considering the number of targets that are thought to have been available in the Bay.'[82]

EG 9's return to Plymouth on 28 August marked the end of Canadian support group operations in the bay. One final act remained in the Allied inshore anti-submarine campaign during the summer 1944, however, and it fell to EG 9 to complete it. On 31 August the group sailed from Plymouth to patrol the now-familiar waters off Land's End. They were no sooner on station than C-in-C,

Plymouth, ordered EG 9 to move northward, to the area off Trevose Head (on the north Cornish coast), in response to an aircraft radar report. At 1845Z, as EG 9 passed Land's End, *Saint John* obtained a good contact fifteen miles due east of Wolf Rock. Layard detached *Monnow, Stormont,* and *Meon* northward to begin the ordered search: *Swansea, Port Colborne,* and *Saint John* stayed behind to pursue the contact. Difficult tide conditions made it hard to hold, but by 2115Z contact was firm enough to attack with both depth charges and hedgehog, which produced slight traces of oil. By 2300Z it was too dark to follow the oil traces and the contact was completely lost.

It was assumed that the U-boat was moving southwest, so Layard organized his three ships in line abreast, and at 2400Z they began to sweep from the last known position of the contact across its likely path. The sweep went first to westward between Land's End and Wolf Rock, then around south of Wolf Rock, and eastward back towards the datum. The group had just turned back to the east south of Wolf Rock when at 0155Z on 1 September *Saint John* regained contact three miles from Wolf Rock light. This time the target was on the bottom in forty-two fathoms. *Saint John* classified it with an echo trace, which showed a clear spike in the centre of what appeared to be a hull: unmistakable evidence of a U-boat. The frigate then carried out two depth-charge attacks, both of which produced oil and tearing explosions. C-in-C, Plymouth, ordered the group to remain on the contact, which it did throughout the rest of the night, although *Saint John* again lost contact and *Swansea* was never able to gain it at all. Only the expanding oil slick gave continuous evidence of the location of the target.

At daylight Layard took the three ships off to search the area of the previous day's contact, which produced nothing. When the group returned to the oil slick south of Wolf Rock light, *Saint John* regained contact and obtained a very good echo-sounder trace. The latter revealed that the U-boat was already heavily damaged. *Saint John* then opened the range and at 1404Z on 1 September dropped five depth charges on the target, directed by her echo sounder. This attack opened up the U-boat on the bottom, littering the sea with wreckage that included an engine-room log, charts of swept channels, clothing, equipment, and internal fittings from the submarine – ample evidence of the destruction of *U247*.

The British were fulsome in their praise for EG 9's persistence off Wolf Rock, particularly that of *Saint John*. The sinking of *U247* demonstrated 'the skill and efficiency of the Commanding Officer [A/Lt Cdr W.R. Stacey, RCNR] and A/S team of H.M.C.S. *Saint John* and the expert and firm control of operations maintained and exercised throughout the varying phases of the hunt by the operating authority, Commander in Chief, Plymouth.' The story of persistence under difficult asdic conditions and a certain lack of faith on the part of others in the group (only *Saint John* ever had a good contact) merited a discussion in the October issue of the *Monthly Anti-Submarine Report*.[83]

The sinking of *U247* off Wolfe Rock on the night of 31 August – 1 September was a fitting reward to EG 9 for its months of labours inshore. With a little luck they might have sunk a few more U-boats, but in many respects their luck was no worse than that of most other support groups. Only EG 6 ended the summer campaign with nothing to distinguish its efforts. Here, too, luck was a factor, but the change in senior officers in July may have had an impact on the group's deployments. EG 6 spent the whole of August in the Spout, providing regular support for the cross-channel traffic. Ships came and went for fuel and stores and to allow for occasional leave, but the action in August was elsewhere and EG 6 was never a part of it.

The destruction of *U247* is taken to mark the end of the inshore campaign that began with the invasion of France on 6 June. Only thirty schnorkel-equipped U-boats operated in the channel, making forty-five individual sorties – a far cry from the numbers the Allies had anticipated. On the whole, the Germans found it difficult to get into the central channel, although once there they found operational conditions good. Conditions in the U-boats were, none the less, appalling. *U218*, which laid mines off Land's End on 20 June, suffered damage to her diesels that resulted in the venting of some exhaust gases into the submarine. Her log on 20 June reported, 'Several men suddenly taken ill during the forenoon. By noon two-thirds of the crew are suffering from severe headaches and stomach-ache, nausea and retching and are no longer fit for duty. The remainder ... keep things going. There are several cases of fainting due to over exertion and carbon monoxide poisoning.'[84] Navigation, too, was difficult from a submarine that spent all its time submerged. Charts were inaccurate, the coast was hard to discern precisely through a periscope, and echo sounders were difficult to use without giving away positions. U-boats struck bottom with some regularity. The navigation of *U763* was so far off that she found herself – unwittingly – in Spithead, the actual entrance to Portsmouth harbour. In the end the Allies took a heavy toll: twenty-two U-boats and 1,000 crewmen lost between Ushant and Dover in the summer of 1944 – 'all scatter'd in the bottom of the sea.' In exchange for this sacrifice, twelve Allied ships (of all types) were sunk and five damaged.[85]

Over June, July, and August the Allies sank thirty-six U-boats in the English Channel and the Bay of Biscay. Ten fell exclusively to aircraft and two to mines; seven were accounted for by convoy escorts and fleet destroyers (*Haida*'s share of *U971* is included here). Seventeen U-boats were wholly or partly claimed by the A/S support groups. In the cases of three of them, *U385* and *U608* shared partially by EG 2 and *U618* shared by EG 3, naval forces finished off U-boats damaged by aircraft. Of the fourteen U-boats awarded unreservedly to the support groups, the distribution of kills was fairly even. Only EG 12, the surface strike force specialists who sank six enemy ships, and EG 6 destroyed no U-boats. EGs 1, 9, 5, and 14 sank one U-boat each, while EG 2 and EG 3 each independently accounted for two. The extent to which EG 2's role was affected by the sudden death of Captain Walker in June is unclear. Only EG 11 could claim three: the best score inshore in the summer of 1944.[86]

Although EG 11's score was not made official until after the war, Prentice's sailors knew they had sunk at least three U-boats and that they were second to none in the summer of 1944. As *Chaudière*'s commanding officer observed, 'At the time my 210 mostly young shipmates felt an awareness and pride in these stunning successes: in fact at ship's company reunions the surviving members still do.'[87] It was a tremendous accomplishment but one that went utterly unheralded.

Nothing used by the Germans at sea in the summer of 1944 – the midget submarines, one-man torpedoes, glider bombs, aerial torpedo attacks, mines, coast artillery, small craft – affected the Allied campaign ashore. In that sense, the efforts of naval and air forces in the narrow seas were entirely successful. But at the tactical level and among A/S specialists, the experiences of the summer were only a qualified success. As the *Monthly Anti-Submarine Report* for December 1944 noted when summing up the experience, 'We had been disappointed of a holocaust when U-boats failed to storm up the English Channel in the first week of June,' and 'we were again disappointed when they were flushed out of their French bases' in August. Moreover, as the bulk of the U-boat fleet escaped successfully to Norway in August, the first serious attacks on cross-channel shipping occurred in which four ships were sunk and two damaged. The attackers got clean away. One of them, *U480*, which sank the Canadian corvette *Alberni*, the British minesweeper *Loyalty*, and one freighter south of the Isle of Wight, was completely covered in a special rubber coating designed to absorb sound waves from asdics.[88] Whether that new 'Alberich' skin saved her or, like *U989*, which scored success in the same area at the same time, survival was primarily due to the skill of her captain is open to speculation. The fact remained that U-boats had struck and escaped retribution. As the *Monthly Anti-Submarine Report* concluded, 'The loss to our resources was small but the skill with which the U-Boats attacked and escaped unscathed was disquieting.' The summer campaign inshore made it evident that there was much to learn about A/S operations against an enemy who was now never seen. The ocean environment and schnorkel gave submariners new life by late 1944. If that feature could be matched by high submerged speeds, a successful renewal of the German campaign in the Atlantic was likely.

5
A Sea of Troubles

The harvest is past, the summer is ended, and we are not saved.

Jeremiah 8:20

On 16 August 1944 *U482*, commanded by Count von Matushka, sailed from Bergen for the North Channel. Matushka's U-boat was an early type VIIC, but she carried some of the latest improvements including schnorkel, the new radar detector 'TUNIS', which could locate 10-cm transmissions, and the first successful U-boat radar. *U482* carried the usual mix of acoustic and pattern-steering torpedoes, as well as 'submarine bubble-target' (SBT) asdic decoys. Matushka intended to operate in less than thirty fathoms and to use the inshore environment to his advantage.

The passage through Allied patrols between Norway and northern Britain was largely uneventful. Three aircraft and three antisubmarine vessels were sighted, but *U482* remained undetected and located two ships on its own radar. By 28 August Matushka was in the approaches to the North Channel, the first U-boat along that coast in some time. There he remained for the next two weeks. Navigation, often difficult from a schnorkelling U-boat, proved easy because radio beacons and lighthouses were 'still operating as in peacetime.' Indeed, during his whole cruise, *U482* surfaced only once to fix its position. Naval forces were all around, their FOXERS and CATs creating a regular din, but Matushka was not intimidated. He kept his battery charged by schnorkelling briefly at night and resting on the bottom during the day, and then he allowed targets to come to him.

The first to do so was the 10,000-ton tanker *Jacksonville*, part of convoy CU 36, on 31 August. The sinking prompted an intense search. Undeterred, the next day a little further west and a little closer to the Irish coast, Matushka destroyed the British corvette *Hurst Castle*, part of the force sent to hunt him. On 3 September *U482* was overrun by ONS 251, from which Matushka sank the Norwegian steamer *Fjordheim*. Finally, on the 8th the tanker *Empire Heritage* and the small steamer *Pinto* were sunk from HXF 305 just fifteen miles from shore. With that action *U482*'s cruise effectively ended. She arrived in Bergen on 26 September having travelled 2,729 miles – all but 256 of them fully submerged. Matushka claimed that he was never located by aircraft and never firmly fixed by asdic, a claim the British official historian accepted as 'well founded'. As Roskill observed, the loss of these ships 'almost on our front doorstep and at our most sensitive spot [where the main transatlantic convoys funnelled into the Irish Sea] ... was an unpleasant shock, the more so because all the victims were sailing in convoy.'[1]

The sinkings by *U482* suggested that the teeth of the U-boat fleet were anything but drawn by the late summer of 1944. Max Horton, Commander-in-Chief, Western Approaches, was particularly anxious that this old U-boat fleet be decisively beaten and its morale crushed before the long-anticipated wave of radically new type XXI and XXIII U-boats entered service.[2] The new U-boats were already behind schedule. In August and September there seemed little likelihood that they would enter service before the Allies crossed the Rhine and ended the war.

Events during the fall of 1944 served only to deepen Horton's anxiety over the situation at sea. The western Allies stalled at the Rhine in late September, and by October it was clear that the war would not end until the spring of 1945 at the earliest. U-boat construction continued unabated, and the new types poured from yards. The Germans were, it is true, finding it increasingly difficult to man the new fleet, and it became necessary to impress air force personnel into the U-boat service. Sabotage and shortages of skilled workers and material delayed completion of submarines, while their training areas in the Baltic were under continuous pressure from the Russians in the east and Allied aerial mining operations. The clock was winding down on Dönitz's plan to renew the offensive. None the less, the very low rate of U-boat losses through the last half of 1944 helped to preserve the U-boat fleet's expanding expertise in inshore operations. Few submarines were now sunk by A/S aircraft, and the sinkings by warships over the fall proved disappointingly low for the Allies. Senior officers at Western Approaches Command were anxious, lest the expertise in inshore and fully submerged operations developing within the existing U-boat fleet was passed on to the crews of the new U-boats. They also understood that if they could not decisively crush the old U-boat fleet, how could they expect to cope with the radically new types already working up in the Baltic? The fortunes of the RCN's support groups, and indeed the whole shape of the anti-submarine campaign in the last months of the war, were inextricably linked to these fundamental problems.

For Canadians the issue of dealing with the old U-boats inshore remained the most vexing. The major German incursion into Canadian waters in 1942 spurred the RCN to push the scientific exploration of the influence of sea conditions on ASW. U-boats operating inshore escaped all underwater detection inshore during 1942, despite the closeness of escorting vessels to scenes of attack. In the same year the RCN received charts from the USN, based on data accumulated largely by the Woods Hole Oceanographic Institute, that proported to show the 'assured' asdic ranges for ocean areas. These estimates were based on the new science of 'bathythermography' (BT), the changing temperature of the ocean mass in relation to depth. It was understood, albeit still vaguely, that the movement of sound in water was affected by changes in temperature. If the temperature profile of the water could be determined, it would be possible to estimate how the shape and range of asdic sound beams were affected. With that knowledge A/S ships could be spaced so that the assured ranges of the asdic sets at any given time overlapped, and so to prevent a U-boat from slipping between searchers.

While the National Research Council, which did the RCN's scientific research and development, set off to learn more about BT, the RCN's director of Anti-Submarine, A.R. Pressy, pushed the Naval Staff at the end of 1942 to acquire an oceanographic vessel for the east coast. That campaign drifted through 1943, and it was not until April 1944 that the Royal Society's schooner *Culver*, lying at Bermuda since 1939, was acquired. In the meantime, the anti-submarine officer based at Gaspé conducted some rudimentary surveys of asdic operators' experience in the Gulf of St Lawrence in the fall of 1943. His entirely unscientific results told the navy what it already knew: asdic conditions inside the 100-fathom line were dreadful for all but a few months in spring and fall. The staff officer (A/S) at Halifax, Cdr P.M. Bliss, RN, flatly rejected the value of such attempts to analyse the problems of sound propogation. 'It appears to me,' Bliss wrote in December 1943, 'that no practical results are at present being aimed at and in addition I am extremely doubtful if any practical results can ever be achieved.' Bliss was unenthusiastic over the prospects of proper BT surveys of Canadian waters in 1944.[3]

The NRC moved ahead in any event, sending one of their scientists, J.P. Tully, to study American use of BT. The results of his visit in early 1944 probably did little to shake the scepticism of Bliss and perhaps others. While it was true, Tully reported, that the USN used BT data 'for immediate tactical purposes,' American efforts were directed at producing deep ocean range prediction charts. Such charts, Tully warned, were not directly 'applicable to the coastal areas where bottom effects and fresh water run off exert considerable influence.'[4] To tackle that problem, a new 'Atlantic Oceanographic Research Group' was established in St Andrews, NB, in May 1944 to begin serious work on sound propagation in Canadian waters. It was not until August, however, when the group released its first report (of which more later), that Canadian scientists concluded that BT would be useful inshore.[5]

Meanwhile, operations over the winter of 1944 confirmed the extent of the problem in the Canadian northwest Atlantic. While in the eastern Atlantic escorts tracked U-boats at extreme depths, those in the western Atlantic seldom – if ever – made asdic contact on known U-boats. The natural tendency was to attribute this failure to poor training, but that was not necessarily the case. The hunts for *U539* and *U854* in February on the Grand Banks revealed that standard hull-mounted asdics were unable to penetrate deeper than 200 feet. Unlike the more complex problem of shallow water with its bottom influences, this was a straightforward problem of temperature. Arctic air masses moving across North America in winter lowered the temperature of the top layer of the sea, producing an *increase* in temperature with depth (a 'positive' temperature gradient). Since sound travels faster in warm water than in cold, the warmer, deep water then bent – or 'refracted' – the sound beam of the main asdic set towards the surface. As long as these conditions prevailed, as they did through the winter months, U-boats off Canada and Newfoundland below roughly 200 feet were – to use Chavasse's words – 'safe as houses'. Only in the spring and fall were conditions off the coast uniform (isothermal) in temperature, which produced little refraction error. The summer months, when the surface of the sea was warmed by continental air masses, provided perhaps the worst conditions of all. The sharp temperature break (thermocline) between the warm surface layer and the deep, much colder water split the asdic beam in two. One portion, trapped in the warm surface layer, could give occasionally very long ranges on U-boats near the surface. Below the thermocline sound was bent sharply towards the bottom, giving very short assured ranges below the surface layer. In fact, surface warming in the summer over a cold sea split the asdic beam like a 'Y,' and cast a large 'shadow' in the theoretical area of the asdic beam into which no sound actually penetrated (see figure 2).

In the spring of 1944 the apparent solution to these problems was to fit asdic type 147B and Q attachments to the main sets. Both directed sound downward more effectively and represented the best hope of penetrating the complex layers of inshore waters. Supplies of Q became available in May,[6] but the acquisition of 147B languished appallingly. Attempts to acquire the equipment from the British in late 1943 failed because of production delays in the United Kingdom. Recourse in 1944 to Canadian production – ordered and administered by the British – proved little more suc-

a. Warm surface layer over a negative gradient

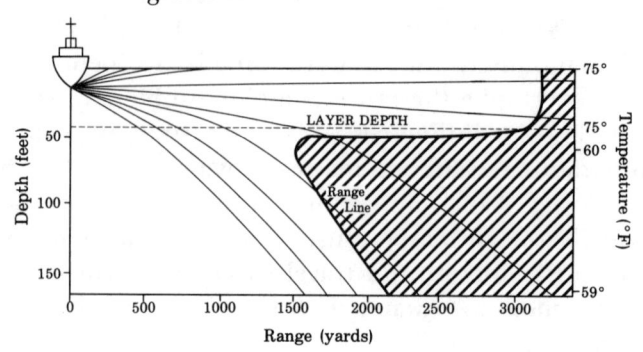

b. Negative temperature gradient

c. Positive temperature gradient

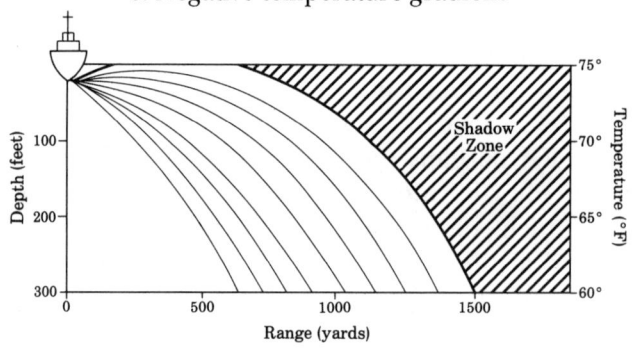

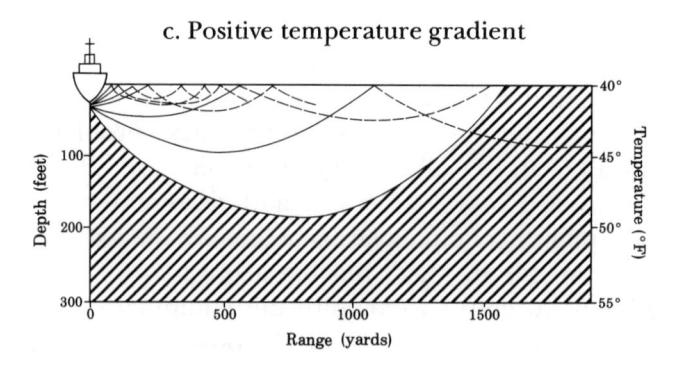

Figure 2 Variations in temperature gradient of sea water

A Sea of Troubles 181

cessful, since it did not begin until August. As a result, few RCN escorts were fitted with type 147B before the end of the war. This was precisely the kind of delay in equipment acquisition that precipitated the firing of Nelles in 1943; the new regime appears to have had no better luck.[7]

Despite the absence of type 147B sets in Canada, the RCN was actually putting the equipment into service in some of its newest ships, the Castle class corvettes. These ships began to appear in the fleet in May with the assignment of *Orangeville* to C 1. The RN was also forthcoming in its allocation of frigates to the RCN, agreeing that the last three transferred would be Lochs, which were assigned to EG 6 and EG 9 because of the 'importance of having in each support group at least one ship fitted with type 147B to enable the depth of U-Boat to be predicted.'[8]

The Canadian debate over type 147B in the spring of 1944 centred on its ability to predict the depth of U-boats in the open ocean, a problem that also forced the RCN to reverse its position on squid. It would be opportune to digress briefly to settle the fate of that weapon in the RCN's inventory. In June 1944 the navy's operational research scientists reported on the likelihood of increased U-boat sinkings from a crash program of fitting squid to River class destroyers and twenty-six frigates. The report was not optimistic. Without ready access to the equipment, time, and dockyard space, and given that the existing ships would require extensive adaptation to carry it, the RCN could not begin to fit squid until late 1944 or early 1945. Even then it could manage only two ships per month. Thus by the end of 1945 the RCN might have an additional fourteen squid-fitted ships available, which, considering time involved in the work, training, and the like, averaged about four operational vessels over the year. The anticipated result was perhaps one additional U-boat sinking by the end of 1945. The whole program, completed by the end of 1946 – assuming the war lasted that long – promised better results: perhaps eight additional U-boat sinkings. As the report observed, no appreciable returns were likely before 1946; results depended upon how long the war was likely to last.[9] In fact, few people were anticipating that the war in Europe would last into 1946, and the RCN did not go ahead with the refitting. The quickest and easiest way to get squid into service was to fit it on ships built to carry it, and the RCN had all but abandoned those late-war types in the fall of 1943.

Concentration on deep-ocean ASW in the spring of 1944 reflected both a contemporary preference in the solution to inshore ASW and a portent of postwar trends. By this time it was evident to the RCN that the best way of dealing with inshore U-boats was to sink them while they were still offshore. The ability of the RCN to take effective action against transiting U-boats was determined by the availability of trained support groups, the effectiveness of cooperation with the RCAF, and the ranges from bases at which the operations were conducted. Problems of poor RCN-RCAF coordination at the tactical and operational level endured for the balance of the war, although in the last months it showed signs of significant improvement. Good naval-air liaison was hampered by the simple fact that the northwest Atlantic was inhospitable to fliers. During the winter there were seldom more than ten days in a month when effective air support could be provided from Newfoundland bases.[10] When weather was not poor at the bases, it was often unsuitable over the Grand Banks, where the mixture of the arctic Labrador Current and the warm Gulf Stream combined to produce almost perpetual fog to a distance of about 400 miles from shore. Operations beyond the fog belt depended upon long-range warships, such as frigates, and very long-range aircraft, such as the Liberator, both of which were in short supply in early 1944.

The alternative was a hunter-killer group based on a small aircraft carrier. Such USN groups operating to the east and northeast of Flemish Cap sank five U-boats in the fall 1943 campaign and another southeast of the Grand Banks in early January 1944. *U490* was claimed by the USN southeast of Flemish Cap in early June. The USN's hunter-killer groups also patrolled routinely in the southern approaches to Canada, south and east of Sable Island. There they sank *U856* in early April and *U233*, which was on its way to mine the approaches to Halifax, in early July.[11] Most of these kills took place in the larger 'Canadian' zone north of 40 and west of 40, looked upon by the RCN as its natural frontier and within which its searches were conducted and intelligence estimates were promulgated.

A Sea of Troubles 183

The success of USN operations in the northwest Atlantic was not lost on senior RCN officers. The USN was helping to solve the inshore A/S problem by maintaining a de facto barrier patrol in deep water – a concept that matured after the war into a major component of ASW. For the Canadians in early 1944, the need to maintain a permanent frigate support group in home waters to do the same was paramount.[12] On 10 July the British were informed of the new group's existence and of NSHQ's intention to use it 'to operate against U-Boats west of longitude 35 degrees west as they approach Canadian coast or are on passage to southward.' To make this modest barrier force more effective, NSHQ asked the British to assign an escort carrier 'under operational control of CinC, CNA to work with this support group.' The timing of the request probably owed something to the recent commencement of operations by two RCN-manned auxiliary carriers, *Nabob* and *Puncher*, the fruits of the 1943 decisions on a fleet air arm for the RCN. Conceivably, the RCN was now looking for a quid pro quo.[13]

The Admiralty, aware that the USN could handle any major threat in the western Atlantic, simply refused to countenance the idea. Not only was their reply to the RCN for the loan of a carrier dismissed summarily, but they seriously questioned the need for a support group in the western Atlantic at all. As the Admiralty observed, 'in view of present U-boat disposition we do not think that a group should yet be spared solely as a hunting group in the West Atlantic.'[14] The attempt to obtain the use of a carrier by a coup de main thus failed – and the Admiralty set its sights on Murray's new group.

The RCN was not eager to cast away its new support group (*Springhill, Orkney, Stettler, Magog, Charlottetown,* and *Toronto*), now designated EG 16. NSHQ informed the Admiralty that it might consider doing so, but for the moment the group was to remain in Canada. In true Nelsonic tradition, the Admiralty put its 'telescope to its blind eye' and operated for the next week on the assumption that EG 16 would be assigned to the eastern Atlantic. When the Canadian staff in London prodded NSHQ on 2 August to get EG 16 to Scapa Flow by 16 August for the next Russian convoy, NSHQ had to remind the Admiralty again that the group was to be retained for

Murray's use. In the event, British anxiety to use the RCN's burgeoning fleet of frigates did not end there.[15]

In the euphoria over the German collapse in France and the race to the Rhine, there was a sense in the Allied camp generally through August and early September that the war with Germany might soon be over. Throughout the late summer of 1944 the attention of most naval staffs, the Canadian staff among them, focused on the Pacific. The great outpouring of new construction, especially frigates, for the first time offered a surfeit of resources. At the end of May 1944 the RCN had thirty-five frigates in commission, about half of them operational. Four months later the number in commission was nearly double – sixty-three in all, with a further seven due before the end of the year.[16] The RCN admitted – after five years of war – that its principal commitments in terms of escort ships were finally met. Just what the navy planned to do with the surplus, apart from bolstering some of its existing forces, is not entirely clear. Some frigates were designated for conversion for anti-aircraft service in the Pacific war, there were refit and training schedules to be met, and some had already been used to form EG 16.

The British, anxious to redeploy their forces to the far east, also had their eye on the burgeoning Canadian fleet in the summer of 1944 and were not shy about making suggestions for its employment. On 25 July 1944 the director of Operations Division (Home) at the Admiralty, Captain C.T.M. Pizey, RN, outlined his proposals for the new ships to the Canadian Naval Mission in London. Pizey suggested that WEF be strengthened, that frigates be sent to the far east, that the RCN Castle class corvettes be transferred to the Arctic, and that other RCN surplus escorts be assigned to overseas as needed. The frigates in particular, Pizey felt, would be useful as fast escorts in the Indian Ocean.[17] Pizey's memo was provided 'informally' to the CNMO, where it was reviewed by Lt Cdr Todd, the naval assistant (Policy and Plans). Todd noted in his own memo to Admiral P.W. Nelles, who was now senior Canadian naval officer (London), on the 28th that the decision to send RCN forces to the far east was really the Admiralty's second choice. Their first was that the RCN take over responsibility for A/S work in the north

A Sea of Troubles   185

Atlantic. This plan was impossible, however, because the RN's U.S.-built Captain class frigates could not be supported in the far east.[18]

The gist of Pizey's plans were passed to Ottawa on 6 August. Four days later NSHQ replied that after a new C group was formed for MOEF to cope with the planned resumption of SC convoys in September, 'proposals will be made with regard to allocation of new construction in excess of RCN operational requirements.'[19] Before the Canadians could articulate their plan, however, C-in-C, WA, entered the fray with his own proposal as part of a broader scheme to inflict a major defeat on U-boats on the run from their bases in France. In a signal to the Admiralty on 13 August Admiral Horton stressed the urgent need for more support groups. He wanted all fast escorts removed from MOEF to form hunting forces. Western Approaches could then find enough ships for another two or three support groups, and he hoped the RCN could provide at least one new group. In addition, Horton wanted all planned dispatch of escorts to the far east cancelled.

The Admiralty accepted Horton's plan and forwarded it to NSHQ on the 17th, noting that subject to Canadian approval it intended to adopt the scheme. The next day a copy of Pizey's memo was also sent. NSHQ replied on the 24th, pointing out that the situation of the Canadian forces was not as rosy as the British believed. The RCN could jostle enough escorts to form C 6, C 7 and C 8, the additional groups needed for MOEF, and W 8 and W 9 for WEF, to meet the forthcoming increase in convoy sailings and to meet refit schedules. To honour these commitments it would be necessary to use EG 16 for close escort, employ more frigates in MOEF, and run both MOEF and WEF below minimum strength. If these steps were taken, the RCN's Castle class corvettes could be freed for Arctic service. As for the frigates per se, the RCN planned to form new support groups with them but offered no commitment on where they might serve.[20]

The RCN had every right to be nervous about the threat of U-boat activity in the Canadian northwest Atlantic in the late summer of 1944. Reliance on schnorkelling coupled with dependence on bases in Norway severely limited the operational range of the existing U-boat fleet. As the U-boat commander-in-chief's War Diary noted on 15 September, 'the US coast, the Newfoundland area and also the St Lawrence' were the only overseas stations where type IX U-boats could operate with any hope of success. The smaller type VII U-boats were now considered so limited in range (in part because of habitability problems), that even the English Channel was thought beyond reach from Norwegian bases. The British expected the Germans to concentrate their efforts west of Ireland again, in the open ocean, in a repeat of the winter 1943-4 campaign. The Germans adapted quickly to inshore conditions, however, and preferred to patrol where the Allies had difficulty in finding them. For that reason U-boat operations in September and October 1944 concentrated in the waters off northern Britain and eastern Canada.[21]

While the debate over the future of the RCN's latest warships went on, EG 16 operated against a wave of U-boats en route for the North American coast. The group became operational at the end of July under Lt Cdr W.C. Halliday, RCNR, in *Springhill*. Halliday had been identified before the war as a keen and capable officer and was given a command almost immediately when war broke out (the armed yacht *Sans Peur*). After a short spell as the first lieutenant of the destroyer *St Clair*, Halliday took command of the corvette *Kamsack* in May 1943. Prentice's assessments of him were glowing, and he was recommended for the command of both a frigate and a group; he got both in 1944.

It was some time before Halliday's group was fully effective, and its radar, which in all but *Springhill* and *Orkney* was the faulty Canadian RX/C, severely hampered the group's efficiency. The burden of barrier operations in the late summer of 1944 therefore fell to the USN. In mid-August two U-boats, *U802* and *U541*, were en route to the Gulf of St Lawrence, while *U1229* was on the way to land spies in Maine. All were boldly making time on the surface as they approached the Canadian coast. On 19 August aircraft from the USS *Bogue* obtained radar contact east of Flemish Cap with *U802* and pursued it for several days. On one occasion the submarine nearly torpedoed the carrier, while the latter's aircraft attacked and nearly sank the U-boat. But *U802* escaped. Now extremely cautious and running entirely on its schnorkel, it could not be located by

RCAF air patrols. *U-802* arrived in the Cabot Strait about 30 August. Meanwhile, the USN group, redeploying in an attempt to re-establish contact as the U-boat emerged from a cold front, actually located *U1229* south of the Grand Banks on the 20th. EG 16 was searching just a few miles to the north, hampered by poor weather and frequent radar breakdowns. But if EG 16 was ever to sink a U-boat, now was surely the time. The captain of *U1229*, Korvettenkapitän Armin Zinke, was rated by one American historian as 'high among the more inept German U-boat commanders'. His submarine was equipped with schnorkel, but Zinke consistently refused to use it and had been on the surface for about a week when aircraft from *Bogue* found and sank *U1229* on 20 August.[22]

EG 16 had another chance against *U541*, which, following behind *U802* towards the gulf, made regular surface runs at night in poor flying weather. But no one gained contact with *U541*, and she arrived in the Cabot Strait about the same time as *U802* – a full day ahead of intelligence estimates. The presence of U-boats in the strait was anticipated, and local convoys began before their arrival.[23] None the less, *U541* found the independently routed small British steamer *Livingston*, which she sank off Cape Breton on 3 September; the 1944 inshore campaign in Canadians waters, the first since 1942, had begun.[24]

Thus, there were at least two U-boats loose in Canadian waters when the scramble for escorts began in early September. By then the British had shifted their interest from attempts to crush U-boats fleeing French ports or support for the Russian convoys to fear of a major inshore campaign in U.K. waters by schnorkel-fitted U-boats. On 2 September the Admiralty informed NSHQ and senior operational staffs of the 'urgent' need for additional support groups. The strength of MOEF groups was to be reduced to five ships, all fast ships were to be withdrawn from B groups, and escorts for the Sierra Leone and Mediterranean convoys were to be reduced to three corvettes (since the French coast was now clear, these convoys were well away from any major threat). A separate signal was sent to NSHQ later the same day asking the RCN to conform to the plan, an agreement that, the Admiralty hoped, would produce three more RCN support groups. Two support groups could then be left in Canadian waters (counting EG 16), while the other two were to be assigned to British waters. NSHQ concurred. The first two of the new groups were to go to Horton, while the third remained in Canadian waters alongside EG 16.[25] Such was the genesis of EGs 25, 26, and, somewhat later, 27.

The British plea for Canadian assistance in U.K. waters was strengthened by the faltering operational reliability of the RCN's two destroyer support groups, EGs 11 and 12. Following EG 12's second action off Ushant, in which *Qu'Appelle* and *Skeena* collided, NSHQ proposed to the Admiralty that the two groups be amalgamated. The Admiralty accepted on 1 September; EG 12 lapsed and its remaining ships were passed to EG 11. Prentice transferred to *Qu'Appelle*, but neither he nor his ships were in a fit state to continue much longer. Prentice was exhausted by August and had begun to loose his grip. Several incidents in the handling of EG 11's new flagship in September drew attention to his plight. When he asked to be relieved of his command, Commodore (D), Western Approaches, recommended at least two months' rest before reassignment.[26] The group itself officially was withdrawn from service in early November after *Skeena* was driven ashore in Iceland during a gale and lost. Only *Assiniboine* stayed in service through the winter, seconded to EG 14.[27]

The collapse of Prentice's health was the result of long years of wartime service without a break. He had arrived in Britain in May 1944 exhausted from sixteen months as Captain (D), Halifax, and ought to have taken extended leave then. Instead, he took EG 11 on one of the most intense periods of wartime service experienced by the RCN, conditions shared by the other support groups who carried the burden of the A/S war. Exhaustion 'was a serious problem,' Captain Denis Jermain, RN, recalled, especially for the command team of a ship. The captain, navigator, A/S officer, and its best asdic operators 'needed to be on constant duty, checking the myriad bottom contacts any one of which could be a U-boat.' After a prolonged hunt Jermain's group typically sailed seaward for eight hours of sleep, and a 'stretch off the card.' 'The person who bore the greatest load was the senior officer of the group,' Jermain observed, since he had to command both his ship and the group itself. To overcome this overload, the RN eventually adopted the American practice of assigning command of the group separate from command of a ship.[28]

The two-watch system seems to have been practised throughout most Canadian support groups for all the crew, not just the specialists. Typically, warships' crews stood watch four of every twelve hours (a one-in-three system), with stand-to action stations at dawn and dusk, in the event of a battle, or if the threat of action was imminent. These conditions certainly obtained in the early phases of the inshore campaign of 1944, but four-on/four-off was a debilitating routine. RN veterans recall that they tried to maintain a one-in-three watch system.[29] As much as possible the duty watch 'fought' the ship. The same was generally true in the RCN groups, but the Canadians seemed to have maintained the two-watch system right to the end of the war. A.B. Sanderson, a young lieutenant aboard *Monnow* of EG 9, recalled that the two-watch system devastated the health of the ship's crew and contributed to the death of three of *Monnow*'s wardroom and her captain within two years of the end of the war.[30] Even the groups formed in late 1944 were affected. EG 26 was on a two-watch system – at least for its key personnel – in early 1945 when, as group telegraphist Ernie Doctor remembers, her much loved senior officer was forcibly removed from duty and sent to hospital suffering from nervous exhaustion.[31] Whether the RCN's choice of shipboard routine was a factor in its declining performance in British waters during the last months of the war remains open to speculation.

Although there was a definite slackening of the intensity of the U-boat war through September and October, few of those at sea had time to notice. Given the overall shortage of support groups and the shifting of German efforts northward, the British used close escort groups occasionally throughout August for searches in the approaches to the North Channel (Operation CW). It was while on one of these patrols that *Dunver* and *Hespeler* of the MOEF group C 5 apparently gained contact with *U484* off Malin Head on the night of 9 September and sank it after a prolonged series of attacks. At the time the Admiralty, well aware from intelligence that a submarine had gone missing in the North Channel, assessed the attack as a U-boat 'probably sunk' – and possibly by an RCAF aircraft from 423 Squadron. It was only after the war that the destruction of *U484* was finally awarded.[32] On the same day (9 September) two ships from another close escort group, *Portchester Castle* and *Helmsdale*, were credited with sinking *U743* further westward.

Controversy still clouds the loss of both *U484* and *U743*. Recent reassessment of these losses by a German historian, Axel Niestle, suggests that *U484* was the U-boat sunk by *Helmsdale* and *Porchester Castle*. The British ships' attacks produced wreckage and human remains, and according to Niestle, *U743* was not ordered into the area until 11 September, two days later. No firm evidence was obtained in the attacks by *Hespeler* and *Denver*, so Niestle concludes that *U484* was destroyed by the two British ships. What the Canadians acted on is unclear. *U743* is believed to have been lost to unknown causes some time between 21 September and 10 October. If Niestle is correct, then only one U-boat was sunk by British and Canadian ships in the main north Atlantic during September 1944 (other than *Saint John*'s kill on the night of 31 August – 1 September).[33] The low kill rate was no measure of the scale of the German effort. In fact, some twenty U-boats operated in northern British waters throughout September, several for very long periods. It is true that few ships were sunk, but the U-boats survived to fight another day and to pass along their experience – just what Horton wanted to avoid.

Support groups from the Channel and Biscay operations began redeploying northwards at the end of August. By early September only four (including all three RCN groups) of the thirteen support groups operating in British waters, EGs 1, 6, 9, and 11, remained under Plymouth command; the rest were either in the northwest approaches or off north Scotland.[34] Here the areas to be covered were much greater than in the channel (hence the appeal for *more* groups in August), and it took time to become familiar with all the non-subs and other local conditions. One saving grace for the channel veterans was that the northern areas contained far fewer wrecks. The English Channel averaged eight non-sub features every ten square miles, or one every mile and a half. From Land's End to Mull (around the west of Ireland) there were nearly as many wrecks as in the channel, some 650, but the vast area reduced their concentration to about 2.5 per ten-square-mile area. West of Scotland the incidence of non-sub features dwindled to 0.3 for every ten square miles before rising sharply again along the eastern coast of Britain.[35]

The lower incidence of non-sub features in northern waters was offset by the absence of the radio navigation aids and shore-based radars. Wrecks and non-sub contacts were reported and buoyed, but not with the same accuracy as they were further south. Problems of accurate fixing of non-subs prompted the Admiralty to appeal to the Air Ministry in November for the hasty establishment of a series of GEE stations throughout northern Britain. These posts were considered essential so 'that escort vessels should not waste their time hunting and attacking charted wrecks in a mistake for bottomed U-Boats.' The 'considerable success achieved in the Channel during OVERLORD by ships fitted with GEE in differentiating between sunken wrecks and bottomed U-boats' necessitated the extension of the system in both the north and over the southwestern approaches as well. The Air Ministry responded, and by the end of 1944 the approaches to the North Channel and the southwestern approaches to the English Channel were covered. Through 1945 GEE stations were also sited to cover waters north of Scotland.[36]

To reduce the problem posed by U-boats on the bottom, the Admiralty also began a large-scale inshore deep-mine-laying scheme, which continued to the end of the war. The likely areas for profitable U-boat operations were laid with deep fields of moored contact mines, which posed no serious danger to surface vessels. Indeed, ships of EG 25 often put themselves in the middle of such minefields when a rest from operations and some relief from GNATs was needed.[37] The first deep mine fields were laid in St George's Channel, the southern entrance to the Irish Sea, in early October.

Mining St George's Channel was part of the plan to change the main entrance to Britain for ocean convoys from the North Channel – now clearly within reach of Norwegian-based U-boats – to the southern approaches. The first ocean convoy to use St George's Channel since June 1940, HX 306, passed through in mid-September. The area was already well covered by support groups, since several, including EGs 9 and 11, remained in the area throughout September in response to the last deployment of U-boats from Biscay. These submarines, and those assigned to north U.K. waters, were primarily intended to draw the Allies away from the U-boats in transit to Norwegian bases.[38]

Meanwhile, much of the A/S action – or at least Allied activity – was concentrated north of Britain, where an attempt was being made to establish a surface and air barrier to stop the movement of U-boats in and out of their bases. This effort was almost entirely futile. Bomber Command sank more U-boats (four) in a single raid on Bergen on 4 October than A/S forces achieved in two months of patrols. The best chances came in September, when the traffic was highest, and two U-boats in transit were sunk by Coastal Command aircraft. But October was a dismal month for the A/S specialists and the northern barrier forces. Most of the U-boat fleet was regrouping for a major offensive and there were few U-boats at sea. Of the six sunk at sea in the north Atlantic during the month, the Germans themselves accounted for four: two were lost in collisions with other German ships, one was lost by a German mine off Norway, and one was believed lost in a schnorkelling accident. Of the two sunk by Allied forces one went to the Fleet Air Arm and the other – the only kill of October by a support group in the north Atlantic – was by EG 6.

EG 6 had moved north, along with EG 18, at the end of September to participate in 'Operation SJ,' an attempt by C-in-C, Rosyth, to achieve, 'The destruction of U-boats on passage to and from Norwegian ports.' The two support groups cooperated with 18 Group, RAF Coastal Command, and were provided with a covering force of two auxiliary carriers screened by Home Fleet destroyers. The support was needed because the plan called for pushing both air and sea A/S patrols well inshore off the coast at Bergen.[39]

Operation SJ was singularly unsuccessful and provided little deterrence to the passage of U-boats. By mid-October the patrols settled in to maintain a barrier between the Shetland and Faroe islands. The support groups' only firm contact occurred on 16 October, when EG 6 was engaged in 'gamma searches' designed to locate a U-boat in an area where continuous air patrols made it impossible to surface. Asdic conditions were good, and many whales were classified – both visually and on the group's asdic sets. None the less, at 1752Z *Annan* had a contact worth attacking. When *Annan* failed to regain contact, it was considered that she had attacked a school of fish or a number of whales and had 'dispersed' them with her charges. In fact, she had badly shaken *U1006*.

The U-boat, a type VIIC/41, commanded by Oberleutnant zur See Horst Voigt, left Bergen on her maiden voyage on 9 October headed for British waters. On 16 October 1944 *U1006* was on course and on time, travelling at slow speed on batteries at sixty metres' depth, when screw noises were heard, followed by depth-charge explosions all around. Several other depth-charge patterns followed. By the time the last one exploded, the U-boat's hydroplanes were wrecked, the forward bulkhead was buckled, and water was flooding in so fast that the pumps could not handle it.

When *U1006* broke the surface, *Annan* and EG 6 were already far away, out of sight in the darkness. Voigt might have escaped but for the excellent type 272 radar perched atop the high lattice mast of *Loch Achanalt*, another newcomer to EG 6. Her radar operator detected a small blip on his screen that sent *Annan*, still the closest escort, back to investigate. *Annan* soon picked up the contact on her 271Q set. When it was clear that the target was a surfaced submarine CAT gear was streamed, a fortunate move, since Voigt turned to fire a GNAT when he saw the frigate. After plotting the U-boat's movements for a few minutes by radar, *Annan* illuminated the scene with rockets, and a gunnery duel broke out. *U1006* returned a brisk fire, putting the frigate's asdic and radar out of action and wounding eight men. As the range closed, however, *Annan*'s superior firepower quickly told. *U1006*'s gunners were swept from their positions, and *Annan* laid two depth charges set to shallow right on top of the wallowing U-boat (one actually hit the U-boat's forecastle and rolled off before exploding). *U1006* began to sink and its crew abandoned ship; forty-six of a crew of fifty-two were recovered.[40] There was nothing fancy about *Annan*'s victory: it was a combination of good luck, basic skills, and the powerful radar of a Loch class ship. A year after the great crisis of 1943 over the failure of the RCN to sink any submarines, the navy could at least claim that it had accounted for all the U-boats destroyed by warships in the waters it operated in during the month of October 1944.

The larger picture, however, was not good by mid-October. Antisubmarine operations fell well short of Horton's earnest expectations. The northern barrier was a failure, and so too were other offensive operations. A report by the director of Operational Research on 1 December condemned A/S efforts in the fall of 1944 as ineffectual. Between 25 August and 17 October no less than fifteen support groups hunted for U-boats north of Ireland, totalling 750 ship-days at sea, supported by over 2,000 hours of flying by Coastal Command. Apart from the two U-boats sunk on 9 September, hunting in the approaches to the North Channel produced nothing. True, the U-boats were suppressed; but this inaction was put down to their generally passive nature.[41]

There was no gainsaying that something had to be done to increase the chances of killing submarines, but Horton was also anxious to keep the Germans from sinking Allied. These were not mutually exclusive goals. He instructed support groups in September that 'During the present inshore threat to our shipping in the Western Approaches,' their duty was to concentrate on 'the safe passage of the convoy through the local area.'[42] That signal was reinforced a month later when a new set of patrol instructions for support groups in the British Isles was issued. Called 'Operation CE,' these orders designated specific patrol areas and assigned 'task force' numbers to the forces operating in them (see map, p. 196). The duties of the task forces were simple: 'The safe passage of shipping through the focal areas. When no shipping is present, the object is the destruction of U-boats operating in these areas.'[43] Operation CE was expanded in early 1945 to include the Irish Sea, and the orders remained the principle guidance for A/S hunting operations in north and western U.K. waters to the end of the war. It is noteworthy as well that the primary duty for Task Forces 32, 33, and 34 remained 'the safe passage of shipping entrusted to their charge' through to May 1945.[44] It was a concept of operations that the director of Warfare and Training in Ottawa, now Captain D.K. Laidlaw, wanted to emulate in the northwest Atlantic.[45]

On the face of it, Operation CE was a retreat from Horton's expressed plan to crush the U-boat fleet.[46] However, Operation CE was well in keeping with one consistent WAC tenet: the best place to find U-boats was around the targets they sought. Whether this return to a previous norm was the direct result of frustrations from months of fruitless searches in wreck-strewn waters remains obscure. One major postwar British study attributed this shift in emphasis back to close support of convoy movements to a noticeable increase in the number of U-boat kills by early 1945.[47] There were other mitigating factors, however, not least of which were increased U-boat activity and the improving skills of the Allies.

Concentration of hunting around convoys was also a much better way to use very limited resources. With some 300 destroyers and escort vessels slated for departure to the far east,[48] the RN was hard pressed to respond. Canada somewhat eased the RN's problems with the establishment of new frigate support groups, two of which were formed during September. EG 25 became operational on the 25th at Halifax and spent the next month off the Nova Scotia coast. It included two new ships, *Ste Thérèse* and *Thetford Mines*. *Joliette* was transferred from C 1, and *La Hulloise* and *Orkney* were assigned from EG 16. With *Orkney* came Acting Commander Victor Browne, RCNVR, as senior officer. Like the other VRs who led support groups, Browne was highly recommended. He had specialized in A/S early in the war, serving as A/S officer of *Saguenay* in 1941 before receiving command of the corvette *Mayflower* in May 1942. Browne was considered an 'outstanding corvette commanding officer' and was recommended for a frigate command, which he gained in June 1944 when he took over *Orkney*. The new senior officer of EG 25 was very young, but also was experienced and qualified. His efforts to mould EG 25 into an efficient unit did not go to waste.

The decision to keep EG 25 close to home for a while probably owed something to the desire of Captain (D), Halifax, now Captain W.L. Puxley, RN, to get new groups to sea for a few operations before they received their final alongside servicing and post-com-

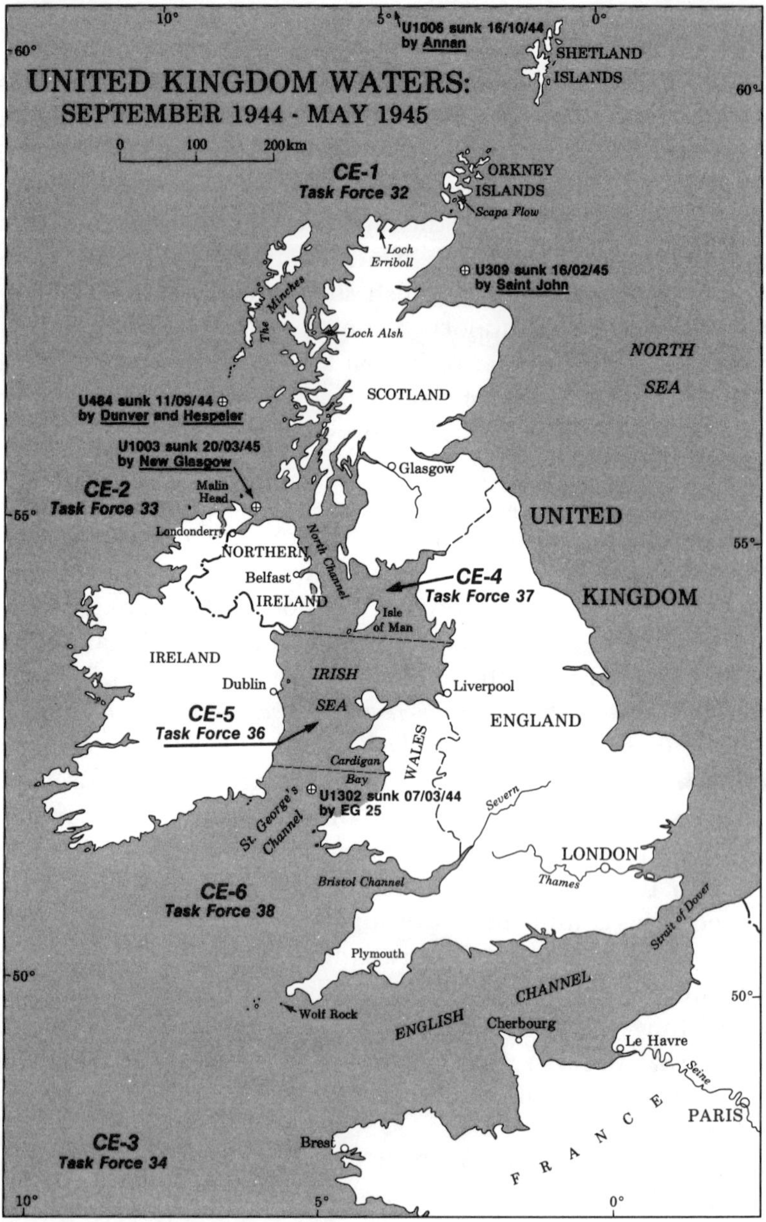
UNITED KINGDOM WATERS:
SEPTEMBER 1944 - MAY 1945
0 100 200km
U1006 sunk 16/10/44
by Annan
SHETLAND
ISLANDS
ORKNEY
ISLANDS
Scapa Flow
CE-1
Task Force 32
Loch
Erriboll
U309 sunk 16/02/45
by Saint John
The Minches
Loch Alsh
NORTH
SEA
SCOTLAND
U484 sunk 11/09/44
by Dunver and Hespeler
U1003 sunk 20/03/45
by New Glasgow
Glasgow
CE-2
Task Force 33
Malin
Head
Londonderry
NORTHERN
IRELAND
Belfast
North Channel
UNITED
KINGDOM
CE-4
Task Force 37
Isle
of Man
IRELAND
Dublin
IRISH
SEA
Liverpool
ENGLAND
CE-5
Task Force 36
WALES
Cardigan
Bay
U1302 sunk 07/03/44
by EG 25
St. George's
Channel
Severn
LONDON
Thames
Bristol Channel
CE-6
Task Force 38
Strait of Dover
Plymouth
CHANNEL
ENGLISH
Wolf Rock
Cherbourg
Le Havre
Seine
PARIS
FRANCE
Brest
CE-3
Task Force 34
10°
5°
0°
60°
55°
50°

missioning additions and alterations.[49] EG 25 was also needed to respond to the U-boat campaign that developed in September in the Gulf of St Lawrence, and it was not sent overseas until Murray's second designated support group, EG 27, was formed. EG 25 finally sailed for the United Kingdom on 30 October.

The other group formed in September was EG 26. Of its numbers only *Beacon Hill* was new construction. The balance was drawn directly from MOEF: *New Glasgow* from C 1, *Jonquière* from C 2, and *Montreal* and *Ribble* from C 4. The group came into service in Britain on 1 October. Its arrival raised the number of support groups controlled by Western Approaches Command to eighteen.[50] EG 26's senior officer was another reservist, A/Cdr E.T. Simmons, RCNVR, who was also CO of *Beacon Hill*. Simmons began his north Atlantic career under Prentice, and his first experience with U-boats was leading *Chambly*'s boarding party onto *U501* in September 1941. He got a second crack at a submarine in January 1943 when his own corvette *Port Arthur* sank the Italian submarine *Tritone* off Algeria. Simmons's career was championed by Prentice, who, as Captain (D), Halifax, considered him 'group leader material.' But Simmons did not need to rely on the enthusiasm of his old mentor to gain command of a group; all his senior officers recommended him highly. After some intense training EG 26 sailed for operations on 24 October: the fastest of the frigate groups to enter service.

The third and final group of the September plan was EG 27, officially formed on 20 October. It comprised *Lévis* and *LaSalle*, both new construction, *Meon* from EG 9, and *Ettrick* from C 3. In November *Coaticook* joined from the shipyards, and group composition remained stable until April 1945. The group's steady composition was matched by the permanence of its senior officer, A/Cdr St Clair Balfour, RCNVR, aboard *Meon*, who remained with the group until the end of the war. Like the others, Balfour was an experienced and highly recommended officer. After a stint at sea in the destroyer *St Laurent* in 1941 he had spent a year on the west coast before taking command of the corvette *Lethbridge*, from which he graduated to *Meon* in February 1944. His transfer back to Canada in the fall of 1944 brought some much needed expertise to Murray's support group operations.

By the end of October 1944 the RCN fielded seven support groups, EGs 6, 9, 11, 16, 25, 26, and 27, with one – EG 11 – soon to be withdrawn from service. It was a credible team, although many members were deficient in both experience and equipment. Many of the frigates assigned to EGs 25, 26, and 27 were still fitted with RX/C radar, which limited their utility until it could be replaced by 271Q.[51] The reliance on RCNVR acting commanders as senior officers was also a mixed blessing. There was no question that the VRs had the experience and the leadership skills to do the job effectively. Their appointments marked a significant departure from the previous reliance on RNR officers – including Prentice. By late 1944 the weight of command in the Canadian navy's offensive against the U-boats shifted to the young reservists in the fleet.

Most RCNVR captains of frigates were qualified only to lieutenant commander rank, the normal grade for command of a such a ship. In September 1944 the Naval Staff reluctantly agreed to appoint all support group senior officers to the rank of acting commander.[52] This change made them equal to their RN counterparts, in form at least. But British support groups – virtually without exception – were commanded by professional RN officers. Professional officers took precedence over reservist officers of the same rank regardless of seniority, just as RNR and RCNR officers took precedence over VRs. Thus, when RN and RCN support groups were operating together, the British officer was invariably 'in command.' Except for very brief periods of acting command during the short existence of EGs 11 and 12, no 'RCN' officer commanded a Canadian A/S support group. The RCN's A/S war, therefore, was left in the capable – but somewhat disadvantaged – hands of the 'hostilities only' officers. RCNVR officers understood the system of seniority very well and generally they accepted it. They also knew that their professional RN counterparts were jockeying for postwar careers. Unfortunately, the system of precedence also meant that sometimes Canadian support groups were elbowed out of the way.[53]

The assignment of two new frigate groups to British waters displayed considerable courage and faith on the part of the RCN, since September and October were anything but quiet in Canadian

A Sea of Troubles 199

waters. The torpedoing of the *Livingston* on 3 September in the Cabot Strait signalled the start of a campaign in Canadian waters that would last to the end of the war, and Murray never had enough forces to deal with the intruders. His only support group at the time, EG 16, arrived the morning after *Livingston* was lost – far too late to accomplish much.[54] Tardy arrival on the scene of action was a recurring problem in the Canadian northwest Atlantic in the last year of the war.

Searches in the wake of the loss of *Livingston* were also hampered by a muddled intelligence picture which remained obscure for the next few days. Like the British, the Canadians were now reduced to locating U-boats largely on the basis of their attacks. Authorities refused to countenance attacks on *U802* between Cape Gaspé and Anticosti Island by motor launches on 6 September, because no U-boat had been seen nor had torpedoes beed fired. The presence of at least one U-boat off Anticosti Island was confirmed only on the night of 7–8 September when the corvette *Norsyd* attacked *U541* on the surface. The U-boat was saved by a jam in the corvette's main gun, which limited *Norsyd* to raking *U541*'s conning tower with 20-mm gunfire just before she dived to safety. *Norsyd* never established asdic contact, but there was now little doubt that a U-boat lay in the mouth of the river.[55]

In fact, there were two: *U541* and *U802*. Murray immediately ordered a Salmon operation. Poor weather curtailed flying, but the navy responded with commendable speed. EG 16, still searching in the Cabot Strait, was ordered to the area within minutes of *Norsyd*'s report and arrived early the next morning (about ten hours later). Murray also dispatched an ad hoc group, W 13, composed of the frigates *Ste Thérèse*, with her captain Lt Cdr J.E. Mitchell, RCNVR, as senior officer, *Thetford Mines* (both designated members of EG 25), and *Stettler*, while *Magog* joined the next day. Along with the motor launches already on the scene, this was a large force, but numbers alone confer no particular advantage in ASW, and the chances of catching either U-boat were slim.

The Germans entered the gulf in 1944 anxious about what the local conditions held in store. They were pleasantly surprised by the complexity and usefulness of the water mass. *U802* found that in the mouth of the river itself it was possible to sit comfortably with motors off on a water layer at periscope depth and, totally silent, watch all that passed. As Michael Hadley recounted in *U-Boats against Canada,* Petersen of *U541* found the entrance to the St Lawrence equally forgiving.

We didn't know the water conditions in the St Lawrence, but very quickly ascertained that conditions obtained here which greatly favoured our sojourn ... Water density increased markedly from the surface to greater depths. On the top was the fresh water of the river, and below the salty Atlantic water had probably been pushed underneath it by the tides ... And in order to lie dead with stopped engines at any particular depth of water, we didn't need to employ the so-called hover gear which would keep us at a pre-set depth by automatically flooding and pumping a few litres of water. The gear was used in operational areas in order to conserve energy and make as little noise as possible when we were, so to speak, at 'lurking stations.' Here however water layerings bore us: they also had the great advantage that A/S vessels could detect us either not at all, or only with great difficulty, for the asdic sound waves deflected at varying depths. Thus we felt as secure as in the bosom of Abraham.[56]

Canadian escorts were only dimly aware of what they were up against, and they lacked both the weapons and the training – not to mention the experience – needed to find and destroy *U802* and *U541*. Although all the frigates and *Norsyd* were equipped with type 144 asdic, apparently none was yet fitted with Q attachments or type 147B, so the chances of penetrating the dense layers were slight. Moreover, apart from *Norsyd, Springhill,* and probably the Norwegian escort *King Haakon,* none of the ships had type 271 radar. Since both U-boats schnorkelled to recharge batteries, however, the hunting forces were not likely to encounter a target big enough for even their 10-cm sets to detect. The Salmon was called off on 10 September without any contact.

The whole affair might have ended without further incident had W 13 not stumbled on *U802* four days later off the western tip of Anticosti Island. *Ste Thérèse* made the initial contact, which was then picked up by *Stettler*. The captain of *U802* watched the attack

A Sea of Troubles 201

develop through his periscope as he attempted to pass the patrol line to find the convoy he believed lay behind it. As *Stettler* raced in to attack, *U802* plunged to safety in the pronounced layer at 170 metres, firing a GNAT that exploded in the frigate's wake. W 13 then conducted a series of attacks well away from *U802*, which crossed the river and, finding the Gaspé Current (of which more shortly), slipped silently eastward carried by the flow.[57] EG 16 worked its way along the north coast of the Gulf of St Lawrence and ended the month searching just east of the Strait of Belle Isle, probably on the assumption that a northern exit from the gulf would put the U-boats closer to Norway. Thus no forces interfered with *U802* and *U541* on their homeward passage of the Cabot Strait.[58]

Both U-boats were gone from the gulf by 25 September, but the intelligence picture remained murky. Five U-boats were estimated to be in the Canadian area (north of 40 and west of 40): three well offshore on the move, one off Halifax, and one in the gulf. Actually there was none in the gulf at the first of the month, although *U1223* was accurately plotted en route. Its arrival off Cape Breton matched intelligence plots, but poor flying conditions again precluded effective searches, and the U-boat reached the mouth of the St Lawrence River unannounced. 'The paucity of [U-boat] transmissions originating from the area during the month,' Murray's War Diary for October opined, 'made estimating difficult and at times the situation was not clear.' Special intelligence, 'Ultra,' was of little help other than in determining where the U-boat was supposed to go. But until she arrived and demonstrated her presence, there was little to do except sweep along predicted lines of advance and hope for the best. 'The U-Boat in the Gulf,' the War Dairy noted laconically, 'made no signals, but the torpedoing of H.M.C.S. *Magog* on the 14th confirmed its whereabouts.'

Indeed it did. The U-boat was deep in the river itself, off Pointe des Monts, when it encountered the gulf section of convoy ONS 33, supported by *Stettler* and *Magog* of EG 16. Ironically, it was the second encounter between ONS 33 and a U-boat. Ten days earlier north of the Azores *U1227* attacked and blew the stern off the escorting frigate HMCS *Chebogue*. She was towed to port, but was declared a total loss. *U1223* blew the stern off *Magog* at the end of the convoy's journey. As the captain of *U1223* watched through his periscope on 14 October, *Magog* turned slowly across in front of his sights. One of a spread of two GNATs detonated under her stern, destroying sixty feet of the ship. Remarkably, *Magog* remained afloat and only three men were killed outright. As salvage operations began, escorts nearby streamed CAT and attacked non-submarine contacts. *U1223* slipped quietly away, and again no asdic contact was ever made on her. *Magog* was towed into Quebec City where she, too, was declared a total loss.

The search for *U1223* lasted until the 18th, and naval intelligence estimated that she remained in the gulf – without further success – until the end of the month. Then the area was declared clear of submarines. The decision, as it transpired, was premature. On 2 November, not far from where *Magog* had been hit, *U1223* torpedoed the steamer *Fort Thompson*. The attack prompted renewed searches, largely by aircraft, until 18 November, by which time the U-boat was long gone. The impunity with which the schnorkelling U-boats could attack was reinforced towards the end of the month when the corvette *Shawinigan*, patrolling alone in the Cabot Strait, disappeared when a GNAT from *U1228* exploded under her. EG 16 picked up six bodies and other ships recovered some wreckage; all that was ever found. The events of the fall of 1944 in the gulf were a source of great frustration for the RCN. Patrols, particularly those by radar equipped aircraft, suppressed the intruders, as they had done in 1942. Also as in 1942, there were no compensating U-boat kills.

In 1942 it had been possible to catch U-boats on the surface. The advent of schnorkel all but eliminated that possibility. Airborne 10cm radar could detect a schnorkel, but only under good conditions, at one-tenth of the range possible on fully surfaced U-boats, and if the radar operator had some inkling of what to look for in the clutter of his screen. The situation was better, but only just, for 10-cm sets fitted to warships. Trials conducted on a mock-up schnorkel fitted to the Italian submarine *Galatea* in the Mediterranean under 'good sea conditions' revealed that the schnorkel could be detected, but only at comparatively short ranges. The type

271P set, the one most commonly fitted by the RCN by the fall of 1944, could detect a schnorkel at 3,000 yards, if the operator knew the submarine was there, and could pick up the contact in any event at about 2,000 yards. The type 271Q fared better, picking up the schnorkel at 5–7,000 yards. However, a practical maximum range under moderate conditions for the 271Q set was closer to 4,000 yards.[59] The Canadian RX/C, when it worked, fell somewhere in between the two sets in terms of performance, but it was seriously handicapped because its information was displayed on 'A' scan screens.[60] The likelihood of an RX/C-equipped ship's finding a schnorkel was considered remote. Ships equipped with type 272 and type 277, as some of the Lochs and Castle class ships were, stood a much better chance.

By early October it was clear that the best solution to the schnorkel was 3-cm radar, and trials were ordered against dummy schnorkels fitted to two British submarines.[61] Some of 3-cm sets were already at sea. Canadian scientists working for the British had developed a small and highly efficient 3-cm set, known as the type 268, for use on MTBs. It became one of the best British radars of the war. The RCN knew of the type 268, but its small antenna and resulting short range limited its value for escort work. The British were aware of the problem and had developed a larger, more powerful antenna. The new radar, known as the type 972, was, according to David Zimmerman, 'intended for use on frigates to enable them to detect the new schnorkel device.'[62] However, the number of type 972 sets in service before the war ended was small. The RN's American-built Captain and Colony (a version of the River) class frigates were fitted with the USN's 3-cm SL radar. By all accounts, this was an excellent set and one fully capable of finding a schnorkel. Peter Elliott claimed – with some hyperbole – that the SL was so good that 'even planks with nails in them, floating in the water, were easily detected at a mile or two.'[63]

The Canadians, despite their indifference to the home-grown type 972, were not unaware of the need for better radar. With the collapse of the RX/C project in June, the Naval Staff agreed on 21 August 1944 to fit all frigates with the new American SU radar, an improved SL set designed specifically for ASW. Sixty sets were ordered, but the program was slow to bear fruit and had little impact before the war ended.[64] The RCN's commitment to SU was confirmed in January 1945, when Canadian Naval Mission Overseas raised the issue of a new RN 'schnorkel-hunting' type 972. NSHQ advised CNMO of its intention to complete the fitting of SU and to let the 972 be for the moment.[65]

The Canadian solution to the schnorkel problem in late 1944 was to concentrate on improving the effectiveness of air patrols. This approach included fitting new 3-cm radar to aircraft and providing additional binoculars and visual search procedure training. Successful air location of a schnorkelling U-boat would facilitate a more intensive underwater search by naval vessels. By November the training submarine *Unseen*, based at Digby, Nova Scotia, was equipped with a dummy schnorkel and was reserved almost exclusively for RCAF exercises. The emphasis in naval training was still on how to tackle the contact once it was made.[66]

Whether the RCN could have fitted the Canadian-built type 972 on its ships in late 1944 is an interesting question. The type 972 was a wholly British project. Neither the RCN nor the government was in any position to intercept the sets as they left the factory. Perhaps more importantly, by 1944 the RCN was weary of trying to pry equipment out of the British. Many officers at NSHQ, including Jones (the CNS) and the director of the Signals Division (which included radar), Captain Sam Worth, were by now ardently anti-British.[67] The decision to opt for American radar was taken, in part, to secure better access to the latest equipment, and it represented a clear trend within the RCN at the end of the war towards American electronics. The move was not entirely successful, since few RCN escorts were fitted with SU radar before the war ended. The best hope for initial detection of schnorkelling submarines was 3-cm radar. In the last winter of the war the Americans and British had it; the Canadians did not.

Even without the latest radar, the RCN came very close to sinking both *U802* and *U541* in Canadian waters in the fall of 1944. However, the submarines slipped easily from the RCN's grasp once they submerged. Captain (D), Halifax, tended to blame poor plotting for the failure of A/S searches in mid-November. 'Experience has recently shown,' Puxley wrote, 'that ships have spent time, energy and explosives on contacts which were near but could not possibly have been near enough to the original evidence.'[68] Good plotting, the key to success in British waters, was in large part a function of high standards of training and leadership: qualities that could not be expected from the small, harried, and often ad hoc forces at Murray's disposal.

It would take more than good plotting, however, to solve the riddle of inshore ASW. Improvements in radio navigation aids were forthcoming on the Canadian east coast in late 1944, but they never reached the same quality or quantity as those present in British waters. LORAN was in operation, but unlike DECCA or QH, it was not accurate enough for tactical purposes. A well-developed radar system was also in place for air warning and to assist flying. No provision was made to use it to coordinate naval efforts. The RCAF had plans to establish eight microwave early warning radar stations throughout the gulf to detect surfaced submarines. These posts held promise for control of warships, but the scheme was behind schedule. Only one station, at Cape Ray, Newfoundland, on the Cabot Strait operated in 1944. Two more opened in 1945, including one on St Paul's Island in the middle of the Cabot Strait; none could detect schnorkels. As the RCAF official history notes, 'The remaining sets were never installed because the navy did not develop the facilities to plot the thousands of contacts made.'[69] The RCN evidently saw no value in using the sets to control its own forces.

Closing the Cabot Strait exercised the minds of many senior RCN and RCAF planners throughout 1944, and some innovative suggestions were advanced. Apart from mining, they included laying seabed asdics, an extension of harbour defence asdics already in widespread use. A variation on this theme, a line of moored sonobuoys, was eventually tried off Ireland by the RN in 1945, with little success. The concept evolved after the war into an extensive system of seabed sensors known as SOSSUS. Blimps equipped with magnetic anomaly detectors were considered, and USN specialists toured Atlantic Canada to select sites and comment on feasibility. Nothing came of the effort. Perhaps the most ingenious idea was

206   The U-Boat Hunters

sowing the Cabot Strait with acoustic homing torpedoes, an idea that would have to wait for the invention of the microchip to permit the necessary discrimination between friend and foe.[70] With few resources at their disposal, the Canadians could not implement these more ambitious schemes before the war ended.

What stood in the way of an effective response to schnorkelling U-boats inshore was the ocean environment itself. In August 1944 the new Atlantic Oceanographic Research Group (AORG) at St Andrews, NB, produced its first report, 'Asdic Ranging Conditions in the Halifax Approaches.' The conclusions confirmed conventional wisdom. As the report explained, the USN made use of bathythermography (BT) readings in deep waters to obtain an 'assured range' for the asdic set. There was no question that USN hunter-killer operations in deep water off the Canadian coast were highly effective. The question on Canadian minds was whether BT would work inshore. The first report from St Andrews was not optimistic. 'Assured ranges are zero during the winter months' off Halifax, it noted, 'and deep submarines could approach to close quarters without detection.' Things were little better in summer months when, because of the warming of the surface layer, 'assured ranges are very short.' The best conditions, as sailors already knew, were in the spring and fall. None the less, the report concluded, BT readings would be 'of great importance in the operations of A/S flotillas.'[71]

The RCN went ahead with BT trials in the fall of 1944. *Springhill* of EG 16 was fitted with the equipment, and during September and October daily readings were taken by her A/S officer, Lt R.A. Nairn. It is conceivable that the contact with *U802* on 14 October owed something to use of BT, but a 'daily' reading was useless for tactical purposes, and there is, in any event, no indication that EG 16 sought to apply it.[72] The reports were passed directly to St Andrews for evaluation.

The complexity of water conditions in the river and gulf were outlined in the AORG's second report, 'Asdic Ranging Conditions in the River and Gulf of St Lawrence in Late Summer,' issued on 17 November. This report drew directly on Nairn's readings, and what it revealed was probably worse than anyone in the RCN had

A Sea of Troubles 207

expected. At least three currents distinct in temperature and salinity swirled throughout the area. Tides from the Atlantic pushed sea water through a deep submerged channel (at 100 fathoms and deeper, really the ancient river bed) from the edge of the continental shelf to the mouth of the Saguenay River. The great mass of water lying over this dense sea water was from the Labrador Current. This chilled body of arctic water, as cold as 30° F, poured into the gulf through the Strait of Belle Isle, dominating the depths from 50 to 500 feet. The thin surface layer was a mixture of three water masses: the Atlantic waters, which welled up at the Saguenay, mixed with the fresh waters of the St Lawrence River and then blended with the cold but less saline Labrador Current. This clash of water bodies also produced a distinctive pattern of water movement, the principal element being the Gaspé Current. This current, about 100 feet deep, followed the southern shore of the river, 'rounding the Gaspé peninsula in considerable volume' and then moving around Cape Breton, where it was known as the Cabot Stream. Not surprisingly, the report concluded that 'for operational purposes, with such variations in Asdic Ranging conditions in the area dealt with, it is of prime importance to know the ranging conditions at a given time and place.'[73]

Although the scientific reports on asdic ranging and the influence of sea conditions upon it were tentative, they were sound in principle and marked the RCN's initial integration of science into its A/S operations. Nairn was removed from *Springhill* and appointed to Halifax as Murray's BT officer. All senior officers' ships were subsequently equipped with two deep and one shallow BT instruments, and in November the RCN's first BT manual was issued. Senior officers were instructed to take daily readings 'and utilize their tactical value.'[74]

The RCN might well have done much more in this regard. In September it was decided not to take the Royal Society's schooner *Culver* into service as the east coast oceanographic vessel – largely because of the cost of making her seaworthy.[75] There is no indication that the RCN considered assigning one of its own vessels to support oceanographic research in the Atlantic (research was being done on the west coast by *Ekholi*). The navy seemed content to feed scientists information from operational ships for the balance of the war.

Canadian scientists entered the tactical debate at the end of November in their third report, on the influence of bottom conditions on sound propagation, by suggesting that it was often better to listen for U-boats than to fill the water with noise that would reverberate in all directions.[76] The suggestion that passive detection might be superior to active was especially valid for the approaches to Halifax, which is rocky and boulder strewn. These features simply bounced asdic echoes in all directions and made it hard to distinguish contacts. Comments on some reports of A/S searches of Halifax in the last winter of the war speculated that genuine 'metallic' contact made occasionally by ships was probably on their own, or their consorts', hull. Passive searches inshore were best in theory, but for the moment naval officers preferred to harass U-boat commanders with active searches in hopes that it would get them moving and eventually into the asdic's search beam.[77] This debate would endure for the balance of the war.

The RCN's application of oceanographic science to ASW reflected the growing influence of the USN on the Canadian home establishment. Both the interest and the expertise in BT work came from the Americans, whose BT readings on every watch and highly successful hunter-killer operations offshore seemed a logical cause-and-effect relationship. The British, too, were no strangers to the problem of layers and their impact on sound propagation. They had noticed the effect of daily surface warming on asdic ranges in the Mediterranean before the war (the same phenomenon noticed by the USN in the Caribbean, which had led to the development of the BT), and had published a confidential book on the subject (CB 1835[2]) in 1931.[78] However, the RN was content to apply the knowledge to its own submarine fleet. The BT equipment initially requested by the British from the United States in late 1943 was destined for use by submariners. By mid-September 1944 the Admiralty reported that 'all the bathythermographs in the United Kingdom have been fitted or allocated to submarines.'[79]

What seems to have prompted RN interest in BT use was not the problems in U.K. waters but rather those experienced on the arctic route to Russia. Asdic conditions for the Russian convoys were comparable to those off Canada and were considered so poor that Captain Walker believed that only aircraft could kill U-boats effectively in that area.[80] The problem was better defined by BT readings done by HMS *Saumarez* during a Russian convoy passage in September 1944. These readings revealed frequent sharp negative drops in temperature at roughly 150–200 feet: the deep, arctic current that lay under the vestiges of the Gulf Stream. As *Saumarez*'s captain, Cdr P.G.L. Cazalet, RN, reported, this finding 'confirmed, in the main, what many of us suspected, i.e., that the submarine at 200 feet or more was almost impossible to detect.'[81] In other areas BT readings revealed uniform temperature with depth, which produced good asdic conditions. Cazalet wanted to use BT to determine when conditions were good for asdic in order that time not be wasted hunting for a U-boat that could never be found. It was probably the problem of layering in arctic waters that prompted the British to request, in August, that RCN Castle class corvettes with their type 147B asdics be released for the Russian convoys. The problem of U-boats along the arctic routes mounted in early 1945 as German submariners sought to take advantage of predictable convoy routes, difficult asdic conditions, proximity to Norwegian bases, and a desire to strike a blow in support of the campaign on the Russian front.

It is doubtful, given the state of equipment and scientific knowledge, if science did much to improve Canadian or British A/S performance inshore before the end of the war. The problems of sound propagation in the sea, particularly inshore, remain almost as vexing fifty years later. Asdic operators and officers already knew conditions were difficult; what they wanted were solutions.

In fairness, the British were preoccupied with the looming threat of the type XXI U-boat in the fall of 1944. British intelligence kept a close eye on the progress of the new submarines and had a fair idea of their size, speed, endurance, and general characteristics. The displacement was estimated at 1,200 tons (400 tons too light), a surface speed of perhaps twenty knots (five knots too fast), and a maximum submerged speed of roughly fifteen to eighteen knots (actually 17.2 knots).[82] It was not until October – when the western front had stablized and it was clear the Germans would last until 1945 – that naval intelligence obtained a good picture of the number of new U-boats built and planned, and the operational status of those already in commission. Of the estimated sixty completed, it was believed that perhaps fifteen were ready for operations.[83] By early November the estimates had grown to twenty-four ready for operations and a further seventy-one under construction or fitting out. The figures for the smaller, inshore type XXIII U-boat were sixteen and twenty-six, respectively.[84] As the intelligence report for mid-November indicated, none of the assembly and fitting yards for these U-boats had 'yet been the primary target for air attack.' None the less, the delay in their arrival was at least partially due to the general impact of Allied strategic bombing. The requirement to delay them even further by targeting their construction facilities and mining their training areas became a recurrent theme over the winter of 1944–5. The new submarines were also delayed by Dönitz's decision to go straight into series production before the problems evidenced in the first production models were solved. As a result, the first few type XXIs failed to reach their designated capabilities and were relegated to training.

As he had been with the GNAT, Dönitz was anxious to introduce the new weapon en masse for maximum effect. His excitement about the type XXI was not misplaced. Schnorkel and improved acoustics and weapons made the type XXI a more potent weapon than the Allies realized. The type XXI had to schnorkel for only three hours per day to keep its batteries fully charged. Running on batteries it could cover 300 miles (at five knots) before needing to raise the schnorkel: enough to clear most barrier operations. Increased submerged efficiency gave the new type XXI enough range to go as far as Cape Town and back without refuelling. It was also equipped with a sophisticated passive sonar array capable of tracking fifty individual ships up to 7,000 metres away and sending targeting data to the torpedoes. Most of the latter were new pattern-running torpedoes that could be fired into a convoy with a theoretical hit probability of 95–9 per cent – a result actually achieved on trials. Moreover, with the new passive sensor, the U-boat did not need to surface to fire. And to tackle escorts, a new

GNAT, the 'T 11,' was developed, with an improved homing head impervious to the FOXER and CAT gears then in service.[85]

In the fall of 1944 the British were less concerned with the offensive capabilities of the type XXI than with how they were going to find and destroy it. Operational scientists estimated that it would be five times harder to find than U-boats currently in service and, once found, extremely hard to attack because of its high submerged speed. The problems of target movement at great depth had complicated A/S tactics in the winter of 1943–4, but no weapon system yet in service – or even contemplated – could handle a target changing its position in three dimensions faster than the surface escort could move. Even Allied A/S ships were built to handle a slow target. Apart from destroyers, only the Captain class frigates could outrun a submerged type XXI with certainty. The others, the Lochs, Colonys, and Rivers, had a slender margin of speed over a submerged type XXI, but that advantage depended greatly on sea conditions – which would not affect a submarine at depth. Moreover, simply keeping up was not enough. All existing asdic domes were also designed for comparatively slow speeds, and escorts travelling faster than twelve to fifteen knots quickly drowned out their asdics in the rush of water passing the dome. In short, high-speed submarines threatened to render obsolete the existing ASW system – ships, sensors, and weapons – with no clear solutions in sight.

In fairness, the problem did not develop overnight. Much work was done on responses to high submerged speeds in early 1944 in anticipation of the Walter boats.[86] In June 1944 the British began modifying the submarine *Seraph* for high submerged speed trials by streamlining the hull and increasing the battery storage. *Seraph* was capable of twelve knots for up to forty-five minutes – not as fast or as prolonged as the type XXI, but enough for trials and training. Exercises conducted between 10 and 30 October concentrated on the counter-attack problems of the escorts. At four knots or less *Seraph* represented a normal asdic target, although the streamlined hull made it a noticeably poor target at sharp inclinations. Active contact could be maintained at over nine knots. At high speeds the submarine itself made enough noise that it could be followed by listening (known as the hydrophone effect, or HE, listening with the transducer of the asdic set) at ranges from 4–5,000 yards. Such ranges could be maintained by HE on *Seraph* at her best speed, provided FOXERs were not used, since the FOXER had a serious impact on active searches and 'almost completely destroys that of H.E. detection.' Contact could be kept quite easily by two ships working together, although it was easily lost by a single hunter.[87]

Based on the *Seraph* trials, RN Operational Scientists placed heavy emphasis on developing passive means for tracking the type XXI. Using more VLR aircraft, extensively employing sonobuoys from aircraft, flooding the U-boat's line of advance with ships and aircraft, and maintaining a listening contact with warships, the plan resorted to a much larger and more complex hunt to exhaustion. With its batteries exhausted, the type XXI might become simply a very large, and with luck, quite stationary target. The scheme might have worked, but it appears that the high level of noise generated by *Seraph* in the nine-to-twelve-knot range was not characteristic of the type XXI. American trials after the war with *U2513* concluded that, 'at 12 knots submerged speed she is quieter than our best fleet submarines at six.' Speeds above ten knots caused some noise by cavitation, 'but up to ten knots she is extremely quiet.' The USN found that it required at least seven ships to conduct an effective search. However, 'The prospects for successful search and detection by aircraft,' the report concluded, 'present a pessimistic picture. The prospects for successful attack are not much better.'[88]

The final results of the *Seraph* trials were signalled to senior commanders in mid-December. They were advised that current equipment could handle the problem but were warned that 'A high speed submarine does not allow time to correct small errors in ship-handling and operating,' and therefore the very highest of training standards would be needed to cope with the type XXI.[89] The RCN reached a similar conclusion in its own evaluation of the type XXI problem in the fall of 1944. In late September the Naval Staff instructed Captain (D), Halifax, to prepare an assessment of the implications for convoy defence of a U-boat capable of a submerged speed of eighteen knots for up to ninety minutes. The report, prepared by Cdr J. Plomer, director of the Tactical Unit, postulated a silent approach to the convoy (three to six knots), the higher speed being used to escape following the attack. Provided the U-boat did not simply tuck itself underneath the convoy and ride along, the escort would probably have to rely on listening for the submarine astern of the convoy. As for attacking the target once found, Plomer was optimistic that a slow and deliberate hedgehog attack would be successful. Plomer agreed that there would be problems, but that with more ships and adherence to and perfection of existing attack procedures, the high-speed U-boat could be dealt with.[90] There is no evidence, however, that any effort was made by the RCN to improve training standards in anticipation of this more challenging enemy. If anything, training in home waters was about to go into its winter decline. Ice and arctic weather closed down Pictou and sharply curtailed training at Digby, while group training under captain (D), Halifax, virtually came to a halt in December. Only the base at Bermuda remained unaffected, and it would not be until the spring of 1945 that the RCN training situation – like the weather – saw any improvement.[91] Murray's only apparent reaction to the threat of the type XXI was to keep the River class destroyers close to home to 'deal with the brutes.'[92]

The British could not afford to be so sanguine in the late fall of 1944. It was understood that the quiet of October would not last, and that the schnorkel boats – followed by the new types some time in the new year – would soon come out in force. British Ultra intercepts revealed German progress towards a new offensive. As early as 9 October naval intelligence warned that Germany 'planned to renew her U-boat offensive with new types of U-boats at the end of the year.'[93] Further intercepts confirmed this warning. German agents in Madrid were instructed that information on sailings from North America was 'most urgently desired from December,' and plans for U-boat cooperation with long-range aircraft were deciphered. At the end of October a special appreciation issued by the RN Operational Intelligence Centre concluded that an 'all out U-boat campaign' would begin in the near future: probably sometime between mid-November and mid-December.

Operational scientists – by the fall of 1944, like A.A. Milne's 'gloomy old Eeyore,' a source of eternal pessimism – painted a grim picture of what the new U-boat offensive held in store. The sombre scientists' estimate of the probable shipping losses was submitted on 17 November and was generally accepted. The existing fleet of old U-boats would soon demonstrate a marked increase in activity and effectiveness, they warned, and would likely sink seventy-five ships in the first quarter of 1945. The new type XXI submarines were expected to make fifty operational cruises over the same period, accounting for 100 ships. The smaller, inshore type XXIII would sink enough to bring the losses to all types of U-boats between January and March 1945 to perhaps 200 ships, a figure not approached since the great crisis of the Atlantic war two years before.[94]

The U-boat fleet remained a potent force and the only military arm of the Reich that showed real growth in fighting potential. Its crews were now experienced and confident in inshore operations, and new equipment like the TUNIS radar detector and a more efficient schnorkel design introduced in October greatly improved operational performance and survival rates. By the fall of 1944 the loss rate among U-boats was down to its 1942 average, a factor that not only sustained operations well but had a positive influence on morale. Throughout the fall of 1944 the director of Naval Operational Studies took pains to demonstrate with mathematical precision that A/S searches were ineffective. The single kill by a surface vessel in October was scarcely bettered in November, when only two U-boats were sunk in the north Atlantic (both in British waters by RN vessels).[95] Moreover, despite the virtual collapse of the German air defences and the presence of Allied armies on the border, U-boat construction reached its wartime peak in December 1944 with thirty-one submarines launched; twenty-one of them were type XXI.[96]

The Germans also discovered that they could get more range out of their schnorkel-equipped U-boats than they had previously thought. *Annan's* U-boat, *U1006*, was one of four type VIICs dispatched in October to the English Channel; eight more followed in late November. The movement of U-boats back into the channel eventually drew all the RCN U.K.-based groups south again by

December. They had had a busy fall, but apart from EG 6's sinking of *U-1006*, there was little excitement. EG 6 itself was among the first to be reassigned to Plymouth command in November, where it remained until mid-December when it went to Larne, Scotland, for 'special duty' – a euphemism for training with *Seraph*. In the meantime, EG 9 ranged far and wide. In October the group was operated briefly by Flag Officer, Gibraltar and Mediterranean Approaches, and was involved in the search for the U-boat that hit *Chebogue* north of the Azores. November and early December were spent supporting Russian convoys, and on 12 December *Montreal* and *Port Colborne* shot down two Ju 88 aircraft. By the end of December EG 9 was on A/S patrol off the Cornish coast.

EG 25 spent November off north Scotland before being assigned in early December to Portsmouth command. EG 26 also undertook its first patrol in November in the approaches to the North Channel, where it learned exactly what it meant to be the junior group. While working with EGs 1 and 20, its task was to form the outer ring of the search area and pass on promising contacts to the British groups.[97] In December EG 26 also went to Larne for 'special duty' before being sent south. Thus by December all the RCN support groups were in the channel, along with four or five RN groups, supporting convoys and waiting for the offensive to begin.

The deployment was timed to meet the arrival of the eight U-boats that had set out for the channel in November. Not all of them made it. Two were accounted for in transit by British support groups in mid-December. Another, *U1209*, was the object of searches by EG 26 off Land's End. Several promising contacts attacked by *Ribble* and *Montreal* were classified at the time as wrecks. Shortly thereafter, however, the group rescued U-boat crewmen adrift in life rafts off Wolf Rock. According to the Germans, their U-boat had struck bottom while navigating in the tide-swept waters, suffering enough damage to warrant abandoning. Both *Ribble* and *Montreal* argued that their attacks contributed to the sinking of *U1209*. British authorities would have none of it. The Admiralty concluded that EG 26 had attacked a wreck and that 'there is no evidence for and N.I.D. [naval intelligence division] evidence against this U-Boat being attacked by surfaces at all.'[98] The Canadian claim was not pressed, but the Wolf Rock incident would resurface, impishly, in the assessment of the RCN's last U-boat kill of the war.

It was the German army that struck first at the end of 1944, smashing through the Ardennes forest on 16 December. The attack, know later as the 'Battle of the Bulge,' took the Allies completely unawares and threatened to drive a wedge between the British and the American armies. On 18 December the steamer SS *Silverlaurel* was sunk in the channel, the first loss in those waters for weeks. Panzers were still rolling westward into Belgium on 19 December when the British Anti-U-Boat Committee sat for its regular meeting. Not surprisingly, Churchill was in a 'petulant' mood, 'inclined to lash out at all & sundry.' He could not understand why U-boats could not be found inshore.[99] His mood was about to darken even more. On the 23rd *U772* sank two ships off the Cherbourg Peninsula, and late in the afternoon of Christmas Eve *U486* found and sank the 11,000-ton American troopship *Leopoldville* off Cherbourg. Initial estimates of the loss of life were very high, perhaps 2,500 men (Roskill records 800).

U-boat activity also increased 3,000 miles away. On 21 December an explosion rocked the Canadian freighter *Samtucky* in the approaches to Halifax harbour. She was beached, and it was thought she had struck a mine. It was only later that pieces of a torpedo were found in her hold. The Canadian response, on the eve of the last Christmas of a very long war, was almost one of disinterest. A few air patrols were flown and some of the local defence forces were instructed to 'sanitize' the approaches for a convoy from Boston. At about the time *U486* sank the *Leopoldville*, the presence of a U-boat off Halifax was confirmed when a GNAT from *U806* shattered the minesweeper *Clayoquot*. Her quarterdeck peeled back like the lid of a sardine can, the tiny ship sank quickly. Eight men were killed outright; another died in the water when *Clayoquot*'s depth charges exploded. On Christmas Eve 1944 the war that would not end was heating up; it was now a race to see if it could be ended before the type XXI arrived.

6
Victory ... of a Sort

> We never gained a firm and final mastery over the U-boats.
>
> Captain S.W. Roskill, RN

During the first major victory over the old U-boat fleet in 1943 the RCN was pushed aside almost entirely. In contrast, the RCN was part of the victory of early 1945. Between mid-December 1944 and the end of the war, Canadian warships sank four U-boats. This number, however, represents a small portion of the thirty-five submarines sunk by naval forces in the main Atlantic theatres. It indicates that during this final phase of the anti-submarine offensive the RCN slipped behind the pace. The reasons for this are obscure and complex. In general they mirror those of the earlier crisis of 1943, not least of which was British anxiety over a major threat to shipping. As the old U-boat fleet attacked in 1945, the British and Americans shifted resources to crush them. Once again the RCN was overcommitted and less well equipped than its larger Allies. Training standards in Canadian waters remained low, and there was a pressing need to refit and update the fleet. By the spring of 1945 the RCN's participation in the offensive against the U-boats was again being marginalized. For the RCN, as for the Allies in general, it was fortuitous that the war ended when it did.

The final campaign of the Atlantic war was one that everyone on the Allied side believed should never have happened. The collapse of the German army in France in August was followed by a victory euphoria which faded only slowly as the winter lines along the Rhine River coalesced. Meanwhile, the Allies began major force reductions and redirection towards the Pacific. In Canada thousands of newly trained aircrew were sent home on indefinite leave, no longer needed for the mills of the air war. The RCN's personnel strength, which had risen steadily for five years, began to contract. No new ships were on order, and the last of the wartime construction was fitting out. The commander of the USN's Eastern Sea Frontier talked in his August 1944 report of downscaling operations now that the end of the war was imminent.[1] The British were particularly keen to shift forces to the far east and to the new British Pacific Fleet. Canadian scientists continued to work on oceanographic problems, and the RCN remained attuned to changes in tactics. However, the professional RCN was on the verge of achieving its long cherished goal of a balanced fleet with cruisers and, now, light carriers. Its attention was focused on the postwar, especially the development of a fleet air arm. The major problems in anti-submarine warfare evident in late 1944 were being addressed systematically only by the RN and the USN: navies with a long-term stake in mastering the submarine menace.

It was enough, therefore, that the European war had not ended by Christmas 1944. The Ardennes and U-boat offensives gave proof that much hard fighting lay ahead. In mid-December the RCN got wind of the deteriorating situation from its staff at the Canadian Naval Mission Overseas (CNMO) in London. On the 17th a 'Cdr Church' from the Admiralty phoned the duty officer at CNMO to discuss a proposed signal to NSHQ regarding the need for yet more support groups. C-in-C, Western Approaches' forces were stretched; the RAF complained that its U-boat sightings were not quickly pursued by naval vessels, the Russian convoys were under attack again, the RN was five support groups short of the twenty-three estimated minimum required, and the number of U-boats in U.K. waters was rising. More support groups were also needed for the mid-ocean, where Dönitz was likely to strike first with his new U-boats, and for the Canadian zone. What prompted Church's call was RCN's 28 November 'Weekly State'; the Admiralty wanted to know specifically what the RCN had planned for the eighteen frigates listed as 'unallocated.'[2] The gist of the Admiralty's case was passed to NSHQ on the 19th. CNMO warned that sinkings of Allied shipping had not increased, but the aim of British anti-submarine policy was to ensure that they did not for fear it would enable U-boat commanders 'to regain confidence and get their tails up again.'[3]

In fact, the first blows of the final offensive had already fallen: one ship was sunk in St George's Channel on the 10th and another off Cornwall on the 18th, despite the presence of eight RN support groups in the area. The attacks prompted the Admiralty to order south the Canadian groups EG 6 and EG 25: EG 6 to St George's Channel and EG 25 to the English Channel.[4] The losses also occasioned a stormy meeting of the Anti-U-Boat Committee on the 19th. Captain C.D. Howard-Johnston, director of the Admiralty's Anti-U-Boat Division, recalled later that Churchill baited the First Sea Lord, Sir Andrew Cunningham, over the recent losses and the inability of the navy to stop them. Cunningham, who was not an anti-submarine specialist, deferred to Admiral Horton. According to Howard-Johnston, Horton lost his nerve in the face of Churchill's anger, opining first that the channel was 'not my area' and second that 'The asdic has let us down.' All eyes then turned to Howard-Johnston, a comparatively junior officer, who was left to mollify an angry prime minister who had just bullied two of his key advisers into silence. Churchill goaded Howard-Johnston by asking, 'Are you trying to tell me that the sea is any different to what it was in 1914?' 'No Sir!' Howard-Johnston replied, 'but the conditions are totally different.' 'There are,' he went on, 'thirty U-Boats hiding in terror on the bottom and this is the first ship torpedoed in two weeks.'[5]

According to Howard-Johnston, Churchill accepted his explanation. But Cunningham, who was trying to get on with the Pacific war, was unimpressed with his anti-submarine experts. At the end of October, during an earlier stormy Anti-U-Boat Committee meeting (when Churchill baited the air force for failing to bomb the U-boat pens at Bergen and Trondheim), Cunningham noted in his diary that 'Max Horton [was] thoroughly mixed-up about wrecks &

Howard-Johnston endeavouring to correct him got himself engulfed in a wave of technicalities.'[6] It seems that the First Sea Lord was caught between an angry prime minister and a complex subject he little understood. Yet Cunningham had to carry the burden of the war that would not end. His presentation to the cabinet meeting of 27 December was 'rather a sorry tale of losses.' Three ships had been sunk and two escorts damaged off Cherbourg since the confrontation with Churchill on the 19th. After a further three ships were sunk in the Channel on 28–29 December, Cunningham recorded in his diary, 'This is getting serious.' On 30 December he advised Churchill that he was dispatching immediately two additional support groups to the channel; another was on the way; fourteen ships earmarked for the Pacific were to be retained for the channel; and two escort groups off west Africa were being moved north to permit two Gibraltar-based groups to shift to the channel.[7]

Canadians made a modest contribution to checking the U-boat menace at the end of December 1944. Only four U-boats were sunk by anti-submarine forces during the month, a slight increase from November, but half of that number were accounted for by Canadian forces. One sinking was in the channel by an RCAF Wellington of 407 Squadron, which swept in on *U772* off Portland Bill, illuminated it with the aircraft's powerful searchlight, and promptly sank it. It was, Roskill observed, 'a rare success for Coastal Command during this period.'[8] The same could be said for sinkings by naval vessels. The RCN made a contribution towards the end of December when the Castle class corvette *St Thomas* of the mid-ocean group C 3 sank *U877*.

A large type IXC/42 under Kapitan Leutnant Findeisen, *U877*, reached her patrol area west of Ireland in mid-December. Her initial task was to transmit weather reports crucial to the German Ardennes Offensive which began on the 15th. *U877* was unable to transmit, however, owing to a faulty wireless set, and by late December BdU feared she might be lost. Assuming the best – that she had radio problems – Findeisen was given discretion to operate at will and he headed for North America. The U-boat was westbound when the radar warning set gave the alarm on 27 December, and *U877* crash dived.[9]

U877's faulty wireless nearly saved her, since there was no indication on Allied plots of a U-boat in her area, and the radars that tripped her detector were never close enough to obtain a return echo. The escorts discovered *U877* by pure luck on asdic, when *Edmundston* reported a contact that she soon abandoned as doubtful. *St Thomas*, coming up astern, obtained the contact and immediately fired one bomb from her squid mortar to 'keep his head down' and then set about proper classification. Leading Seaman G.A. Elsey, at the asdic set, was convinced he had a U-boat, but his A/S officer believed it to be a school of fish. *St Thomas*'s captain, Lt Cdr L.P. Denny, RCNR, listened for a while and then asked Elsey what he thought. Elsey was adamant, noting that the target had doppler. This effect indicated movement, which was confirmed by the plot. Once the course of the target was determined, Denny placed *St Thomas* directly astern at a slow but closing speed, lined up the squid and allowed it to fire automatically when the U-boat was in the right position. Once again a single bomb was fired, but this time it exploded directly over *U877*'s stern. Water poured into the U-boat, causing her to sink to nearly 1,200 feet before ballast was blown and full speed brought the U-boat back to about 500 feet. Trim could not be maintained, however, so all tanks were blown and the U-boat surfaced.

Both *St Thomas* and *Sea Cliff*, who had joined to help, opened fire, but there was little fight left in *U877*. By the time the two escorts closed, all the U-boat's crew were safely in their life rafts, and all were rescued. One German submariner felt compelled to compliment *St Thomas* on the accuracy of her attack by pointing skyward and exclaiming, 'Vasser Bomb, goot Vasser Bomb!!!' As Denny said, 'With Squid it was like Duck Soup... difficult to miss if you were careful.'[10] *St Thomas*'s kill marked the end of a very successful year: the best of the war for the RCN's A/S efforts. The next few months were not so kind.

On the day *U877* was sunk, CNMO forwarded to NSHQ the news from the latest British cabinet Anti-U-Boat Committee meeting. The London staff reported a delay of at least two months in the expected arrival of the new German type XXI and XXIII U-boats but none the less painted a gloomy picture of their potential. Even

Bomber Command was being brought onside to stem the rising tide of new high-speed submarines. Admiral Nelles, in one of his final acts as a serving officer, appealed personally to NSHQ on 4 January, stressing the gravity of the situation, the forces needed to cope with it, his belief that the British concern was justified, and that the RCN should do what it could to help.[11]

The RCN was aware that a final confrontation with the U-boat fleet was looming.[12] An upsurge in U-boat activity in Canadian waters was anticipated, but not enough to forestall sending more escorts overseas. Admiral Jones, the CNS, supported Nelles's appeal, and on 5 January he passed along the CNMO's signals to the War Committee of the Canadian cabinet with his recommendation that the RCN comply. The Canadians were well prepared, therefore, when the official Admiralty request arrived on 8 January. Noting that with the new types available for service the Germans would have at sea by March 1945 more submarines than they had during the crisis of March 1943, the British raised the spectre of a 'two-front' war – major campaigns inshore and in the mid-ocean. This was, in fact, just what Dönitz hoped to achieve. The British wanted five additional groups for C-in-C, WA, from all sources and hoped that the Canadians would find a few of them.[13]

NSHQ responded on the 12 January, with less enthusiasm than might be expected. The demand for more support groups coincided – typically – with a plan to shorten the convoy cycle that meant that both MOEF and WEF required an additional group, and other escorts were needed to support the feeder convoys, which would also sail more frequently. Again the RCN was being pulled in several directions, just as it was trying to refit some of the existing fleet for the Pacific. The best NSHQ could offer was a new frigate group (designated EG 28) planned for mid-February and EG 11, which was expected back in service by mid-March. Apart from those two support groups, the balance of new construction was needed to keep existing groups – which were already running at a minimum – up to strength, and to allow for a 'large number' of badly needed refits.[14] The Canadian offer was accepted, and the British found themselves enough escorts to form three additional support groups in January (EGs 22, 23, and 24), the five new groups needed.[15] At least one, EG 24, was composed of vessels already slated for departure to the far east.

The delay in building up the British Pacific Fleet occasioned by the intensification of the U-boat campaign weighed heavily on Cunningham, and he felt compelled to explain the situation at length to Admiral Bruce Fraser, commander of the British Pacific Fleet, on 19 January. Cunningham's letter summarizes the mood of the moment. The First Sea Lord wrote,

> We are ... having a difficult time with the U-boats. There is no doubt this schnorkel has given them a greater advantage that we first reckoned on. I believe some of them havedone as much as six weeks at sea without surfacing. The scientists have not yet caught up and the air are about 90 per cent out of business. The asdic is also failing us. The U-boat has found out that, if he conducts his patrol close inshore or in confined waters where there is a good tidal stream the tide rips will to a great extent defeat the asdic – so it is about 50 per cent, I imagine, out of business. As you probably know, they have been active and caused us losses in the last few weeks in the Channel, off Holyhead and one especially bold one penetrated into the Firth of Clyde. Science having failed us for the moment we are entering on an extensive deep mining programme, starting by mining them out of the Irish Sea and the English Channel and then extending the minefields to the seaward to try and keep them in deep water again. This, of course, will take time but we hope to finish in the early summer.[16]

Cunningham's gloomy estimation reflected his own ignorance, badgering by Churchill and the need to explain to Fraser why his ships would not be coming. The air force was indeed largely out of the business of killing U-boats. But it was airpower that kept the submarines down. As for the asdic, it would work well enough in British waters once the U-boat was localized, and increased German attacks on shipping would provide that information for waiting hunters.

The Canadian offer in January 1945 of two more support groups was a generous one, particularly in view of the attacks then under way off Halifax. The RCN was aware of the presence of U-boats off the Nova Scotia coast in December, but since they no longer engaged in routine radio traffic, it was largely impossible to conduct pre-emptive operations against them. There was no support group available, then, for an immediate hunt after *U806* sank *Clayoquot* off Halifax on Christmas Eve. EG 27 was far to the northeast supporting convoys, and EG 16 was dispersed throughout Murray's command doing numerous tasks. Murray responded with an ad hoc group designated W 12 – motor launches, corvettes, minesweepers, and the frigate *Kirkland Lake*, twenty-one ships in all – and led by his training commander, Cdr R. Aubrey, RN, who was carried to the scene in a motor launch four hours after the attack. The haste with which this ragged assembly of vessels poured forth from Halifax had a Keystone Cops flavour, but the searches were sensibly organized under dreadful asdic conditions.

The searches for *U806* were also entirely without effect. The U-boat quickly bottomed – in the middle of the swept channel – turned off all machinery, and lay there, listening to the racing of ships' propellers, the pounding of depth charges, the buzz-saw of CAT gear, and the searching asdics. By Claus Hornbostel's reckoning, vessels passed over his U-boat repeatedly without ever gaining contact. *U806*'s crew enjoyed a cold Christmas dinner in safety. The RCN expected him to 'clear the area tonight on schnorkel,'[17] but Hornbostel waited fourteen hours before lifting *U806* off the bottom and heading south on her batteries. He hugged the sea bed until well clear and only raised *U806*'s schnorkel forty hours after *Clayoquot* went down.[18] Hornbostel was unimpressed with the RCN response to his intrusion, noting that the Canadian ships were 'unpractised, for the U-boat wasn't even detected a single time.'[19] It was a view that senior naval officers shared, but the problem admitted of no easy solution.

The USN, accustomed to operations in deeper waters with better asdic conditions, watched the Anglo-Canadian struggle with inshore waters with interest in 1944. Since September their highly successful carrier-based hunter-killer groups had been without targets in the mid-Atlantic. In the fall of 1944 the USN kept one task unit of four destroyer escorts at Argentia in readiness to forestall westward U-boat movements. Admiral Murray asked if he might have the use of the group and was advised in early December that it would be possible 'on specific request.'[20] Apart from simple kindness, C-in-C, U.S. Atlantic Fleet, may well have been interested in supporting Murray's operations for other reasons. His area contained the only significant U-boat activity in the western Atlantic and the direct U.S.-Cherbourg convoys (the UC-CU series) operating in support of U.S. armies in Europe passed immediately south of the Canadian zone. The RCN also provided a useful conduit for tactical ideas straight from the British inshore campaign. In mid-December two of the three task units now waiting in Argentia were warned that they might be sent into Canadian waters to 'familiarize themselves with every phase of Canadian anti-submarine warfare technique.'[21]

Murray, obviously strapped for resources, was not about to turn down the offer. By late December Task Units 27.1.2 and 27.1.3, each composed of four destroyer escorts (virtually identical to the British Captain class), were searching the comparatively deep and favourable waters south of Sable Island under Canadian operational control. But the USN gained a much better appreciation of the Canadian problem in January, when American units moved inshore to hunt for U-boats in the approaches to Halifax. The first instance occurred on 4 January, when *U1232* under the enterprising Kapitan zur See Kurt Dobratz, attacked the small Sydney to Halifax convoy SH-194 off Egg Island, east of Halifax. The escort of a single minesweeper and one aircraft was no deterrent for Dobratz. His first torpedo shattered the small tanker *Nipiwan Park*, while his second, aimed at the leading ship, missed. Dobratz then cooly repositioned the U-boat for another attempt and ten minutes later hit the tiny *Polarland*, which 'literally flew into the air and burst.' The remaining ship of SH-194, *Perast*, took violent evasive action, while the minesweeper *Kentville*, according to *U1232*'s log, was too concerned with streaming its CAT and with self-survival to attempt any effective counter-attack.

Kentville passed over the U-boat several times, but no contact was ever made. The same was true of the ships sent to conduct the hunt. EG 16 arrived two hours after the attack, followed by an array of ships from Halifax. They too passed over *U1232* without detecting it, and Dobratz eventually moved off northeast along the coast, skimming the bottom. Two USN task groups, 22.9 and 22.10, arrived two days later and searched until 8 January without result.[22]

By the time the USN forces had abandoned the search for *U1232*, Dobratz was watching shipping patterns in the entrance of Halifax harbour. There, Dobratz wrote in *U1232*'s War Diary, the ships 'have to maintain a tight and rigid route between shoals and banks and cannot escape.'[23] Such was the case with the Boston to Halifax convoy BX 141 on 14 January. Its slender escort of two minesweepers was supported by EG 27 (recently reduced in size following the groundings of *Lévis* and *Lasalle*) which stayed with the convoy to help shepherd it through the long, narrow, swept channel that led into Halifax. There was, in fact, little effective protection that four warships (*Nipigon* was well astern with a straggler) could offer to nineteen ships strung out in single file along the channel. *Meon* led the convoy, *Ettrick* took station on the inshore side with *Coaticook* on the seaward side, while the minesweeper *Westmount* brought up the rear. Dobratz was able easily to torpedo the third ship, *British Freedom*, at 1041Z. It was a few minutes before anyone knew what was happening. As Louis Audette, the commanding officer of *Coaticook*, later remarked, 'The torpedoings, ugly and effective as they might be, were not spectacular: the usual puff of smoke from the victim's funnel and the muffled explosion heard only by the nearer ships.'[24] When it became clear what had happened, St Clair Balfour, EG 27's senior officer, ordered his group to join him at the head of the column to commence a search. Meanwhile, Dobratz hit the fourth ship in line, the *Martin van Buren* with a GNAT. This second attack caused the ships astern, in Michael Hadley's words, 'to pile up like a traffic jam.' One of them, the convoy commodore's ship, *Athelviking*, broke from the pack and crossed *U1232*'s sights long enough for Dobratz to hit her, too, with a GNAT at 1052Z. As the stricken ships stopped and began to settle, Balfour ordered EG 27 to 'adopt scare tactics,' by dropping random depth charges in likely firing positions. Although rash, the tactic saved two other ships and nearly accounted for *U1232*. Dobratz had turned his atten-

Victory ... of a Sort   227

tions to other targets and had nearly completed firing when he noticed *Ettrick* bearing down on him. Dobratz fired at his next intended victim and then dived in haste – and none too soon. *Ettrick* was simply dropping random depth charges and was completely unaware of *U1232*'s presence when she nearly rammed the U-boat. The frigate's magazine crew heard a dull boom, which sounded like a glancing blow on a submerged rock. *Ettrick*'s captain, Lt Cdr E.M. More, and others felt the ship heave 'as if over a mud bank.' Divers later found damage to one propeller and hull plates. The assessment of the staff in Halifax was that *Ettrick* had struck 'a submerged object,' and indeed she had: *U1232*. The U-boat was struck a glancing blow, bending the attack periscope, smashing fittings on the conning tower, pushing the submarine over sharply, and eventually snagging the U-boat with the CAT gear.[25] *U1232* missed certain destruction by a matter of inches.

The search for *U1232* in the aftermath of the attacks failed to find her, but it was not for want of trying or the application of sound principles to the search plan. Asdic conditions were so poor that contact could not be made on the stern of *British Freedom*, which lay on the bottom as her bows pointed skyward in the crisp winter air. Visibility was also obscured immediately by a dense fog, followed by blizzard conditions and a rough sea. Balfour contented himself with a broad front sweep at maximum theoretical asdic range between ships (allowing for some overlap) of 3,000 yards along *U1232*'s anticipated line of retreat: towards the difficult asdic conditions on nearby Sambro Ledges. According to Doug Maclean, Balfour's search plan 'succeeded in placing the ships [of EG 27] in a position where a detection of their prey could have occurred' some time between noon and 1300 hours. That *U1232* was never found Maclean attributes to 'poor asdic conditions.'[26]

Searches around the area of the attack for a bottomed U-boat were organized by *Westmount*, and salvage operations on the three ships failed under dreadful conditions. By the time EG 27 returned to the scene late in the day, the 'probability area' for the attacker was very large and it was dark and snowing heavily. Balfour organized sweeps for a bottomed U-boat throughout the night. EG 16 joined on 15 January, sweeping to the northeast, while C 5 swept down from the north. Finally, on the 17th the USN's Task Group 22.9 joined the hunt for a U-boat now long gone.

The success of *U1232* on the RCN's very doorstep was a shattering experience, and the next BX convoy to pass through the swept channel was heavily screened. As EG 27's report of the passage of BX 142 observed, 'This convoy was indeed well supported with practically all types of Canadian A/S vessels present. Two minesweepers majestically covered the rear of the convoy with a well placed smoke screen.'[27] The Canadians came to rely less on smoke than on the maintenance of standing patrols in the approaches to Halifax harbour. On 19 January two patrol areas on either side of a line running at 145 degrees from Chebucto Head to a depth of thirty miles (the axis of the swept channel) were promulgated: area 'X' to the west of the line, area 'Y' to the east. These two areas provided the central focus of RCN support group operations in Canadian waters until the end of the war. Convoys passing into and out of Halifax were directly supported, as were some more distant convoy movements, but the groups invariably returned to the standing patrol areas X and Y.[28] It was, in fact, the RCN's modest equivalent of the RN's plan for U.K. waters, Operation CE.

The search for *U1232* also demonstrated the nagging problem of poor training among Halifax-based ships. Virtually all training during late 1944 was devoted to working up new ships, with little time given over to operational training of ships and groups already in service.[29] Captain (D), Halifax, complained, and the number of hours devoted to support of operational ships by both his staff and that of the Tactical Unit increased, but proper training for Halifax-based support groups was never obtained. When EG 16 transferred to the United Kingdom in March, it had not received sea training since the previous July.[30]

The contrast between training standards, facilities, and emphasis in Canada and the United Kingdom was marked, and it contributed to the comparative failure of RCN A/S searches in Canadian waters. But even RN standards of A/S training had slipped by late 1944. Escorts were worrying more about anti-aircraft problems than about A/S, and the Western Approaches Tactical Unit had returned to an earlier, pre-October 1942 emphasis on defence of trade against surface attacks, both in preparation for the Pacific war.[31] The RN's operational scientists, who looked at the 'failure' of ASW during the fall of 1944, were critical of the state of the training. Escorts were being taught how to destroy a contact, when the real problem was one of initial classification. Senior staff at the Admiralty concurred. Howard-Johnston, the DAUD, confided to the scientists, 'I think you have pointed out a weak spot in our plan of training.'[32] Reviews of Western Approaches' training establishment in late 1944 were undertaken with an eye on the apparent decline in A/S effectiveness. The surviving evidence of training, including the pattern of operations evident in reports of proceedings and war diaries, suggests an increased emphasis on A/S training of U.K.-based support groups throughout the last months of the war. As CNMO warned Ottawa on 17 January 1945, 'The outstanding fact is that Admiralty's policy is to finish the war with the weapons and control equipment in existence at present. Therefore, training is crucial.'[33] The RN's training establishments were fortunate that they never had to cope with the brutal arctic conditions of a Canadian winter.

The inability of the searching forces to obtain any firm contact on *U1232* also highlighted the problem of asdic conditions inshore. The RCN was content, for the time being, to maintain a policy of active asdic search despite the fact that assured ranges off Halifax were virtually zero. On 15 January Murray signalled his policy on employment of bathythermographs and how to deal with the marked positive (increasing temperature with depth) temperature gradient off Canada's coast in winter. This temperature gradient caused sound to be bent back towards the surface and made detection of bottomed U-boats 'most difficult'. Vessels equipped with BT were to use it to determine the temperature profile and sweep with Q if a positive gradient was found. Bottomed contacts were to be classified by echo sounders and depth charges set to fire on the bottom.[34]

The RCN at home was actually moving towards something of a blend of its own, British, and American procedures for handling the inshore problem. Despite the order on 15 January for BT-equipped ships to use that equipment tactically, the director of Warfare and training at NSHQ observed at the end of January that 'there is not as yet enough data on its performance to prove its value.'[35] The RCN therefore retained the asdic procedures as promulgated in the fall by C-in-C, Western Approaches: active searches during the day and 'transmitting listening watches' at night. The latter called for alternate active and passive use of the asdic in hopes that the sound of a schnorkelling U-boat might be heard. Trials conducted later in the spring revealed that under good conditions ships travelling at four knots could obtain passive contact with schnorkelling U-boats at 2,500 yards, and at 1,500 yards at speeds up to eight knots.[36] Murray's groups used BT to modify this British practice by indicating when a passive search might be better during daylight hours as well. EG 27 seems to have made particularly good use of BT, reporting in March that its use had 'justified its inclusion as equipment in all Groups.' and had 'led to a greater appreciation of the A/S problem.'[37]

A further perspective on the intractibility of the inshore problem was gained at the end of January when the USN Task Group 22.9, helping in the search for *U1232*, entered Halifax for a conference. The bulk of discussion concerned procedures to be followed in the current search. In line with the practice in British waters, it was agreed that the senior officer present should take charge of the search, in this case Cdr V.A. Isaacs, USN, commander TG 22.9. This practice was followed for the remainder of the war. It was also agreed that EG 27 would adopt 'the fundamental doctrines of FTP 223A,' the latest USN Fleet tactical pamphlet, 'without in any way conflicting with the doctrines set by the Canadian Escort Group.'[38] This cooperation marked the start of a new relationship: in February the RCN adopted the search plans of 'FTP 223A' as the basis of its ASW in home waters.

Isaacs also reported candidly to the C-in-C, U.S. Atlantic Fleet, on 21 January about the problems his groups were having adjusting to conditions inshore. Current USN procedures and doctrine, designed primarily for deep-ocean operations, were inappropriate and would have to be substantially altered.[39] A more detailed American assessment of the difference between inshore and offshore ASW was produced by the commander of TG 22.10, who arrived with his group in Halifax at the end of January to work with Murray's forces. Hunting inshore, the report noted, was affected by 'the proximity of land and shoals, the positions of two swept channels, numerous aids to navigation, including two lightships, the presence of several small A/S vessels assigned continuous patrol in the same area, the presence of a similar task group (EG) in an adjacent area, and the constant flow of merchant shipping.' 'One regular encounter,' the report noted with some pique, 'was with a minesweeping group which maintained the right of way whether we had a good contact or not.' This was a new experience. After spending February off the Canadian coast, Cdr E.H. Headland, the senior officer of TG 22.10, went to Block Island Sound to conduct trials on locating a bottomed submarine.[40]

The problem of attacks off the Canadian coast also prompted the RCN to return to an old idea: sinking U-boats while they were still in deep water. Perhaps hoping to trade on its willingness to provide maximum support for operations in British waters, NSHQ once again prodded the Admiralty at the end of January 1945 over the matter of an auxiliary carrier. Again the plan was to use it to intercept U-boats in transit to the Canadian zone. CNMO reported a few days later that the British were unwilling to part with a carrier, when they had only six operational ones. They also continued to doubt if U-boats in transit could be located.[41]

February was a quiet month in Canadian waters, so there was little basis for a Canadian appeal on the carrier issue. Moreover, Murray continued to receive direct help from the USN, in the form of Task Group 22.10, which operated in the Canadian zone under Murray's command throughout the month. EGs 16 and 27 patrolled areas X and Y south of Halifax, shifting as need be to direct support of convoys. In this approach the RCN was consistent with the practice in British waters of concentrating support group operations around the U-boats' likely targets – the convoys. Meanwhile, TG 22.10 pursued possible submarine contacts exclusively. RCN support groups sometimes were on hand to help, but trying to squeeze in on a promising search under the direction of a more senior officer was no easy task. EG 16 found this out on 11 February off Halifax when TG 22.10 declined direct help, and the Canadians had to content themselves with patrolling to the south and east of the search to prevent 'the U-boat' from escaping.[42]

The number of support groups in the Canadian zone increased during February with the addition of EG 28 on the 17th, the last RCN support group formed during the war. It was composed of the new frigates *Prestonian, Buckingham, Fort Erie,* and *Inch Arran,* and also *Ste Thérèse* which was recalled from EG 25. *Ste Thérèse* brought with her an experienced crew and Lt Cdr J.E. Mitchell, RCNVR, to serve as senior officer. Mitchell's overseas training and experience is plainly evident in his excellent 'Reports of Proceedings,' and his choice as senior officer for EG 28 proved a good one. Moreover, his group was quite well equipped. All ships were fitted with asdic type 144Q, and two also had 147B (a third fit the set in March). Three of the frigates were equipped with type 271Q radar, not the best for locating schnorkels, and *Ste Thérèse* laboured on with an RX/C set. But *Prestonian,* in addition to her 271P radar, also carried the American 3-cm type SU set, one of the first fitted to a Canadian ship.[43] One of the earliest reports from EG 28 talked glowingly of the 'superiority of Prestonian's SU radar ... again and again throughout the period.'[44]

The formation of EG 28 freed EG 16 for service in U.K. waters, and the group transferred to the eastern Atlantic in March. The other promised group, EG 11, was delayed. As the destroyers completed refits, Murray kept them close to home, typically escorting fast troopships in the approaches to Halifax harbour. He may have been hedging his bets against the type XXI submarines. It was not until May that a restored EG 11 was back in British waters.

The drama off Halifax in January had its counterpart overseas, when U-boats penetrated the Irish Sea for the first time since 1939 and sank shipping. Six merchant ships were sunk, and two others and two escorts were damaged in the Irish Sea, an area long considered safe. Three new task forces were promulgated under Operation CE: TF 37 covering the North Channel and the area north of the Isle of Man, TF 36 from the Isle of Man to the northern edge of Cardigan Bay, and TF 38 covering St George's Channel (see map, p. 196). Unlike the other areas of Operation CE, forces committed to the Irish Sea were tasked first and foremost with destruction of the enemy.[45] A major redeployment of forces took place during January, and up to six support groups were assigned to these new duties, EG 6 among them.[46] The increase in U-boat activity in the eastern Atlantic brought only a slight rise in the number of kills by naval A/S forces during January: four, up one from December. All fell to RN warships.

At the end of January 1945 the Allies experienced a marked increase in shipping losses with little improvement in the rate of U-boat kills. This was, fortunately, a transitory phase. According to the RN's official historian, by early February there were fifty-one U-boats on patrol off or on passage to and from the British Isles – a far cry from the handful off Canada. Eight allied merchant ships were lost in British waters during February 1945, three in the Irish Sea and five in the channel, and two warships went down. A/S forces destroyed nine submarines, however, a marked improvement over January and one usually ascribed to increased support for convoy movements. That was only true in part. Three of the sinkings were accomplished by EG 10 (whose senior officer, P.W. Burnett, had commanded C 2 the year before) north of the Shetlands during an epic two-week offensive patrol. In all instances the U-boats were discovered on asdic by the same operator in the same ship, HMS *Bayntun*. The first, *U1279*, was located on 3 February in deep water (600 fathoms, so no chance of bottoming) and was destroyed by the squid of *Loch Eck*. EG 9 joined briefly to help but was long gone when *Bayntun*'s asdic operator got a second firm contact on *U989* on St Valentine's Day. The U-boat was fatally damaged by squid bombs from *Loch Dunvegan*. EG 10's final success came three days later when *Bayntun* located *U1278* and sank her with the first hedgehog attack. For the staff at Western Approaches Command, EG 10's kills demonstrated 'the very great hitting power of the Squid.'[47]

Four other U-boats were sunk in the channel during February, where the action was concentrated. One was by an RAF aircraft, one was shared by various forces, and one each was claimed by EGs 2 and 3. In the channel support group operations *were* tied to convoy movements. Many of the large and important oceanic convoys were heavily supported, with advance screens to push them through any wayward attacker. But much of the action off Cornwall centred around the small coastal BTC-TBC convoys. They were often 'supported' by taking station three to four miles *behind* and waiting to pounce on any U-boat rash enough to attack. This 'trolling' for U-boats suggests a relationship between higher shipping and higher U-boat losses. If such a relationship existed, it did not work to the Canadians' advantage. EGs 6, 25, and 26 operated in the English Channel and the Irish Sea during the month without luck, EG 26 ending February 'trailing convoys' between Eddystone Light and Start Point.

EG 26 might well have been successful in its trailing operations off Cornwall. U-boats were present, as EG 3 discovered, when, on 22 February, the group attempted to coordinate a large force of destroyers, frigates, minesweepers, and motor launches in search of *U1004*, which had just attacked BTC 76. Although the search was considered well organized, the U-boat escaped. As the senior officer of EG 3 wrote afterwards, it was 'difficult in inshore anti-submarine operations for an assorted force to co-ordinate in close harmony as a team.' This difficulty, he went on, was not due to unwillingness on the part of the escorts sent to help, but because the job required 'experience, training and the best equipment.' Ships not equipped or trained to recognize non-sub contacts simply wasted ammunition and time and confused the search 'by unreliable reports.' This was exactly the kind of problem that plagued searches off Halifax. In its next contact with a U-boat on 24 February EG 3 restricted its help to EG 15, when they found and sank *U480* after another attack on a BTC convoy.[48]

The only RCN support group to sink a U-boat in February was, in fact, about as far away from the scene of action as it was possible to be in British waters. After two days of operational training, EG 9 sailed for Cape Wrath in early February. Between 9 and 13 February it worked with EG 10 searching between the Faroes and the Shetland Islands, and then on the 13th it was ordered south to support convoy movements across northern Scotland.

On 15 February EG 9 joined WN 74 (seven ships escorted by three trawlers) and supported its passage from Loch Ewe to Moray

Firth. Early in the evening of the 16th, when the convoy was well inside Moray Firth, *Saint John* obtained a contact on the bottom. The target showed no doppler, which indicated it was stationary, and a fix with QH revealed no match with any known wreck. *Saint John* attacked with five depth charges set to explode on the bottom (225 feet), which produced some light oil. A further classification of the target with echo sounder followed, and then two hedgehog attacks were made, again producing only oil. The fourth attack, made with depth charges, produced both oil and a quantity of wreckage. Charts of the Caribbean, cork, splinters, biscuits, signal books, and other miscellaneous material was recovered, enough to convince Cdr Layard that *Saint John* had destroyed a U-boat. Repeated attacks on the 16th and 17th failed to release evidence 'of a large and conclusive nature such as bodies.' In the end EG 9 was credited with a 'probable,' which was confirmed after the war as the sinking of *U309*.[49]

British staff officers were annoyed and mystified by Layard's failure to use the 147B and squid of *Loch Alvie* to attack *U309* (and earlier contacts on the patrol).[50] This criticism was probably a reflection of the recent success of EG 10 with squid in deep water. Burnett had two ships with squid and was operating where U-boats could not bottom; Layard had one such ship and a myriad of bottom contacts to handle on a daily basis. The simple fact was that *Saint John*'s captain, A/Lt Cdr W.R. Stacey, RCNR, had handled the contact superbly and destroyed the U-boat after a thorough and efficient classification and attack.

If Layard failed to respond with the fullest energy and enthusiasm every time one of his ships obtained a contact it was with some justification. He had joined EG 9 a year earlier, and had been at sea constantly ever since. So too had many of EG 9's ships, and they were in desperate need of repair by February 1945. Not only had they been run steadily for months on end, since the spring of 1944 EG 9 had operated largely inshore, attacking and classifying an endless stream of contacts with high explosives. Layard estimated that his ships had expended, on average, sixty to eighty depth charges per month over the previous eight months. In many cases this figure was much greater: during one ten-day period *Saint John* dropped no less than 120 depth charges, and during the twenty-four hours she was in contact with *U309*, the frigate dropped sixty-five. 'In recent months the 9th EG has been [on] operation almost entirely in shallow coastal waters,' Layard wrote in his report, 'making daily and sometimes hourly attacks on bottomed contacts, and the average monthly expenditure of D/C's and H/H has been equivalent to the normal annual expenditure by ships operating in deep water.' All of these attacks, he noted, were made at the proper speed (fifteen knots), but the cumulative effect on the ships of such regular pounding in shallow water effectively eliminated EG 9 from the war by February 1945. When the group returned to Londonderry on 18 February, only *Loch Alvie* was fit for operations. *Saint John*, leaking fore and aft, was shipping thirty tons of water a day and could keep some compartments clear only by continuous pumping. *Port Colborne* had leaks in her distiller machinery, which Layard attributed to the pounding of explosives inshore. She was in Scapa Flow, when *U309* was sunk, trying to get the problems fixed, and she was sent back to Londonderry as unfit for operations. *Nene*, too, was leaking and the staff at Londonderry declared her unfit. *Monnow* leaked badly enough to require immediate docking 'before any serious damage was done.'[51] *Monnow* was docked, and she and *Loch Alvie* carried on for the remainder the war: the rump of EG 9.

The collapse of EG 9 in February 1945 was indicative of the strain placed on both men and ships by the new A/S war. That *Monnow* was repaired quickly and sent back to sea may well account for Lt A.B. Sanderson's comment, noted earlier, that the ship's captain and several of her wardroom died of exhaustion shortly after the war. Clearly the process of classifying targets with high explosives was hard on the ships, and may well explain – as much as age – why EGs 11 and 12 had collapsed at the end of the previous summer. Whether pounding from squid charges would have had the same effect remains uncertain. EG 9 ceased to exist as a viable support group by the end of February. While its veterans refitted and new ships were assigned, the rump conducted special escort duties in March. Then in April, when *Nene* rejoined, it was sent to the channel, where it appears to have been used largely for close escort duty as well. The group was re-

Victory ... of a Sort   237

established in early May, and its final operation would mark a fitting end to the RCN's Atlantic war.

EG 9's modest contribution to sinking U-boats was part of a brightening picture overall during February, achieved in part through the efforts of A/S forces. U-boat kills were up, and the type XXI submarine was now being delayed by bomber strikes. Cunningham had warned the British chiefs of staff in early January to expect first-quarter shipping losses on a scale not seen since 1943. The strategic bombing directive of 15 January specified U-boat production as a marginal priority item. That priority was changed by the Allied combined chiefs of staff, meeting in Malta on 2 February, who directed that type XXI production be directly targeted.[52]

In fact, the type XXI was already behind schedule, thanks to the general disruption of communications caused by Allied bombing attacks and the mining of training areas. By mid-February British operational intelligence estimated that the sixty-four already in commission were plagued by problems. According to deciphered signals, the Germans did not expect the type XXIs to become operational in numbers before June. Only one, *U2511*, was well advanced, and the intelligence people watched her progress with great interest.[53] To a considerable extent, then, by February the pressure had eased. Cunningham reflected this sentiment in his diary entry for 19 February, 'The S/M position which I had explained to me by Captain Wynne [Captain Rodger Winn, head of the Operational Intelligence Center] is rather better, but we are losing escort vessels & some merchant ships. Still some eight S/Ms have been sunk this month & the asdics are working better.' The delay in the arrival of the type XXI also meant that the Allies could concentrate efforts on defeating the old U-boats and thereby avoid the spectre of a 'two-front' war. Everything was now to be sacrificed to holding, or defeating, the existing U-boat threat for the next six months and, if possible, driving them out into deep water.[54] As the *Monthly Anti-Submarine Report* for February observed, better training and increased experience were finally beginning to show results.

That victory was achieved over the next month. During March

fifteen U-boats were sunk in British waters, for the loss of ten merchant ships and three warships. As Cunningham recorded later, 'The Germans were starting to totter' on all fronts. Destroyers were recalled from the Mediterranean, and Cunningham collected all 'the escort vessels I could lay my hands upon,' even to the point of battling Churchill over the release of trawlers to the fishing fleet. He ordered that 'all refits were to be postponed, and that even ships with only one propeller working were to continue to run.'[55] The result was the best score of U-boat kills by A/S forces for over a year. Six U-boats were sunk in the English Channel: two on mines, one by grounding, and three by British support groups. By the end of March the channel was largely abandoned by the Germans. The only RCN groups to spend any time in the area during the month, EG 26 'trailing' coastal convoys between the Lizard and Start point from the 3rd to the 9th and EG 6, which supported oceanic convoys between the 8th and 17th, took no part in the kills. The RN also accounted for all three U-boats patrolling the Minches (the channel between the mainland and the Outer Hebrides), all of them succumbing to the hedgehog and depth charges of EG 21's Captain class frigates. One U-boat operating off Scotland was sunk by the South African frigate *Natal*.

It was in the Irish Sea and in the North Channel where the RCN made its contribution to this final victory, sinking two of the three U-boats accounted for in those waters during March. The first fell to EG 25. In early March command of the group had fallen, temporarily, to Lt Cdr Howard Quinn, RCNVR, of *Strathadam*.[56] Quinn was well up to the task (he later joined the RCN and retired as a commodore). When EG 25 returned from a trip to Gibraltar on 5 March, it was refuelled and immediately sent to sea again as part of Task Force 38: St George's Channel. The group arrived on the morning of 6 March and was instructed by the senior officer present, in EG 5, to patrol and support local convoys. The next day, now under the control of the senior officer of EG 18, who Quinn remembers as particularly pushy, the group was patrolling the narrowest portion of St George's Channel. A contact earlier in the day, which Quinn judged to be 'the real thing,' was attacked repeatedly before EG 18 arrived to take it over. EG 25 then received orders

Victory ... of a Sort 239

from Western Approaches to patrol northwest of Ireland. Leaving the RN group to pursue the promising contact, EG 25 increased speed to sixteen knots and set course roughly northeast (033 degrees), searching in line abreast while en route through the Irish Sea to their new station.

A short time later the group slowed to eight knots to investigate a radar contact by *Strathadam*, which turned out to be false. *La Hulloise*, the port-wing ship, was also toying with a radar contact, which was initially dismissed as a buoy. It took the crew a few minutes to realize that the radar contact and the subsequent asdic contact were one and the same. The contact was closed, and when, at a range of about 350 yards, the radar lost contact, the frigate's port 20-inch searchlight was switched on. What was illuminated, about 200 yards away fine on the port bow, was an 'unmistakable periscope and schnorkel.' The light also alerted *U1302*, which was schnorkelling in accordance with practice with her periscope up and manned. The U-boat dived before the frigate was close enough to ram. Since *La Hulloise* was out of hedgehog bombs and travelling too slowly for a hasty depth-charge attack, her captain, Lt Cdr John Brock, RCNVR, prudently contented himself with locating *U1302* on his asdic (conditions were excellent) and waiting for help to arrive.

A 'Flash report' from *La Hulloise* at 2142Z sent *Strathadam* to action stations. She increased speed, streamed her CAT gear, and was in asdic contact by 2157Z. *Thetford Mines* closed at safe speed to within 2,000 yards of the submarine, gained contact and began a square search around the area. In calm seas, with excellent visibility and asdic conditions and the target firmly held, the rest of the action was comparatively straightforward. *Strathadam* made its first hedgehog attack at 2212Z, which produced a definite hit, with a 'spectacular underwater flash' followed by a huge air bubble and wreckage. Moments later the U-boat briefly broke the surface, its black hull illuminated in a searchlight and seen clearly by the quarterdeck crew members. As the attack developed, the senior officer of EG 18, aware of the quality of EG 25's contact, radioed that he was coming to take charge. Quinn, his group's teeth firmly in the U-boat, informed his senior colleague politely over the radio that if he intended to push EG 25 off the contact, 'he had better come in shooting!' Quinn's firm rejection of 'help' was heeded, and EG 25 kept its weapons trained on the Germans. In fact, little more was seen of *U1302*. Subsequent attacks produced oil and miscellaneous wreckage, including books, clothing, wood and 'numerous other small objects.' The contact was finally abandoned at noon the next day, and EG 25 sailed away content that it had destroyed its first U-boat.

And so it had. Quinn was fulsome in his praise for *La Hulloise*'s commanding officer, John Brock, for his 'cool and deliberate stand aside ... after having established contact and sighting the U-boat's schnorkel and periscope.' It was, Quinn concluded, 'Quite the most notable feature of the whole episode.' The steady and accurate reports on the contact received from both *La Hulloise* and *Thetford Mines* 'throughout bore positive evidence of sound group training by S.O. EG 25 [V.E. Browne] and a well developed appreciation of team work.'[57] British staff officers agreed. Commodore (D), Western Approaches, remarked, 'HMCS *La Hulloise*'s decision to direct instead of attacking with depth charges or attempting to ram is considered eminently sound and indicative of the good teamwork displayed generally by the group and particularly in this action.' His sentiment was echoed by the Admiralty's Anti-U-Boat Division, where Brock's action was judged 'outstanding' and the group's efforts as a whole – in the understated British terminology for excellence – as 'most satisfactory.'[58]

Quite apart from the excellence of the action and the accuracy and coolness of *Strathadam*'s asdic and hedgehog teams, the British were especially interested in how the *U1302* was initially detected. It was, Horton reported to the Admiralty on 26 April, 'the first reported instance of an escort approaching a schnorkelling U-Boat and obtaining Asdic echo, Radar, Adsic H.E. and sighting contact in turn.'[59] In fact, Horton's memo appears to have got the order wrong. As *La Hulloise*'s report bears out and the radar officer in the Admiralty's Anti-U-Boat Division observed, 'This was the first sinking for several months for which radar was responsible for the initial contact in inshore operations.'[60] Finding schnorkels by radar was, of course, the impetus behind the fitting of 3-cm radar. The irony of *La Hulloise*'s contact was that it was made with Lloyd Dawson's rebuilt RX/C.

There was an even more marvellous irony in the RCN's other kill of March, and its final one of the war. It began on the night of 20 March as EG 26 sailed from Londonderry following a ten-day layover. The group was on its way to Loch Alsh for training, where the new senior officer, Cdr R. Aubrey, RN – formerly the training commander at Halifax, who had replaced an ill and exhausted Ted Simmons in early March – could familiarize himself with the group and the local situation. EG 26 passed the Foyle buoy (the seaward entrance to Londonderry) at 2230Z, turned north, and formed into line abreast: *Ribble, New Glasgow, Sussexvale*, and *Beacon Hill*, a mile apart and zig-zagging independently along the same mean course, making about fourteen knots and streaming CAT gear. About twenty minutes later the port lookout on the bridge of *New Glasgow* reported, 'Low flying aircraft approaching!' All eyes and glasses turned his way and 'a loud noise, not unlike an aircraft was heard.' As eyes strained to discern objects in the sky, the changing position of the noise drew the port lookout's attention down and he quickly changed his warning to 'Object in the water – Very Close!' The officer of the watch and the captain had time to see it forward of the port bow about seventy-five yards distant and closing on a collision course: it was a periscope and schnorkel, visible in the moonlight and partly obscured by a pall of thick yellow smoke. Actions stations were sounded and a shallow depth-charge pattern was ordered, but it was too late. The U-boat struck *New Glasgow* just below the bridge moments later.

The surprise on the bridge of *New Glasgow* was matched only by that aboard *U1003*. The U-boat, a type VIIC, was on her second war patrol under Oberleutnant zur See Struebing. On the night of 20 March *U1003* was running on her schnorkel at about three knots when her radar search receiver obtained a contact that the operators failed to report. Shortly thereafter the U-boat experienced a violent impact, which pushed it over 30 degrees out of control. Struebing ordered all engines cut, while the engineer ordered full ahead on the electric motors. The result of the collision and conflicting propulsion orders was an equally violent contact with the bottom at sixty metres.

As *U1003* settled on the bottom, *New Glasgow* informed the group of the incident, opened the range and attempted to get asdic contact while waiting for the rest of EG 26 to arrive. A quick damage assessment revealed leaks in *New Glasgow*'s after 4-inch magazine and that the after ballast tank and – perhaps more important to the crew – the spirit room were flooded. C-in-C, Western Approaches, was promptly informed that *New Glasgow* had 'rammed' a U-boat. Divers later found considerable damage to the portside propeller, its support bracket, and hull plates along the length of the ship where the submarine had scraped by.

New Glasgow's asdic was also affected by the 'ramming,' and EG 26 searched, in company with C 4 and EG 25, without ever locating *U1003*. Some of the initial depth charge attacks by EG 26 came close enough to shake the U-boat, but the hunt soon drifted away. An hour after the collision Struebing lifted *U1003* off the bottom and started slowly westward, out of the North Channel. Progress was painfully slow, and without the schnorkel the air in the U-boat soon became foul. Just before dawn on 22 March Struebing brought *U1003* to the surface, forced open the conning tower hatch and assessed the damage. The bridge was wrecked, railings and fairing were smashed in, periscope and schnorkel were buckled, radar and search receivers were torn away, and one of the 20-mm anti-aircraft was guns dismounted. Another radar search receiver was mounted on the bridge, and for about half an hour the U-boat was ventilated, batteries were charged, and water was pumped out. Then the detection of a radar signal sent *U1003* down, where, after about an hour, she settled on the bottom in 260 feet of water.

The hasty dive revealed that the conning tower hatch was no longer watertight. When the conning tower was allowed to flood, water cascaded into the U-boat through a voice pipe, which took some time to plug. *U1003* then lay on the bottom for twenty hours until the batteries weakened and the air grew foul again. Struebing resolved to surface and beach his submarine in the Irish Free State. The U-boat was no sooner on the surface and the radar receiver mounted than strong signals were picked up and it was decided to scuttle *U1003* and take to life rafts. In haste, some crewmen jumped straight into the sea. Others were soaked in the process of getting into the rafts. The cold sea and the wait for rescue took their toll.

The signals detected by *U1003* probably came from EG 25 or EG 26, both of which were in the area. EG 25, it will be recalled, was on its way to patrol northwest of Ireland when it sank *U1302* on the 7th. After having shared briefly in the hunt for *U1003*, EG 25 had finally arrived on station on 22 March. The next day *Thetford Mines* had detached to Londonderry for urgent repairs, when at 0825Z – about six to eight hours after *U1003* was scuttled – her lookouts spotted a cluster of yellow life rafts sixteen miles northwest of Innistrahull. They contained thirty-three survivors of *U1003*, all that remained of the crew of forty-nine. Among them was Struebing, who passed himself off to *Thetford Mines* as 'Striezel,' merely one of the U-boat's officers. The rest of EG 25 arrived to screen the rescue operation, and then *Thetford Mines* continued on her way to Londonderry. Two of the crewmen died en route, but the remaining thirty-one survivors of *U1003* were landed in Northern Ireland at 1600Z. *Thetford Mines* sailed a little over an hour later to rejoin EG 25 on patrol, stopping briefly on the way to commit the remains of two erstwhile enemies to the care of the deep.[61]

There was no doubt that *U1003* was sunk, but the circumstances of its loss presented the British with something of a dilemma – and an occasion for some mirth. EG 26's report of proceedings, and the supplementary reports on the incident itself, all refer to it as a 'ramming' of the U-boat *by New Glasgow,* despite the clear contrary evidence in the reports themselves. It may be that the truth, that *U1003* had rammed *New Glasgow,* was simply inconceivable. More likely, EG 26 was merely trying to get credit for the kill by putting the best possible face on it. Senior British staff officers were clearly sympathetic. They also remembered that EG 26 had tried to claim *U1209* – albeit without pushing too hard – in December, when that U-boat ran aground and was abandoned off Wolf Rock. EG 26 picked up the survivors and claimed that its attacks had driven

U1209 to the reckless manoeuvres that caused its stranding. Now the same group claimed a U-boat kill following a collision.

There were two fundamental problems to be solved. The first was posed by an Admiralty staff officer with a delightful double entendre: 'Does *New Glasgow* get an "A"?!!' An 'A' assessment was that given to a U-boat 'known sunk.' The other meaning, clearly evident from the other staff comments, was what to do with a student who solves the problem but does not use the formula. As one staff officer noted rather puckishly, if *New Glasgow* was awarded credit for *U1003*, 'won't this lead to a claim by [the] Wolf Rock lighthouse keeper?'[62] Further, if *New Glasgow* was awarded the kill, should Hanbury, her captain, receive the customary DSC that went along with it? The solution was to award the kill to *New Glasgow* and say no more.[63]

The sinking of *U1003* and *U1302* accounted for two of the three U-boats sunk in the Irish Sea and the North Channel during March 1945. The total for the month, fifteen, brought the number sunk by A/S forces in U.K. waters since the start of the new year to thirty-five. This, at last, was U-boat sinking on a scale long desired. The significance of this trend was not lost on the Germans. Although their command and control system was collapsing, and few U-boats used radios at sea (only thirty of the 114 U-boats that sailed on operations from February to the end of the war reported their arrival in the Atlantic) BdU was generally aware of the increased level of losses in British inshore waters. They could not be sustained, and at the end of March large-scale deployments to inshore U.K. waters ceased. This was the victory Horton had sought since the previous summer.[64]

U.K. waters were not entirely abandoned during the last weeks of the war, and the RCN might well have increased its score of U-boat kills in April but for a bit of bad luck. The first incident occurred on 6 April, when the rump of EG 9 – *Loch Alvie, Monnow,* and *Nene* – along with the British destroyer *Watchman* and Free French destroyer *L'Escarmouche* – were escorting a convoy of empty troopships (VWP 17) between Le Havre and Southampton. In the middle of the channel *U1195* torpedoed and sank the 12,000-ton steamer *Cuba*, which 'went to the bottom like a stone,' settled on an even keel leaving five feet of her cargo mast above water with a cat stranded on the top. The scene was already a litter of wrecks, all now marked with buoys. According to Lt Sanderson of *Monnow*, EG 9 attacked a number of contacts before *Watchman*'s captain announced he was taking over the hunt. Many years later Sanderson remained bitter about *Watchman*'s intrusion and about the fact that the British destroyer 'was never seen from *Monnow*' but was none the less later credited with destroying *U1195*. It was not the first time, Sanderson recalled, that *Watchman* had horned in on a search. Her captain was regarded as pushy and ambitious and apparently was resented by the Canadians because of these qualities. That may be so, but Layard's report at the time makes it clear that *Watchman* got the kill.[65]

The other Canadian missed opportunity occurred a few days later in the Irish Sea. The action began around convoy HX 346, headed for northern English and Scottish ports. The convoy was escorted by an RCN group, C 5 (the frigate *Runnymede* and the Castle class corvettes *Huntsville* and *Belleville*), and supported by *Strathadam* and *Thetford Mines* of EG 25. As it passed Holyhead, Wales, on 7 April, *U1024* lay squarely in its path.

The U-boat was resting quietly on the bottom, seventy metres down, waiting for the fog to clear when at 1600Z the CAT gear of HX 346's escort was detected. The submarine quickly came to periscope depth, which revealed an empty and utterly flat sea. None the less, over the next fifteen minutes *U1024* fixed the convoy and slipped unseen behind the leading escorts. By then Gutteck, the U-boat's captain, was so close he had to crash dive to avoid being overrun by the first ship in the port column. Once the danger had passed, *U1024* returned to periscope depth and one torpedo was fired.

The torpedo hit the SS *James W. Nesmith*, the fourth ship in the port column, in her port quarter. As the stricken ship slowed to a stop, *Thetford Mines* raced in to lay a buoy one mile north of the scene to mark the attack, while C 5's senior officer, Acting Commander G.H. Stephen, RCNR, in *Runnymede* took charge of the search. *Strathadam* and *Runnymede* conducted a square search around the datum, while *Thetford Mines* swept in the vicinity of the damaged freighter. The assumption was that the torpedo had come from well ahead of the convoy. In fact *U1024* had carried on through south and west of the scene after the attack. The only escort estimated to have crossed the U-boat's path in the aftermath was *Belleville*. She was in station astern of HX 346 because her asdic was out of service; she and her squid were no danger to *U1024*.

HX 346 continued on course for Liverpool, escorted by *Huntsville*. *Belleville* took the *James W. Nesmith* in tow and the three frigates continued the search. Four hours later EG 5 and and the RCN's EG 16 arrived to help, and they expanded the searches southward, EG 16 to the coastal area off Holyhead and EG 5 southwestward towards Ireland. Neither group obtained firm contact with *U1024*, which had altered course westward towards Dublin. EG 5 was, as Gutteck admitted, 'pretty well on our track' and attacked a non-sub contact close enough to shake *U1024*. But EG 5 was distracted by false contacts and *U1024* slipped away. The sounds of EG 5's attacks could still be heard as Gutteck brought his U-boat to schnorkelling depth off Dublin to recharge batteries, to fix his position, and to try to repair a defective torpedo tube.

Searches for *U1024* off Holyhead were abandoned at midday on the 8th, and by the 9th EG 5 had given up. EG 25 departed to support BB 79 as it entered St George's Channel, leaving *U1024* to lie quietly on the bottom northeast of Dublin during the 9th. Gutteck eventually brought *U1024* back across the Irish Sea on the 10th to the convoy routes southwest of the Isle of Man. There the U-boat settled on the bottom at 0630Z on the morning of 11 April.

The pattern of 7 April now repeated itself, at least in part. At roughly 1008Z Gutteck detected the sound of approaching ships. *U1024* was brought to periscope depth, but visibility was poor and nothing could be seen although the sound grew louder and asdic and FOXER or CAT noises could soon be distinguished. At 1200Z *U1024* went to action stations, but nothing happened for the next forty-eight minutes. Gutteck was alerted to a sudden change in the tactical situation when the noises from the CAT gear stopped and a quick sweep by the periscope revealed an escort emerging from the fog only 300 yards away and heading straight for *U1024*. The escort was *Strathadam*.

EG 25 was supporting a Bristol-to-Belfast convoy northward on 11 April, with CAT gear streamed, when just west of the Calf of Man *Strathadam*'s asdic operator reported a contact at maximum range. Thus began a long afternoon of cat and mouse in the Irish Sea. Below, in *U1024*, the sounds of the hunt could be heard plainly. *Strathadam*'s screw noises, 'slow and quiet' could be discerned, as could the 'Asdic noise-like soundings' made by the frigate's echo sounder as she attempted to classify the contact. At 1312Z the CAT was recovered and a series of attacks began. The first hedgehog attack was close, but it exploded on the bottom without damage to *U1024*. Two others produced no better results, probably because Quinn mistakenly believed that the U-boat had bottomed. In fact, Gutteck kept under way at slow speed, just fast enough to make the attack inaccurate but not enough to shake the pursuers. A series of depth charge attacks by *Thetford Mines* starting at 1340Z produced some light oil and one in particular, at 1540Z, shook *U1024* badly. Gutteck described the attack in his log as 'Pretty good aiming', and it was close enough to encourage him to make a break to the south to try to escape.

The two frigates of EG 25 continued the search, launching three more depth charge attacks until 1745Z when Quinn decided to try his hedgehog again. Having made his intentions known to Lt Cdr Wilkinson, EG 25's group A/S officer, Quinn retired to his cabin for supper and the hunt proceeded with the watch on duty. At 1811Z the firing buzzer sounded and the first hedgehog bombs – which were ripple fired – began leaving their mounting. Eleven of the twenty-four bombs were in the air when one of them suddenly detonated thirty feet above *Strathadam*'s foredeck. It was believed that the explosion countermined some of the other bombs. The force of the blast knocked down the rating operating the firing switch and smashed the firing circuit, preventing the remainder of the bombs from leaving the hedgehog mounting. The blast also raked the foredeck and bridge with fragments, killing one rating on the bridge instantly. Fragments also drove down through the deck, killing three men in the forward mess outright and fatally wounding two others. Seven more of *Strathadam*'s crew were wounded, two of them critically, and small fires broke out.

Strathadam suffered no serious structure damage and the fires were quickly extinguished. But Quinn now faced two problems. The severely wounded men required care (most had head injuries), and *Thetford Mines* sent over her doctor to assist until they could be landed. Equally pressing was the problem of the remaining hedgehog bombs which sat, menacingly, on their spigots. The idea of throwing them over the side was quickly abandoned for fear they would detonate upon striking the water. Nor, in view of the incident itself, was the prospect of firing them off a welcome one. In the end Quinn and Wilkinson cautiously replaced the safety pins to disarm the impellers and hoped for the best. Leaving *Thetford Mines* to pursue the contact, *Strathadam* set course for Belfast to see to her wounded and the dangerous ordnance. The result of the incident was the breaking of contact between EG 25 and *U1024*. As the RN's immediate postwar analysis of the whole affair observed, 'it must have been with great relief that Gutteck entered in his log:- "1930. Corvette has been shaken off. She looked pretty practised in the U-Boat hunt."'

Whether EG 25 would have completed the hunt successfully had *Strathadam*'s accident not occurred will never be known. *U1024* lived to fight another day – one in fact – when she attacked convoy BB 80 on 12 April, hitting the SS *Will Rogers*. The attack brought EGs 2 and 8 to the scene, and in a series of attacks the submarine was severely damaged. Finally *Loch Glendhu* fired a squid pattern that wrecked *U-1024*'s machinery and forced her to the surface. Gutteck, the first out, had his hand shot away and then he took his own life. Thirty-seven of *U1024*'s crew were rescued, and the submarine was temporarily taken in tow before foundering. EG 8 was awarded the kill.

British authorities were mildly critical of Quinn's abandonment of the contact on the 11th, believing that, with *Thetford Mines*'s doctor aboard, *Strathadam* ought to have continued. They did not share Quinn's concern for the hedgehog bombs still on their mountings, although the port authorities at Belfast did. *Strathadam* arrived in the dark of night, in a driving rain with dead and wounded aboard. Unable to anchor because the explosion had cut her anchor cables, Quinn brought the ship alongside, where a hor-

Victory ... of a Sort 249

rified officer took one look at the hedgehog mounting and ordered the ship away. *Strathadam* then steamed to Londonderry to obtain help. The bombs were removed for inspection, which proved inconclusive. As Quinn recalled many years later, some 257,000 hedgehog bombs were fired during the war, with only two premature detonations – one of them in *Strathadam*. The final casualty toll from the explosion was six, all of whom died of head wounds. EG 25, less *Strathadam* and now under Cdr J.M. Rowland, RN, formerly Captain (D), Newfoundland, sailed on 23 April and completed the war without incident supporting convoys in the Irish Sea.[66]

In fact, the war ended rather quietly inshore, because by April the U-boats were concentrated 2–300 miles out to sea. Eighteen U-boats operated in the Southwestern Approaches during April and the first week of May, and the Admiralty deployed support groups of Captain class frigates to deal with them. They accounted for three and aircraft sank three more. Total U-boat kills by A/S forces in waters where the RCN operated in April 1945 were fifteen: the same count as for March. None fell to the RCN. The bulk of killing in the last days of the war, roughly twenty-three U-boats sunk at sea from 1 to 10 May, was the work of Allied tactical airpower, as Germany collapsed and U-boats made one final desperate attempt to save themselves by fleeing to Norway.

The qualified victory in the eastern Atlantic was mirrored in the mid-ocean and off eastern North America, where U-boat operations also continued uninterrupted until the end of the war. One U-boat was sunk within the Canadian zone in March by warships under Canadian operational control, but the ships were American. Task Group 22.14, composed of four destroyer escorts, was loaned to Murray to operate against an inbound U-boat passing through the Canadian zone south of Sable Island. The submarine was just one of a wave of three detected by operational intelligence that were thought to constitute the start of the spring offensive in Canadian waters. While the RCAF flew patrols to seaward along estimated approach routes, EG 16 and EG 27 along with TG 22.14 began retiring searches along *U866*'s line of advance towards the Gulf of Maine starting around 6 March. On 16 March, as the search moved south of Sable Island, EG 16 detached in support of CU 16 en route to her new deployment in Western Approaches Command. The same day, EG 27 left to support the Halifax section of ON 288. TG 22.14 was therefore on its own when the USS *Lowe* made contact with *U866* a few miles southwest of Sable Island at 1019Z on 18 March.

Lowe attacked with hedgehog, missing with the first pattern. Then the U-boat attempted to hide on the bottom, largely impossible to do on the smooth sandy bottom off Sable Island. The second hedgehog attack was on target, two bombs clearly exploding earlier than the rest, followed by a surge of oil, air, and wreckage to the surface. Samples were recovered, but the commander of Task Group 22.14 was not satisfied. USS *Menges* was sent to help, and the two destroyer escorts then delivered a series of depth charge attacks using the new USN Mk 8 depth charge. This weapon had a dual-action detonator and could be fired either by water pressure, like a standard depth charge, or magnetically in response to a target's magnetic signature. *U866* was plastered with sixty-five of these new weapons until 1623Z when, forty minutes after the latest depth charge attack, an underwater explosion occurred. *Lowe* and *Menges* held the contact until 2155; TG 22.14 returned the next day, attacked again, and obtained more wreckage. When they were done there was little doubt that *U866* lay shattered on the bottom.[67]

U866 was the only submarine sunk by a warship within the official confines of the Canadian northwest Atlantic zone during the war.[68] With luck, it might well have been a Canadian success, but all the CNA War Diary for the month could claim was that, 'T.G. 22.14 was under operational control of this Command at the time of the attack.'[69] It is significant, however, that the kill occurred in comparatively deep water with good bottom conditions. The USN repeated the feat a month later, a few miles south of the spot where *U866* was destroyed. This kill, of *U369*, occurred just outside the official Canadian zone.[70]

American success off Canada suggests that the USN had mastered the intricacies of ASW and offered role model for the RCN at home. To some extent this is true, and Murray's command moved closer

Victory ... of a Sort 251

to the USN in some of its A/S measures in early 1945. The RCN received the U.S. Atlantic Fleet's memo on 'Bottomed Submarines – Finding and Destroying,' based on a series of trials conducted off New London and Casco, with interest, in part because it wrestled with the use of BT inshore. Significantly, the results confirmed the Canadian view that the value of BT very close inshore, where tide, run-off, and currents conspired to change conditions hourly, was marginal.[71] None the less, in March the RCN also adopted the observant and retiring search plans from the USN's new Fleet Tactical Pamphlet 223A (FTP 223A), *Anti-Submarine and Escort of Convoys Instructions*. These instructions were, in fact, the basis of searches by EGs 16 and 27 when they were operating under TG 22.10 during the hunt for *U866*.[72]

The primary focus of RCN operations in home waters remained – to the very end of the war – the safe and timely arrival of merchant shipping, not a concerted effort to find and sink U-boats. Therefore, not all of the tactical procedures worked out by the USN for its A/S groups were applicable to the Canadian situation. For example, the two navies disagreed fundamentally on the matter of spacing ships during searches. This was a subject of some debate in the spring of 1945. Cdr J. Plomer, the director of the Tactical Unit in Halifax, reviewed the problem at the end of February after the Joint RCN-RCAF A/S Committee recommended that the RCN adopt a closer spacing following the American practice. The USN, Plomer noted, kept ships spaced at one and three-quarters the assured asdic range and limited their zig-zagging. The idea was to ensure that the effective ranges of the asdic of searching vessels overlapped slightly and with a minimum of disruption. The theory was sound. But the Canadians were alarmed at the steady, comparatively high-speed courses the American system required. EG 27's report of proceedings for late December 1944 commented that in one night-time sweep with TU 27.1.1 they had steamed a steady fifteen knots on a straight course in excellent visibility, a practice that seemed designed to invite a GNAT. Moreover, such tightly spaced and closely regulated searches called for fair weather, a well-drilled team, and modern radars with plan position indicators. Canadians, in contrast, wanted to prevent the U-boat from side-slipping the sweep by employing a broad spacing (2–3,000 yards), zig-zagging widely, and streaming CAT gear. This method, Plomer noted, produced 'a tortuous path with which the submarine is unable to conform.' Murray supported the concept of broader, noisier, and more movement-intensive searches, and he was authorized to promulgate this as RCN policy in early April.[73] If the RCN could not sink U-boats with certainty, it could at least suppress their activities. Surviving U-boat logs indicate that the doctrine of harassment was largely successful, and shipping losses, although occasionally dramatic, were minimal.

Not everyone in Halifax was content with this 'holding action,' not least Captain (D), Halifax, who wanted his ships to be sinking U-boats. Indeed, Puxley complained to Murray at the end of March that his support groups desperately needed training. From 'one work-up to the next,' Puxley wrote, 'no opportunity has occurred to exercise with a submarine.' Puxley wanted access to the training submarines at Digby and a derelict submarine sunk as a target as soon as possible. Sinking a derelict sub had already been arranged with the Admiralty.[74] Access to subs at Digby came in early May 1945.

Captain (D) may well have been spurred by the destruction of *U866* right under his nose, but he was also under some pressure from Mitchell, the senior officer of EG 28. Mitchell had served overseas and knew how frequently U.K.-based groups were trained. On 11 April he complained to Puxley that his group had trained together as a unit for only two days since its formation in February, and some ships in the group had never exercised together. More specifically, the senior officer of EG 28 wanted time to train on real submarines.[75]

While Mitchell's request for proper training lay on Puxley's desk, *U190* familiarized herself with the approaches to Halifax harbour. Her captain, Oberleutnant zur See Edwin Reith, was under orders to attack shipping in the Sable Island – Halifax area and had followed Dobratz's boldness by pushing well inshore. Just before dawn on 16 April Reith torpedoed the minesweeper *Esquimalt*, of the Local Defence Force, which was on A/S patrol in the eastern approaches to the harbour. Miraculously, most of the crew escaped

Victory ... of a Sort 253

the blast and the suction of the sinking ship to the dubious safety of two small rafts and a chilling sea. *U190* herself then went to the bottom to wait out the inevitable counter-attack, but for most of the day nothing happened. It was only when *Sarnia*, patrolling further east and scheduled to rendezvous with *Esquimalt* shortly after dawn, returned to port seven hours later that *Esquimalt* was noticed missing. EG 28 and a few other ships were sent to search the area, where *Sarnia* recovered twenty-six survivors; most of the rest – forty-four in all – perished from exposure. The actual search for *U190* concentrated to seaward of the site of the attack and found nothing because Reith had taken his submarine further inshore before laying her on the bottom. Like Dobratz before him, Reith stole silently along the coast once the search moved away. On the day the search for *U190* began, Puxley informed Mitchell that arrangements had been made to use one of the A/S school's submarines for training.

Reith remained in the approaches to Halifax harbour until 29 April and then set course for Norway. Meanwhile, the burden of the action shifted seaward, to the mid-ocean, where intelligence plotted a pack of six U-boats sweeping westward along the convoy routes towards North America. Dönitz was staging his last pack operation of the war, but there was fear that the submarines carried missiles for a desperate attack against American cities. That fear accounts for the massive American forces deployed to attack group Seewolf in early April: four escort carriers and over forty destroyer escorts in two successive barriers. Code-named Operation Teardrop, the barriers caught and destroyed the group. The RCN had to wait until the first USN destroyers arrived in St John's to learn of the destruction of four of Seewolf's six U-boats. If the British needed evidence of the potential of carrier operations against U-boats in transit, Operation Teardrop provided it.[76]

The USN followed-up these successes later in April by the sinking of what was believed to be *U879* south of Sable Island on the 19th and *U881* southeast of the Grand Banks on 6 May.[77] The only U-boat to make it into American waters in the spring of 1945, *U853*, revealed herself by torpedoing a ship off Block Island on 6 May. These were the same waters where TG 22.14 had conducted trials on a bottomed submarine two months before. Not surprisingly, the USN made short work of the U-boat.

The success of American forces in the waters east of Canada in April 1945 was just a sample of the enormous power latent in the U.S. Atlantic Fleet. There seems little doubt that the weight of the anti-U-boat offensive in the western Atlantic would have been borne increasingly by the USN had the war dragged on into the summer of 1945. 'The USN had such enormous forces available,' the RCAF official history concluded, 'that in April and May 1945, without denuding American waters, it could and did maintain nearly as many ships and aircraft in the Canadian zone as the entire naval and air strength normally available to the commander-in-chief Canadian Northwest Atlantic.'[78]

USN forces were around to take the surrender of submarines in the western Atlantic as well. On 5 May Dönitz ordered all U-boats to return to base, but the conditions of the cease-fire reached with the Allies later required him to instruct his U-boats to surface, fly a black flag, and surrender to Allied forces. Five did so off eastern Canada and Newfoundland. The USN Task Group 22.8 from Argentia outraced an RCN force sent from St John's and secured *U805* off Cape Race, and the USN nabbed two U-boats, *U858* and *U1228*, south of Sable Island. EG 28 was dispatched to mid-ocean to intercept *U234*, but she too fell to the USN before the Canadians could arrive. In the end the RCN had to be content with *U889* and *U190*.

The A/S forces themselves deserve credit for inflicting a major defeat on Germany's old U-boat fleet in the dying days of the war. It was, none the less, a very close-run thing. *U2511*, the first type XXI to become operational, sailed from Norway on 31 April, and five more XXIs were scheduled to follow in early May. Forbidden to attack until she reached her war station, *U2511* conducted only a mock attack on an unsuspecting British cruiser before receiving the recall signal. The inshore type XXIIIs fared better. Several were on patrol off the British east coast in April and early May, sinking a number of ships. *U2326*, which failed to hear the recall, sank two ships (one for each of its two torpedoes) off Newcastle on 7 May before returning safely. Others operated with impunity in the Thames estuary right to the end of the war. By all accounts, none of the type XXIII or type XXI U-boats to make operational cruises at the end of the war was ever detected by Allied forces.

The extent to which the victory of the spring of 1945 constituted a defeat of the U-boat fleet remained a controversial subject after the war. Only a few weeks following the end of hostilities Captain Howard-Johnston observed, 'the new U-Boat with new propulsion, the new torpedoes, the new W/T gear, has enabled the U-boat arm to complete this war to all intents and purposes undefeated at sea.' His sentiment was widely shared. Captain Stephen Roskill, the RN's official historian, concluded in the penultimate draft of his history that 'We never gained a firm and final mastery over the U-boats.' That conclusion raised objections within the British cabinet in 1960, when the draft was circulated. Germany recently had been re-armed as a member of NATO, and senior British officials wanted to ensure that there were no illusions about its defeat in 1945. Roskill agreed to soften his conclusion, qualifying it by saying, 'we did not gain so high a degree of mastery as would have forced them to withdraw from out coastal waters – as the heavy losses inflicted in the Atlantic in May 1943 forced them to withdraw from that ocean.' The basis of Roskill's position was simply that on 5 May 1945 there were twenty-five U-boats either inshore or on passage to or from inshore U.K. waters and that these submarines were tying down 400 A/S vessels and 800 aircraft.[79]

Epilogue: Loch Eriboll

The end of the war in Europe brought an end to Canada's A/S war. Plans for Pacific operations required a different kind of expertise, largely an anti-aircraft capability. Even Prentice, ever at the cutting edge of the small-ship war and now commander of the work-up base at Bermuda, threw himself into reshaping the fleet for this new challenge,[1] and with reason. In the spring of 1945 the first of the RCN's new cruisers, HMCS *Uganda*, joined the British Pacific Fleet off Okinawa just in time for the introduction of kamikaze attacks. HMCS *Ontario*, the RCN's second modern cruiser, was en route in the Indian Ocean when the bombs fell on Hiroshima and Nagasaki. The core of this new navy, the two light fleet carriers *Warrior* and *Magnificent*, were allocated for transfer to the RCN in April, as was a flotilla of fleet class destroyers. At the same time the RN began to gather Canadians in the Fleet Air Arm together into nascent RCN squadrons. The Pacific war ended long before this new fleet could be assembled, but the direction of the postwar RCN was clear: it was, in no small way, a rejection of the war against the U-boats.

In retrospect, it was not surprising that the professional RCN turned a blind eye to the A/S war in the Atlantic in 1945. It had striven during the war to acquire the key components of a balanced postwar fleet and to qualify its officers for the necessary posts. Fighting U-boats and guarding convoys were suitable work for reservists, and it was left to them. Even the River class frigates, a large and powerful group of ships, was left in the hands of hostilities-only personnel. Of the seventy frigates commissioned into the RCN, none was commanded by a professional RCN officer. Several were commanded by RCN(R) and RN(R), but even the frigates were quite outside the stream of Canadian professional naval officer development. Such was not the case in the RN, where the principal A/S vessels – the Captains, Lochs, Rivers, and Colonys – were commanded by *RN* officers.[2] The traditional argument advanced by the professional RCN for its lack of presence within the A/S fleet was that there were simply too few RCN officers to go around. There is some merit in that position. By December 1944 there were only approximately 1,026 RCN officers in service: a tiny portion of the navy's total personnel strength of over 92,000. Moreover, most of those RCN officers (7–800) were enlisted during the war and were quite young. The overwhelming majority of officers, roughly 8,000, were reservists of some sort.[3]

But it is also true that the professional service funnelled its own people into 'traditional' naval operations, especially cruisers and fleet class destroyers by 1944–5, in order to qualify them for the postwar fleet. The acquisition of fleet class ships late in the war also probably accounts for the fact that the few RCN officers who were qualified either by training or by experience to command support groups late in the war were posted elsewhere. In fact, remarkably few officers had time to take any qualifying courses during the war, particularly the year-long course required of professional officers. Of the 1,000 professional RCN officers in service in 1945, no more than thirty-five took qualifying courses during the war.[4] Not surprisingly, the overwhelming majority of qualified A/S officers were reservists. Of the 104 A/S qualified officers in the fleet in early 1945, ninety-nine of them were RCNVR – all but a few of whom had taken the short course (a few weeks of specialist training).[5] In the six long years of war only five professional RCN officers qualified as A/S specialists by taking the necessary long course. In the spring of 1945 three of these officers were assigned to fleet class destroyers, one was posted to the base in Bermuda and the other two held shore postings. The Canadian navy had to rely on its reservists, or on British officers, to command its front line A/S forces right to the end of the war. When temporary replacements were needed in EGs

258 The U-Boat Hunters

25 and 26 in 1945, the RCN sent two RN officers from Murray's command to fill the slots; Rowland, formerly Captain (D), Newfoundland, and Aubrey, the training commander at Halifax.

Not surprisingly then, the postwar RCN saw little of itself in the A/S and escort war in the Atlantic. That 'sheep dog' navy had been tied to defensive tasks, which it performed stoically if not always well. Its efficiency was passable, as were its standards of training and equipment. Even the navy's operational research scientists saw little value in studying the small-ship war closely. When Dr J.S. Vigder recommended – without much enthusiasm – in 1949 that the RCN undertake a thorough analysis of its A/S war, the recommendation was rejected by the Naval Staff. The answers to the problem of RCN efficiency in ASW seemed straightforward enough. As Cdr J.V. Brock, the director of Naval Plans and Operations, commented on Vigder's report, the RCN had expanded too quickly from too small a base, its equipment and training were poor, and most of the fleet was tied to convoy escort duty. 'It is agreed,' Brock concluded, 'that little purpose can be served by attempting any further analysis of the relative efficiency of the R.C.N. and R.N. in A/S Warfare in the Second World War.' Rear Admiral F.L. Houghton, now the VCNS, concurred, suggesting that the idea 'be dropped for good.'[6]

Rejection of the A/S war extended even to the personnel involved. In the euphoria of victory in 1945 decorations were distributed widely within Allied armed forces, and the RCN had considerable discretion in where its share went. These honours were awarded liberally to personnel in Tribal class destroyers, D-Day minesweepers, and MTBs; none went to the A/S forces. Even the RCN's two U-boat aces – albeit modest ones – Clarence King of *Swansea* and Prentice were given nothing other than the usual wartime awards for sinking submarines. In Prentice's case, this oversight was probably due to professional jealousy. All he ever received from the chief of the Naval Staff, G.C. Jones, was a comment that 'he was fortunate to have had his opportunities.'[7] No awards were ever made for kills confirmed after the war. It is ironic that the wartime experience so shunned by the professional navy in 1945 would become the raison d'être for Cold War expansion of the fleet and the development of world-class capability in ASW.

Epilogue: Loch Eriboll 259

The rejection of the A/S war by the professional RCN was the result of a combination of two main factors. In the first instance, the close escort and A/S war in the Atlantic was never at any time part of the RCN's long-term ambitions. Planning for the postwar fleet began as soon as the war started. Even the British were well aware – at the height of the crisis in the mid-Atlantic during the winter of 1942–3 – that the RCN was busy positioning itself for postwar battles within the Canadian defence establishment. Murray's warning in 1941 that 'The reputation of the RCN in this war depends on the success or failure of the NEF'[8] and by extrapolation the whole expansion fleet, fell on deaf ears. One can only speculate on the extent to which this preoccupation with building for the postwar while the situation at sea drifted along affected British perception of – and actions towards – the RCN during that dreadful year of 1943. As it turned out, the period from late 1942 to the late summer of 1943 was the only time when the problems of the escort fleet dominated the RCN's agenda. That phase passed with the advancement of plans for the acquisition of major fleet units in the fall of 1943.

The second cause of professional rejection of the A/S fleet was that it was, between 1941 and the end of 1943, a major source of embarrassment for the RCN. For a service that aspired to the rigorous standards of conduct and efficiency of the RN, the Canadian escort fleet demonstrated a casual disregard for propriety and an occasional reckless enthusiasm for getting on with the job with a minimum of pomp and circumstance. In no small way, the professional RCN remained more 'Royal' than Canadian, while the sheep dog navy had a distinctly North American flavour. The escort fleet earned a reputation for zealous inefficiency in 1941, a reputation that stayed with it throughout the war. That legacy affected British assessments of Canadian problems in MOEF in late 1942 and coloured their response to Canadian attempts to modernize the fleet in 1943. The depth of British scorn for what they saw as the bungling enthusiasm of the Canadians is best captured in the postwar writings of Captain Donald Macintyre.[9] In this sense it was cruelly ironic that the last RCN U-boat 'kill' of the war was the result of a simple collision. Perhaps for that reason, but perhaps also because of an order issued by the commander of the RN dockyard at Bermuda forbidding Canadian ships under training from entering the inner harbour because they kept ramming the outer mole, the RCN earned the brief nickname 'The Royal Colliding Navy.'

Certainly no other group of escort vessels has undergone the same penetrating investigation into its operational efficiency as has the wartime RCN. And while it is true that the overall standards of small, inshore escort forces off the Canadian coast were not up to those in the RN's Home Fleet, the same might well be said of other secondary forces elsewhere. But what was lost in all the wartime and postwar debate over what was wrong with the escort fleet was that by 1942 the best of the RCN – its escorts in MOEF and those on loan to the British and Americans in 1942–3 – were really quite good. C groups were passably effective at defending convoys in the mid-ocean by 1942 within the severe limits imposed by their lack of equipment, leadership, and modern destroyers, and given the fact that they escorted the bulk of the slowest and most vulnerable convoys. The British, who took over slow convoys in early 1943, fared no better in defence of them until major improvements were made in the allocation of aircraft and support groups the following spring. It is ironic, moreover, that the same escort fleet was remarkably adept at sinking submarines by 1942–3. Half of the U-boat kills claimed by MOEF in the last half of 1942 went to the RCN, even though they represented only some 35 per cent of MOEF forces. In fairness, the slow convoys put Canadians in contact with the Germans more often and for longer periods, so they ought to have sunk more. But RCN corvettes also claimed five U-boat kills in the eastern Atlantic and western Mediterranean in early 1943 as well. The RN staff at Gibraltar thought very highly of its Canadians. What prevented the RCN from participating in the major slaughter of U-boats which characterized the great Allied victory of 1943 was therefore not inefficiency or lack of equipment per se; it was the legacy of 1941–2 – and the British decision to win the Atlantic war on their own terms.

The Washington Conference of March 1943 awarded the RN command and control over the decisive theatre of the Atlantic war, and it set out with ruthless determination to defeat the U-boat fleet. This it did with the best of its own forces, largely to the exclusion of not only the Canadians, but also the Free French, Belgians, Norwegians, and Poles, who also contributed to the mid-Atlantic war. It is fatuous to argue, as the naval minister did in 1943, that the British ought to have tempered their zeal to defeat the U-boat menace enough to accommodate the ambitions of either the RCN or Canadian politicians. There was both a battle and a war to be won, and a clear path to victory which the British took. The result was a beaten U-boat fleet and a somewhat bruised RCN. The extent to which those bruises affected the long-term relations between the RCN and the RN remains largely for historians to measure. It seems that once the late-war senior staff officers – Jones, Worth, Reid, and others – passed from the scene and were replaced by those, like Grant, who served in British ships during the war, the wounds healed quickly.[10]

The fact that the RCN claimed no U-boat kills between April and November 1943 was therefore no reflection of their ability to sink submarines when given a chance; they simply seldom had the chance. The exception to this general pattern was, of course, the battle of ONS 18 / ON 202. It ought to have produced better results, but there were mitigating circumstances. The escorts were warned early on that they were facing a new weapon, and that expectation probably affected their performance. Moreover, the pace of the action left little scope for deliberate and extensive prosecution of contacts.

Operations over the winter of 1943–4 provide a better measure of the competence of the A/S fleet, and it performed very well. This was true not only of the support groups, but also of the close escorts that participated in the hunts. This trend was carried over into the summer, when EG 11 proved to be the best support group of the campaign in the English Channel and the Bay of Biscay. The RCN stayed with the pace through the fall of 1944 when there was actually little pace to keep up with.

In no small way Canadian success in the Atlantic during 1944 stemmed from the fact that, from the British perspective, the war against the U-boats was in a period of stalemate; the RCN could be left holding the fort. That situation changed significantly at the end of 1944, when the type XXI loomed as a major threat. To a considerable extent the pattern of 1943 reasserted itself in early 1945. The British poured resources into the danger area, and the proportion of U-boat kills by the RCN fell off dramatically. The reasons for this decrease are complex, but are less reflective of peculiarly Canadian problems than of the situation two years earlier. In the first instance, the proportion of RCN support groups in U.K. waters declined in early 1945. EG 11 was withdrawn in November 1944 and then EG 9 collapsed in February, leaving only three RCN groups, EGs 6, 25, and 26, until EG 16 became operational in British waters in April. At the same time the number of British support groups rose to twenty-one.[11] The RN therefore responded to the latent crisis of early 1945 in the same determined manner it had responded to the crisis of 1943, by pouring its own best into the thick of the battle. The pattern was being repeated in the western Atlantic, where RCN forces at home were being eclipsed by the USN.

Also contributing, in a marginal fashion, to declining RCN fortunes late in the war was the issue of seniority of support group senior officers. The extent of this problem is hard to measure, but it was enough of a concern to warrant mention by a number of veterans and there is sporadic evidence in surviving reports. The source of the problem, according to Howard Quinn, seems to have been at sea, not at operational commands. RCN groups were not 'ordered' to serve under more senior RN officers, nor is there any particular evidence that Canadian groups were assigned to less promising areas. Indeed, the evidence of operational cycles that can be derived from summaries of dispositions suggests that all groups were worked extremely hard and that there was little scope for excluding some groups from active areas. Once at sea, however, the situation was sometimes different. It was customary for the RN officer present to organize a search by two or more groups. One would expect, under those circumstances, to find the RN officer's group assigned to the most likely search area. On occasion, as with Quinn's problem with the senior officer of EG 8, some of these officers were pushy. Ironically, the same problem of seniority affected RCN operations in home waters whenever a USN group was present.

Epilogue: Loch Eriboll 263

There is some, albeit very limited, evidence that shortfalls in RCN equipment affected tasking in U.K. waters late in the war. Clearly the best of the British support groups in 1945 were those equipped with both squid and 3-cm radar. It would have been entirely logical to put these ships into the most likely spots, and indeed the success of such groups during the final phase is noteworthy: they were either simply better by virtue of their equipment, or their equipment led them to be placed in better hunting areas. The result was much the same. The first groups to be assigned to the new U-boat concentration west of Ireland in April 1945 were Captain class frigates, ships that were, on the whole, considered by some senior RN officers to be better than anything the British built during the war.[12] Problems of equipment shortfalls suggest that had the war extended into the summer of 1945, another, much less dramatic, equipment crisis would have ensued until RCN groups were refitted with 3-cm radar and type 147B asdic. They would still have been forced to operate largely without squid unless the Castle class corvettes from MOEF were assigned to the support groups. It was just as well, therefore, that the war ended when it did. Of the twenty-eight kills made by surface escorts in the last three months of the war, only three (including one that was entirely fortuitous) were accredited to the RCN.

In contrast to the success of RCN A/S forces in the eastern Atlantic, not a single U-boat was sunk by Canadian warships within the Canadian northwest Atlantic between 1943 and 1945. Indeed, no U-boat was sunk anywhere near the Canadian coast by the RCN during the entire war. Immediately after the war the navy's operational research scientists did a comparative study of A/S operations in Canadian waters with those in British waters in an attempt to determine why forces at home were so ineffective at sinking submarines. Their study assessed only actual hunts, not the larger issues of taskings or the influence of seniority during cooperative efforts. They concluded, not surprisingly, that the real reason why the RCN failed to sink a U-boat in Canadian waters was a general lack of 'experience and tactical training.' There were other factors, among them too few support groups for too big an area. Murray really needed to have two support groups consistently at sea in order to reduce reaction times and start hunts while the scent was still hot. The RCN also needed to develop shore-based operational staffs to coordinate searches, much as the RN did in its inshore campaign. The assessment also mentioned – briefly – the impact of poorer asdic conditions in Canadian waters, a subject it might well have dwelt more on. In the end, the RCN's scientists concluded that the navy really ought to have sunk at least one U-boat in home waters during the last years of the war.[13]

With a bit of luck the RCN might have sunk a couple. The jamming of *Norsyd*'s main gun on the night of 7–8 September 1944 undoubtedly saved *U241*, and *U1232* was spared certain destruction at the hands of *Ettrick* on 14 January 1945 by a matter of inches. In a number of other instances RCN hunters were right on top of U-boats, while dreadful asdic conditions kept them as safe as in the 'bosom of Abraham.' It is interesting to contemplate the effect a kill or two in 1942 might have had on the RCN's attitude towards ASW in its own waters. By 1944 the attitude seems to have been one of resignation: nothing worked very well off the Canadian east coast, and it was not worth trying. The overwhelming majority of convoys moved without loss or delay. It was, after all, the RCN's contribution to the close escort of the main transatlantic convoys between North America and Britain that formed the bedrock of the Allied victory generally, and the victory over the U-boats in 1943 and – perhaps – in 1945. In the end, RCN escorts sank, or contributed significantly to the destruction of, eighteen U-boats in the last two years of the war – not bad, considering that the vast majority of its forces were tied to close escort in areas under little threat.

Convoys continued to be escorted and supported through May 1945, in fears of 'rogue' U-boats that had not heard of – or refused to accept – the German defeat. The last A/S hunt by RCN forces in the western Atlantic took place on 30 May, when EG 27 attacked a non-sub. On the other side of the Atlantic most RCN support groups ended the war quietly. EG 25's final report for U.K. waters, submitted on 21 May, concluded by noting that twenty-five convoys were supported, 'all without incident worthy of report.' EG 6, recently transferred to Murray's command, ended the war quietly in Halifax. A reconstituted EG 11 lay in Londonderry waiting to become operational. EG 26 finished the war patrolling in the English Channel, arriving in Portsmouth on 20 May after an uneventful patrol.

It remained for EG 9 to end the RCN's war against the U-boats. The group, now reformed with Layard in *Matane*, along with *Loch Alvie*, *Nene*, and *Monnow*, and now joined by *St Pierre*, sailed from Londonderry on 10 May to help escort convoy JW 67 to Russia. By the 14th the convoy was north of Britain when two U-boats (*U516* and *U1010*), flying black flags and travelling on the surface towards Scotland, were encountered. *St Pierre* and *Matane* escorted them clear of the convoy and then rejoined. Two days later a German convoy was reported east of JW 67, and EG 9 was detached to investigate. What Layard discovered was a formation of depot ships, auxiliaries, and fifteen U-boats in two long columns moving down the Norwegian coast.

A boarding party from *Matane*, led by her executive officer, Lt J.J. Coates, RCNVR, revealed that EG 9 had intercepted the remnants of the German Arctic and Barents Sea command under Kapitan Subren. They were on their way to Trondheim under the conditions of the surrender and as ordered by German Naval High Command. Boarding parties from other frigates which checked a number of the U-boats confirmed that they had complied with the terms of surrender: all ammunition had been landed, no mines were carried, and all the pistols had been removed from the remaining torpedoes. Through Coates Layard then ordered Subren to instruct his U-boat captains not to attempt to scuttle themselves. Such an act would contravene the terms of surrender, Layard informed Subren, and the U-boat 'would be destroyed and no mercy would be shown its crew.' Layard ordered Subren and the surface ships to proceed to Trondheim while EG 9 escorted the fifteen U-boats to Scotland. Subren protested that this action contravened his orders, but Layard was insistent and refused to provide Subren with either his name or his rank. The German captain was given *Matane*'s pendant number and told that the change in plans would be passed through Allied command. With that assurance Subren relented. Command of the U-boats was passed to *U278*, the boarding party –

which had been treated 'with extreme courtesy, but no show of friendliness' – returned, and EG 9 set off for Loch Eriboll with its fifteen prizes.

The 500-mile passage to Scotland was largely uneventful. Layard suspected some fabricated mechanical problems in several of the U-boats and sent out a curt warning that any mischief would be handled by a long postwar detention before repatriation to Germany. This signal, he wrote, 'seemed to have a good effect.' None the less, *St Pierre* was forced to escort two U-boats into the Shetlands for fuel, and so, when, at 1915Z on 19 May, EG 9 arrived at Loch Eriboll, it had thirteen U-boats in tow. The U-boats were handed over to the local senior officer, although several that were unable to anchor were berthed alongside the ships of EG 9.

By all accounts, EG 9's personnel had acted with complete propriety and did not resort to looting either the U-boats or their personnel. This was a point of some pride within EG 9,[14] but the 'sharks' in Scottish ports were not so restrained. As *Monnow* lay in Loch Eriboll in a dense fog, one of 'its' U-boats was boarded by RN personnel from a nearby ship. The RN motor boat then lost its way the fog, and when its engine failed, the boat drifted onto *Monnow*. The boat's crew was anxious to get away, and when they were refused permission because of the fog, they insisted on posting a guard over the boat for the night. The next morning the engine was repaired and the boat set on its way before personnel from *Monnow* discovered that one of its charges had been stripped clean – including the case of brandy that had been duly noted as being aboard. When it was learned later that the wayward motor boat carried the loot, *Monnow*'s people were furious, but by then it was too late to take action.[15]

EG 9 departed that day, 21 May, with five of the U-boats under escort for Loch Alsh, where the submarines were once again handed over to local authorities. By that point the Canadians had been travelling companions with their erstwhile enemies for five days. The local naval officer in charge at Loch Alsh censured the Canadians for fraternization. Layard admitted the charge had merit, but he noted that the German crews accepted the terms of surrender 'willingly and without question,' and that it was 'indeed difficult to prevent English or Canadian personnel from displaying small acts of kindness and goodwill.'[16] Perhaps being charged with showing kindness towards those who had also endured the rigours of the Atlantic campaign was a fitting way for reservists to end their war.

Appendix I
Escort Building and Modernization, 1941–3: The Technical Background to the Crisis of 1943

From 1941 until the end of 1943 the bulk of the RCN's war against the U-boats was carried by the seventy-odd corvettes of the first two RCN wartime building programs. Apart from a small batch of 'Revised Patrol Vessels' commissioned in 1942, which embodied many of the structural improvements needed for ocean escort duty, most RCN corvettes sailed on through the middle years of the war little changed from their original form. Their short forecastle, mercantile style bridge, primitive electronics, and weapons systems were ill suited to modern war. Only their seakeeping and endurance allowed corvettes to operate in the broad reaches of the Atlantic.

By the time the RCN became involved in oceanic escort of convoys in 1941 with the formation of NEF, the British were well advanced in modifying their corvettes for ocean work. This procedure involved a complete rebuilding of the forward portion of the ship to improve seakeeping and habitability, the reconfiguration of the bridge to naval standards, and the addition of modern weapons and sensors. The RCN was aware of these modifications at the end of 1940, while the bulk of its first corvette program still lay in the builder's hands. Whether the navy seriously considered major modifications to its corvettes at that time remains a mystery. Had they done so, the ships would have been better suited to the longer patrols, larger crews, and the more active war that lay ahead. But they would also have been seriously delayed. Not only would it take time to do the actual work, but the complex Canadian system of contracting for warship construction mitigated against changes in design. Interposed between the navy and its contractors lay the government's Department of Munitions and Supply (DMS), which was responsible for the administration of the contracts. Changes in design affecting costs had first to pass the scrutiny of the DMS, who then dealt directly with the shipbuilders. It was not possible under this system for the RCN to respond effectively to changes in design and fittings while the ships were still in the builders' hands.

The issue of making radical changes to corvettes nearing completion or in service was clouded in late 1940 and throughout 1941-2 by the news of newer, better designs, such as the 'Twin Screw Corvette.' They were a direct development from the original Flower class corvettes: 100 feet longer and outfitted with two sets of corvette machinery and double screws. In other respects, including the mercantile method of construction and even auxiliary minesweeping gear, the Twin-Screw corvette was similar to the Flower class. It was designed from the outset for the broad ocean, however, with a speed of nineteen knots, and a length well suited to north Atlantic wave patterns. When it was decided to forgo the minesweeping gear in favour of addition fuel storage, Twin-Screw corvettes obtained a range of 7,200 miles – twice that of the original corvette. So different were these new ships that the chief of the Canadian Naval Staff, Percy Nelles, recommended they be given the new class name of 'frigate,' a term not applied to warships since the age of sail. The term was quickly accepted into use.[1]

By April 1941 NSHQ had enough information to begin exploring the possibilities of frigate construction in Canada. It was soon apparent that the new ship was too large to build on the Great Lakes, where about one-third of the corvettes were built. It was also more expensive, took longer to build and required a larger crew (107, compared with about fifty for the corvettes by early 1941) than a corvette. 'In spite of these impediments,' Tucker concludes, 'the advantages of the frigate were so great as to be decisive.' In May 1941 the Naval Staff decided that when current corvette contracts were completed, all future escort building would be limited to frigates. By the following October twenty were authorized for the 1942 building program,[2] and changes from the original Admiralty design, most notably the abandonment of minesweeping gear in favour of increased fuel storage, were adopted by the RCN.[3]

This was a modest start, to be sure, towards a fleet designed and built for oceanic operations, but the RCN could not make the switch quickly. Much of the building space for large ships was already under contract to the British Ministry of War Transport for merchantship construction, and the British Admiralty Technical Mission (BATM) had tied up other yards with orders for ten frigates for the RN. These foreign contracts could not be set aside. Both the RCN and Canadian shipbuilders needed the BATM's technical assistance, particularly its expertise in ship construction. Moreover, the financing of the BATM projects was structured to help Canada in its balance-of-payments deficit with the United States, which was seriously in arrears by 1941. The British frigates, for example, were purchased from the shipyards by War Supplies Ltd, a Canadian crown corporation, were sold to the U.S. government, and then were handed over to the British under lend-lease agreements.[4] As a consequence, the British frigates were ordered early, had priority, and were the first of their kind completed in Canada. Moreover, merchant shipbuilding had priority over all RCN construction.

Finding room for Canadian naval requirements in Canadian shipyards was therefore not an easy task, and few frigates were ordered during 1941. Contracts were finally let in 1942, when forty were ordered: thirty for the RCN and ten for the RN. The RCN wanted to have twenty frigates in commission by March 1943[5] and all its frigates delivered by November 1943. Had the navy been able to achieve this goal, the story of 1943 might have been radically different. For the reasons mentioned above, however, priority was assigned to the British ships, and the RCN was promised that only five would be forthcoming by the end of 1943. As Tucker observed, this was a most 'distressing situation,' since it would leave the RCN with only five frigates 'three full years after word of the new design had reached N.S.H.Q., and two-and-a-half years after the first orders had been approved by the Naval Staff.'[6] In the end, the RCN ordered many more frigates and moved them into service faster than this dreary forecast promised, but not soon enough to affect the RCN's role in the great battles of 1943.

Frigates were the obvious solution to oceanic operations, and the pursuit of arrangements to build them in 1941 left a noticeable gap in RCN escort orders during that year. Thus there was also a delay in the ordering of corvettes of improved design. By the end of 1941 yards in Canada were idle at a time when the Japanese attack on Pearl Harbor and the outbreak of war between Germany and the United States changed the war into a global conflict. The requirement for warships of all types was high, and it was unthinkable that Canadian shipyards should be left wanting work. The problem lay in deciding what to build. More frigates were ordered, but many shipyards could build only smaller vessels. The issue of building more corvettes was a vexing one. By early 1942 the British had all but abandoned corvette construction in favour of new, larger designs. But as the war spread into the western hemisphere, the need for coastal escorts increased and the Naval Staff authorized the building of additional 'revised, increased endurance' corvettes in early 1942.[7] These corvettes were 'intended primarily for escorting coastal convoys,' and were part of a policy, adopted in 1942 at the height of the Axis tide, 'that every shipbuilding yard should be worked to capacity until we see light ahead.'[8]

The Naval Minister, Angus L. Macdonald, was opposed to renewed corvette construction in early 1942 because he considered the whole class of warships obsolete. But corvettes of the 1942-3 program were anything but obsolete. All of the lessons of wartime experience were pulled together into what came to be known as the 'increased endurance' (IE) corvettes. The improved sheer and flare of the earlier revised corvettes formed the basis of the new ships, but the bridge was a full deck higher and was built from the outset to naval standards. The platform for the 4-inch gun was also raised, to clear spray and provide a better field of fire, and it was connected directly to the wheelhouse level of the bridge by a wide platform. The larger platform not only eased access to the gun and the ready crews to shelter in poor weather, but the new ahead-throwing anti-submarine weapon 'hedgehog' was mounted there as well. By augmenting the fuel storage in the design, the IEs doubled the range of the original design to 7,400 miles at ten knots. This alteration and the hull improvements made these late-war corvettes excellent ocean escorts, and it was in this role that the RCN eventually used them.

The decision to order more corvettes from Canadian yards in early 1942 was followed by a review of shipbuilding capacity, prompted in good measure by the desire to find a way to build as many frigates as possible. The review revealed two basic facts: Canada had more capacity – overall – than anticipated, but frigates could not be built in the Great Lakes yards because they were too long to pass through the existing locks on the St Lawrence River. Concentrating frigate construction in yards outside the Great Lakes left surplus capacity in Ontario, where only small ships could be built. The RCN preferred to concentrate on the new 'Algerine' class minesweeper, a design that incorporated two engines (compared with the corvette's one). Switching some yards engaged in corvette production to Algerines promised lengthy delays, however, and to avoid that situation, to keep these yards busy, and to get ships to sea, a small program of IE corvettes – twelve ships – was ordered from the Great Lakes as the 1943-4 program. Thus the final decisions about corvette acquisition were dictated more by the limitations of Great Lakes yards and St Lawrence locks than by the assessed needs of the navy.

The debate over building programs after the end of 1940 reveals that the RCN recognized that its initial corvette program was inadequate and that the solution to the problem of outdated ships was to build *new* ones. With hindsight it would have been better had the RCN moved directly into building improved corvette designs as soon as the original corvette program was completed. Failure to do so resulted in a major gap in escort building, which left the RCN to struggle through 1943 with precious few modern ships. The other solution open to the navy in the winter of 1941-2, and the one forced upon it hurriedly by 1943, was modernization of the existing fleet. The crisis over modernization was also affected by decisions concerning new building programs, and by bleak Allied fortunes in the early half of 1942.

Appendix 1

Already well behind in modernization, the dilemma faced by the RCN in early 1942 was whether to start the process on seventy ships now considered obsolete or to wait for the new types. Developments at sea complicated the issue. The requirement in early 1942, as commitments multiplied, was for ships of any type to expand the convoys that were the bedrock of Allied trade defence. Whatever the failings of the RCN's senior staff, they understood that the surest way to defend shipping was to organize it into convoys, provide it with a minimum of air and naval escort and route it clear of danger. Their choice in 1942 was unquestionably correct. But the decision not to proceed with modernization in early 1942 and to wait on the arrival of new ships exploded upon the Naval Staff a year later. Senior RCN officers would be vilified by their sailors as short-sighted and unprofessional and charged by their own naval minister with bungling incompetence.

In early 1942 it was simply better to wait for the new ships than to try to fix the old ones. Few of the resources needed to undertake a modernization program were within easy grasp of the RCN. The navy never planned major modifications to its first corvettes, and little had been done to develop the infrastructure needed.[9] The Ontario shipyards that built many of the ships and needed work were inaccessible to a fully equipped and ballasted corvette. The BC yards were many weeks of steaming away. American yards were out of the question: the government restricted all work in the United States to no more than $10,000.00. And of course all work was subject to a tendering process administered by the Department of Munitions and Supply.[10]

The work of modernizing the fleet would have to be done on the east coast and it offered little in early 1942. While new shipyards sprouted like mushrooms around the Great Lakes in 1940–1, none appeared in Atlantic Canada. Only three – *Amherst, Moncton,* and *Sackville* – of the hundreds of escorts built for the RCN during the war were ordered in the Maritimes.[11] Ship-repair resources on the east coast belonged to the director of Ship Repair and Salvage, an arm of the Department of Munitions and Supply. His efforts were devoted entirely to civilian and foreign repair and salvage work, while the Department of Munitions and Supply controlled access to materials, mainly steel and brass. Most of this work was concentrated in the major ports of Halifax and Saint John. Shipyard services outside Halifax and Saint John were understaffed, poorly equipped, and badly coordinated. Small Maritimes yards, it is true, did annual refits of escort vessels, but they lacked the manpower and resources to undertake major structural changes. The RCN could not create, of its own will, the needed modernization facilities on the east coast, nor would it spring fully formed from the existing arrangements had the RCN desired. Canada had built a fleet that fit well into the government's industrial plans, but no one had developed the infrastructure to maintain it properly, let alone completely rebuild it.[12]

If these problems were not enough to dissuade the RCN from pursuing modernization in early 1942, the task itself was. The corvettes were so far behind in both equipment and structure that little improvement was possible short of a major reconstruction. The keys to modern anti-submarine warfare were three: gyro compasses, asdic type 144, and hedgehog. Installation of these pieces of equipment required the complete rebuilding of the bridge and forecastle of the ships and the installation of a low-power system to run the asdic and the firing control, night sight, and illumination circuits of the hedgehog.[13] It was possible to add these weapons without major structural changes, but there was little point in doing so if the ship would then have to be torn apart to extend the forecastle and rebuild the bridge later.[14]

All of these problems confronted staff planners in early 1942 as the war itself spiralled out of control. A request in March for staff policy on modernization languished until the summer. At that time the director of Naval Construction (DNC) was cool to the idea, in large measure because of the complexity of the task, the inability of Canadian yards to cope with the work, and the difficulty of obtaining the needed equipment (most of which had to be obtained from the British). The rest of the Naval Staff lacked enthusiasm as well. Even the director of Anti-Submarine Warfare recommended caution. As the DA/SW pointed out, the new A/S mortar hedgehog – the subject of modernization debates in 1942 and 1943 and about which the Canadians then knew little – was due to be superseded by a better weapon (squid). Far better to wait until British modernization policy was clearer, and meanwhile all the latest equipment was authorized for new construction.[15]

The most the DNC was prepared to authorize in June 1942 for ships already in service were improvements that could be made with little effort. These included converting the type 123A asdic to 123D by adding another standard magnetic compass binnacle and bearing indicator on the open bridge ahead of the asdic hut (so the captain did not have to run inside during an action to get his bearings), and a loudspeaker system so the asdic operator could pass his information to those on the bridge. These improvements had already been carried out on five of the British-owned corvettes operated by the RCN and were incorporated in the designs of the revised corvettes (just entering service). Other simple improvements authorized by the end of the year included the construction of a housing behind the bridge for the new 10-cm radar set, the type 271, and the movement of the radar office from the former officers' head to the bridge. The fitting of 20-mm oerlikon guns on the bridge wings and resiting the mast astern of the bridge where it would not obstruct the forward search of the 271 radar could also be done piecemeal.

Rather typically, the Naval Staff decided to wait to hear from the British before committing the navy to a wholesale program of modernization. An Admiralty anti-submarine mission, which visited Ottawa in early July following a trip to the United States, convinced the RCN of the need to modernize its corvettes. The Naval Staff then approved in principle the acquisition of the equipment necessary to modernize the ships.[16] It was estimated that the program would take two years to complete and cost approximately $2 million.

As an interim measure the Naval Staff further agreed on 20 August 1942 that corvette asdics should be immediately upgraded to type 127DV, 'to enable our corvettes to work in company with Admiralty corvettes.' As a result, asdic – and modernization – policy got into something of a muddle. The type 127DV was a radical improvement, as the staff knew, and a major step towards a complete type 144/145 set. With the type 127DV the training of the asdic dome under the hull was electronic (the type 123 set was turned manually by a handwheel attached to a bowden wire), the set required the addition of a large recorder and bearing plot, and because the oscillator itself was stabilized in azimuth (which held the asdic bearing regardless of the yawing of the ship), the set also required a gyro compass. The RCN's DNC believed that it was possible to add the gyros and modernize the sonar without the addition of a low-power system.[17] Since the LP system would ultimately be needed for the type 145 asdic and the hedgehog, however, it was prudent to proceed with its installation in conjunction with conversion of sonars to type 127DV. There is no evidence in the Naval Staff minutes that the problem of rewiring before structural changes were made was seen as an issue.

The admission that Canadian corvettes, with their poor asdics, could not work in company with other escorts was a considerable indictment of the class and of the qualities of the escort fleet. The Naval Staff's endorsement of the modification of asdic from type 123A to type 127DV, as a piecemeal move towards modernization, suggests a poor understanding of the tasks ahead. The decision in August 1942 to move towards the type 127DV asdic as an interim step threw into question the earlier decision to upgrade the corvette's type 123A set to 123D. Eventually, in early 1943 the staff rescinded its decision to proceed with type 123D conversions, although the fitting of additional binnacles and loudspeakers was allowed.

The type 144 and 145 asdics were really derivatives of the best prewar RN asdic, the type 128, a fully gyro-stabilized set with range recorder and bearing plotter. Apart from refinements in the display of data on the instruments, the major difference between the type 128 and the 144/145 sets was that the latter included fire control for the new 'ahead-throwing weapons' like hedgehog. The new system required a complete rebuilding of the bridge to carry a new sonar hut with sufficient space for the sonar equipment, a plot, the A/S officer, and two operators. Anti-submarine searches would now be controlled from the hut by the A/S officer, who had to have good communications with the conning position – or pilotage – portion of the bridge. A gyro compass system was central to the new equipment and organization. In addition to its role in controlling the asdic set, gyro compass repeaters were to be fitted in the conning position, wheelhouse, captain's position on the bridge, asdic-directing gear compartment, and the radar and direction-finding offices.

Other extensive and important improvements, particularly the fitting of new weapons and radar, were hindered in late 1942 by basic administrative roadblocks. Canadian changes to the original corvette design led the RCN to restrict the adoption of some British improvements, such as heavier secondary weapons, until stability trials could be conducted. Therefore, 'alterations and additions' authorized by the British for their corvettes had first to be authorized for Canadian escorts by the RCN 'A&A' staff in Halifax. Newfoundland-based ships operating against the Wolf Packs in the mid-Atlantic in 1942, the ones most in need of modernization, routinely called in British ports where modifications could be done. However, MOEF ships could obtain permission to undergo the modifications only by first visiting Halifax. Until this administrative roadblock was removed at the end of 1942, MOEF escort captains could obtain improvements only surreptitiously.

Although orders for the equipment needed to modernize the ships were completed by the fall, the Naval Staff still lacked enthusiasm in late 1942. It was only at the end of the year that signals went out to east coast commands to report on the extent of fitting of modern equipment in the fleet, and the full details of the situation were not available to Ottawa until the spring of 1943. Meanwhile, work was authorized to begin on one corvette, as a trial, in order to ascertain details of costs and time involved. This lead ship was to get the full treatment, but finding the time, place, and resources to tackle even one ship proved insurmountable in 1942. The staff hoped to start with *Saskatoon* in August, but the navy, Halifax Shipyards, and the Department of Munitions and Supply were unable to find the small amount of steel need to complete the job and the lead-ship project languished. *Sackville*, the next ship earmarked, went off to her regular scheduled refit in early January 1943 without going in for modernization. It was not until *Edmundston* was taken in hand by HM Dockyard, Halifax, on 4 January 1943 that the modernization program effectively started, six months after the Naval Staff had agreed on the need. The second corvette to begin modernization, *Calgary*, went into a yard at Cardiff, Wales, in early January as well. *Calgary* was chosen because she suffered from serious mechanical defects that required extensive dockyard repair; it made sense to modernize her in the process. As a revised patrol vessel *Calgary* required little structural change and she was back in service by April.

With hindsight it is easy to fault the staff for its vacillation and indecisiveness during 1942, but two considerations weigh in their favour. First, although the expansion fleet was poorly equipped for modern anti-submarine warfare virtually from the day it went to sea, it was none the less adequate for the tasks assigned to it. The convoy system was implemented and it ran effectively. It was numbers of escorts, not their quality, which allowed that to happen. Second, the urgency over modernization in early 1943 was a direct result of the embarrassing removal of Canadian ships from the decisive theatre of the war just as the battle was joined. This crisis arose overtly only at the very end of 1942, when it dawned on senior RCN officers that steps had to be taken to keep its old corvettes in the fight until the new ships arrived.

The result was that 1943 – the year of crisis and victory in the Atlantic – represented for the RCN something of a hiatus in its escort fleet development program. It was bitterly ironic that this gap occurred at the moment when the issue at sea was decided in the most dramatic fashion.

Appendix II
U-boat Kills by the RCN, May 1943 – May 1945*

Date	U-boat	Ships/Units	Position
13 May 1943	U753	Drumheller with 423 Sqn RCAF and HMS Lagan	48.37° N, 22.39° W
20 Nov 1943	U536	Calgary, Snowberry, and HMS Nene	43.50° N, 19.39° W
8 Jan 1944	U757	Camrose and HMS Bayntun	50.33° N, 18.30° W
24 Feb 1944	U257	Waskesiu	47.19° N, 26.00° W
6 Mar 1944	U744	Gatineau, St Catharines, Chilliwack, Fennel, Chaudière, HMS Icarus, and HMS Kenilworth Castle	52.10° N, 22.37° W
10 Mar 1944	U845	St Laurent, Swansea, Owen Sound, and HMS Forester	48.20° N, 20.33° W
13 Mar 1944	U575	Prince Rupert, USS Hobson, USS Haverfield, aircraft from USS Bogue, and 172 and 206 Sqns, RAF	46.18° N, 27.34° W
14 Apr 1944	U448	Swansea and HMS Pelican	46.22° N, 19.35° W
22 Apr 1944	U311	Matane and Swansea	52.09° N, 19.07° W
24 June 1944	U971	Haida, HMS Eskimo, and 311 Sqn (Czechoslovak)	49.10° N, 05.35° W

6 July 1944	*U678*	*Ottawa* II, *Kootenay* and HMS *Statice*	50.32° N, 00.23° W
18 Aug 1944	*U621*	*Ottawa* II, *Kootenay*, and *Chaudière*	45.52° N, 02.36° W
20 Aug 1944	*U984*	*Ottawa* II, *Kootenay*, and *Chaudière*	48.16° N, 05.33° W
31 Aug 1944	*U247*	*Saint John* and *Swansea*	49.54° N, 05.49° W
11 Sept 1944	*U484*	*Dunver* and *Hespeler*	56.30° N, 07.40° W
16 Oct 1944	*U1006*	*Annan*	60.59° N, 04.49° W
27 Dec 1944	*U877*	*St Thomas* and *Edmundston*	46.25° N, 35.38° W
16 Feb 1945	*U309*	*Saint John*	58.09° N, 02.23° W
7 Mar 1945	*U1302*	*Thetford Mines*, *Strathadam*, and *La Hulloise*	52.19° N, 05.23° W
20 Mar 1945	*U1003*	*New Glasgow*	55.25° N, 06.53° W

* Information courtesy the director of history, NDHQ, revised to April 1993.

Notes

PREFACE

1 The sensitivity of this subject is evident in the unpublished memoirs of the captain, G.H. Roberts, the justly famous director of the Western Approaches Tactical Unit. His commentary on the 1944–5 period was limited to *two* sentences, as follows: 'In 1944 and 1945, for a year, every U-boat using the Channel with any effect was sunk, but the problems of finding them on the bottom were numerous and very tricky. These problems are *still* [emphasis in the original] pertinent and little or nothing should be said about them here.' See his papers in the Imperial War Museum (IWM), London, England.

PROLOGUE

1 For a detailed discussion of the inshore campaign in 1942 see Hadley, *U-Boats against Canada*, passim.
2 *The Montreal Gazette*, 1 Feb 1943, clipping in the Mackenzie King Papers, National Archives of Canada (NAC), MG 26, J5, v 73
3 See M. Hadley and R. Sarty, *Tin Pots and Pirate Ships: Canadian Naval Forces and German Sea Raiders, 1880–1918*, pt three.
4 'Defence of Trade,' CNS memorandum, 12 Feb 1937, NAC, MG 27, III, B5, v 37, D-26
5 DNO to ACNS and CNS, 15 Aug 1940, Directorate of History (DHist), National Defence Headquarters, Ottawa, NHS 8000, HMCS *Captor*
6 In fact, the debate over concentration on defensive over offensive measures never wholly subsided, and some heretics were tolerated, such as Captain Johnny Walker. He endured because he could sink submarines, although often at heavy cost to his convoys.

7 For a detailed discussion of the NEF and its operations in 1941 see *North Atlantic Run*, chaps 2-3.
8 *Escort-of-Convoy Instructions*, Nov 1941, part V, Primary Duties of Escorts, Naval Historical Centre (NHC), Operational Archives Branch, Washington, DC
9 Many of J.D. Prentice's ideas about screening, irregular zig-zagging, distant patrols, and the like are mirrored in Lantflt 9A, and – as one might expect – in the orders for USN escort groups. See, for example, the orders for group A 3 in 1942; Annex Baker to COMESCORT Unit 24.1.3 OPPLAN No. 6-42,' NHC, Heinneman Papers, Box 4.
10 The tactics, and Prentice's enthusiasm for corvettes, are outlined in his letter to Tucker of 15 Jan 1947, DHist, NDHQ, NHS-8000, *Chambly*. His tactics are also discussed in *North Atlantic Run*, 75-6.
11 A draft narrative on operational training prepared by the naval historians refers to such training in 1942, but it clearly describes refresher training provided to operational ships in specific areas. From very early on the RCN provided sound basic training. Asdic operators were trained on shore-based simulators and on 'tame' submarines, gunnery specialists spent time on ranges, and so on. But operational training, as understood in the RN and subsequently by Canadians, was never grasped effectively by the RCN during the war.
12 Capt. S.W. Roskill, *War at Sea*, I, 475
13 CO, HMCS *St Laurent* to Captain (D), Newfoundland, 18 Oct 1941, and related correspondence in NAC, RG 24, 11938, NSS 8440-2, v I
14 Prentice to Tucker, 15 Jan 1947
15 HMCS *Chambly*, ship's log, NAC
16 The Canadians, Chalmers wrote, adopted a cowboy form of convoy defence, keeping a portion of the escort to ride 'herd' on the convoy while others operated at a distance to stir up trouble. Prentice easily identified himself in Chalmers's remarks and complained that Chalmers had entirely misunderstood both his background and his tactics. He had indeed been born on a ranch, but he left it in 1912 to join the RN. As for the escorts Chalmers claimed were intended to 'keep 'em well stirred up,' Prentice noted that his practice of keeping ships at visibility distance from the convoy was intended to keep U-boats down. Stirring up trouble Prentice 'considered to be the job of support groups.' See Prentice's letter of Nov 1955 to the Royal United Services Institute *Journal* in DHist, NHS 8280, B, v 1a
17 *North Atlantic Run*, 129
18 For a detailed account of these actions (except that of *Morden*), see *North Atlantic Run*, chaps 4-6.
19 Minute of Captain (D), Newfoundland, of 24 Aug 1943, appended to report

Notes to pages 14-26   285

from the Commanding Officer, HMCS *Assiniboine,* of 9 Aug 1943, NAC, RG 24, 3997, NSS 1057-3-24
20 As quoted in *North Atlantic Run,* 140
21 For a succinct discussion of the tactics of Canadian MOEF groups, in the absence of modern equipment, see M. Milner, 'The Implications of Technological Backwardness: The Royal Canadian Navy, 1939–1945,' *Canadian Defence Quarterly* 19:3, (Winter 1989) 46–52
22 See Hadley and Sarty, *Tin Pots and Pirates,* 292-3.
23 Naval Board Minutes, 5 Oct 1942
24 Secretary of State, External Affairs, to Secretary of State, Dominion Affairs, 5 Dec 1942, NAC, MG 26, J1, v 334
25 The political side of the 1942 inshore campaign is dealt with at length in Michael Hadley's *U-Boats against Canada;* see especially chaps 3 and 4.
26 'Anti-Submarine Warfare Operational Research, Report No. 3,' covering letter dated 21 Apr 1943, NAC, RG 24, 83-84/167, NSS 1670-3-2, v I
27 'OR Report No. 3,' Apr 1943, NAC, RG 24, 83-84/167, NSS 1670-3-2, v 1
28 William Eggleston, *Scientists at War* (Toronto: Oxford University Press 1950), 147
29 Director of A/S memorandum, 22 Dec 1942, in DHist, Naval Auxiliary Ship file, *Culver*
30 'OR Report No. 3,' observed, 'we may reasonably expect intensified U-Boat activity in the area N. of 40N, W. of 40W and especially in the Canadian Coastal Zone.'
31 Secretary of State, Dominion Affairs, to Secretary of State, External Affairs, 17 Dec 1942, NAC, MG 26, J1, v 334, as quoted in *North Atlantic Run,* 194
32 The transfer crisis is dealt with at length in *North Atlantic Run,* chap 7
33 A.L. Macdonald Diary, 23 Dec 1942, Public Archives of Nova Scotia (PANS), Macdonald Papers
34 CNS to FSL, 7 Jan 1943, DHist, NHS 8440-60
35 In fact, the event was entirely unknown until 1979, when the author presented the details in his MA thesis, which formed the basis for *North Atlantic Run.*

CHAPTER 1: THE FORTUNES OF WAR

1 For a more detailed discussion of the Canadian campaign for a separate command see W.G. Lund, 'The Royal Canadian Navy's Quest for Autonomy in the North West Atlantic,' in *The RCN in Retrospect,* ed. J.A. Boutilier (Vancouver: University of British Columbia Press 1982), 229–35; and Douglas, *Creation of a National Air Force,* 548–50.
2 Minutes of the Atlantic Convoy Conference, Washington, DC, 1–10 Mar 1943, DHist

3 Ibid., appendix A to A.C.C. 1, s 5, para f
4 CNA War Diary, July, DHist
5 See M. Milner, *Canadian Naval Force Requirements*, Extra-Mural Paper No. 20, Dec 1981 (Ottawa: Operational Research and Analysis Establishment, NDHQ)
6 'Summary of Naval War Effort,' second quarter of 1943
7 Minutes of the Atlantic Convoy Conference, Washington, DC, 1–10 Mar 1943, DHist
8 The RCN's 'Summary of Naval War Effort' for the first quarter of 1943, prepared at the end of April, records that the British were now to provide MOEF with nine operational groups.
9 As indicated in Roskill, *War at Sea*, II, app J
10 *RCNMR* 16 (April 1943), 66–7
11 The reassessment was done by the British Ministry of Defence; see DHist, 1650-U-boats, 1939–45.
12 'Summary of Naval War Effort,' first quarter of 1943
13 Details of the agreement on support groups in the Minutes of the Conference do not conform entirely with those set forth in the 'Summary of Naval War Effort' for the first quarter of 1943. It is the latter, for example, that mentions W 10.
14 Moore's letter to the First Sea Lord, 8 Mar 1943, as quoted in *North Atlantic Run*, 232
15 For a reprint of the Naval Board memo on staff reorganization, including descriptions of duties, see Tucker, *Naval Service of Canada*, II, app XIII, from which this quote is taken.
16 See the somewhat vague discussion of the organization at NSHQ in ibid., chap 15.
17 See the excellent summary of the gulf issue in Captain (N) Robert H. Thomas, 'The War in the Gulf of St Lawrence: Its Impact on Canadian Trade,' *Canadian Defence Quarterly* 21:5, (Spring 1992) 12–17.
18 Ibid., 15
19 The 'Battle of the St Lawrence' is discussed in detail in Hadley, *U-Boats against Canada*, while the role of the air force is outlined in Douglas, *Creation of a National Air Force*, II, chap 13.
20 Senior Officers met in Ottawa in February to map strategy for the gulf in 1943; see 'Summary of Naval War Effort' for the first quarter of 1943, from which the quote is taken, and the final plan 'Gulf of St Lawrence, GL 43,' 25 May 1943, NAC, RG 24, 11579, NSD 16-59-19.
21 'Gulf of St Lawrence, GL 43'
22 See Flag Officer, Atlantic Coast, War Diary, NAC, RG 24, 11052, NS 30-1-10, v 18.

23 Movement in that direction was supposed to be enhanced by the establishment of the Joint RCN-RCAF Anti-Submarine Warfare Committee in February. But the committee, constituted in haste to help to explain at the forthcoming Washington Conference how the two Canadian services affected cooperation, was composed of comparatively junior officers. It accomplished little until the fall of 1944, when senior RCN and RCAF officers were appointed. See especially the minutes of 19 Sept 1944, NAC, RG 24, 11026, CNA 7-19-2.
24 Douglas, *Creation of a National Air Force*, II, 550–1
25 Graves, 'Royal Canadian Navy and Naval Aviation,' 82–3
26 'Hints on Escort Work,' Captain (D), Halifax, 30 Mar 1943, NAC, RG 24, 11938, 8440-2, V I
27 'Hints on Escort Work - Part IV,' Captain (D), Halifax, 8 June 1943. This part, and by all accounts all parts issued, are found in the Public Record Office (PRO), ADM 1/13749.
28 The RN agreed to promulgate the memoranda and one Minute sheet attached to the British copy described 'Hints' as 'Interesting and Refreshing.' 'The only heresy,' the Minute sheet noted, 'is on page 4,' a reference to Prentice's preference for quick attacks at short range with depth charges; PRO, ADM 1/13749.
29 'Monthly Operational Report for April 1943,' Newfoundland Command War Diary, DHist
30 Nelles reported to the cabinet War Committee on 11 March that the RCN would provide a support group of six corvettes; CWC Minutes, DHist copy.
31 Mackenzie King Papers, NAC, MG 26, J5, V 73
32 Connolly Diary, 7 Apr 1943, NAC, MG32C71, V II. I am grateful to Michael Hennessey for drawing this to my attention.
33 For a detailed account of the passage of ONS 5 see Gretton, *Crisis Convoy*.
34 'Report on operations of support groups,' DA/SW, 15 June 1943, PRO, ADM 199/2020
35 Macdonald Diary, 26 May 1943, PANS
36 *North Atlantic Run*, 239–40
37 'Summary of Naval War Effort,' second quarter of 1943
38 Ibid.
39 Admiralty to CNS, personal 181338Z/4/43, DHist, NHS 8440 Support Groups – General
40 Details on the shipping situation are drawn from the draft of M. Hennessey, 'RCN Modernization, Expansion and Maintenance, 1943–45,' unpublished research report for DHist, 1993.
41 *North Atlantic Run*, 216–17
42 Ibid.

43 As quoted in ibid., 252, which also contains a more detailed discussion of the AA/SSB's activities and their impact on the RCN.
44 'Summary of Naval War Effort,' first quarter of 1943
45 Mansfield to Pound, PRO ADM 1/13746
46 CNA War Diary, June 1943
47 Hennessey, draft, 40
48 Bidwell to Murray, 220644Z/6/43, NAC, RG 24, 11115, 61-1-3, v 2
49 Details of the formation of the CSG taken from the following DHist sources: CNA War Diary, NHS 8440 EG 5/6, Ships' Movement Cards, and NHS 8440 Support Groups – General
50 For a list of support group compositions between March and May 1943 see Roskill, *War at Sea*, II, 367, table 30. Reference to their withdrawal in May comes from 'Summary of Naval War Effort,' second quarter of 1943. See also Admiralty Pink Lists for May 1943, PRO, ADM 187.
51 'The First Year of Canadian Operational Control in the Northwest Atlantic,' Naval Historian's Narrative, DHist. See also Admiralty to NSHQ, 25 May 1943, DHist, NHS 8440 Support Groups – General.
52 C-in-C, WA, to C-in-C, CNA, 071000z/6/43, DHist, NHS 8440 EG 5
53 See Admiralty *Pink Lists* for May-June 1943, PRO, ADM 187.
54 The evidence for this shift remains conflicting. The 5-4 breakdown in favour of the RCN after May 1943 comes from CNA War Dairy. The 'Summary of Naval War Effort' for the third quarter lists RN groups in MOEF as seven, but it does not distinguish between RN close and support groups.
55 See Admiralty Pink Lists for this period, PRO, ADM 187.
56 Roskill, *War at Sea*, III, pt 1, 29
57 See Y'Blood, *Hunter-Killer*, app II, 'Submarine Sinkings by Escort Carrier Groups.'
58 See G. Hessler, *U-Boat War in the Atlantic*, III, 8-21, for a general discussion of the U-boat campaign in the summer of 1943, and for the figures quoted here (save those attributed to Roskill).
59 Roskill, *War at Sea*, III, pt 1, 32
60 For a thorough account of these clandestine operations see Hadley, *U-Boats against Canada*, chap 6.
61 'Summary of Wardroom Grouses,' Lt J. George, RCNVR, [June 1943], DHist, M-11
62 The Piers report is discussed in *North Atlantic Run*, 245-6, the report from Lt Cdr W.E.S. Briggs aboard *Philante* can be found in NAC, RG 24, 11960, CS 34.
63 The signal is referred to in a letter of 9 July 1943 from Captain (D), Newfoundland, A/Capt. J.M. Rowland, to Commodore (D), WA, but an original has

never been found. See NAC, RG 24, 11948. I am grateful to Michael Hennessey for this information.
64 Commanding Officer, HMCS *Assiniboine*, to Captain (D), Newfoundland, 9 Aug 1943, NAC, RG 24, 3997, NSS 1057-3-24
65 Strange to Macdonald as quoted in Macdonald to Nelles, ca 20 Nov 1943, Macdonald Papers, F276/3, PANS
66 Macdonald Diary, 11 Aug 1943, PANS
67 Minister to CNS, 21 Aug 1943, NAC, RG 24, 3995, NSS 1057-1-27
68 Graves, 'Royal Canadian Navy and Naval Aviation,' 84–108
69 Admiralty to NSHQ, 010319Z/7/43 and DOD memo of 21 Aug 1943, DHist, NHS 8440, EG 5
70 'Allocation of 5th Escort Group to C-in-C., W.A.,' DOD memorandum, 21 Aug 1943, DHist, NHS 8440 EG 5
71 Houghton to Bidwell, 18 Aug 1943, NAC, RG 24, 11987, 1292. Again, my thanks to Hennessey.
72 C-in-C, CNA to FONF, 181250Z/8/43, DHist, NHS 8440 EG 9
73 '5th Escort Group – Training Report: Period of Training from 16th to 18th August, 1943,' HMS *Philante*, 23 Aug 1943, DHist, NHS 8440 EG 5
74 *RCNMR* 23 (Nov 1943)
75 Roskill, *War at Sea* III, pt 1, 30
76 L. Burrow and E. Beaudoin, *Unlucky Lady: The Life & Death of* HMCS *Athabaskan 1940–44* (Stittsville, Ont.: Canada's Wings 1982), 42-51 contains a detailed account of the destroyer's action.
77 Roskill, *War at Sea*, III, pt 1, 30

CHAPTER 2: CINDERELLA: ACT II

1 The comment was made in response to a request by Mackenzie King in late December 1943 for a comparison of RCN and RN U-boat kills. See PRO, ADM 199/1937.
2 Admiralty Appreciation of the U-Boat situation, 1 Aug 1943, PRO ADM 223/21, as quoted in Rohwer and Douglas, 'Canada and the Wolf Packs,' 163
3 Edelsten to Cunningham, 11 Sept 1943, Roskill Papers, 5/16
4 Rössler, *The U-boat*, 204–10
5 Hessler, *U-Boat War in the Atlantic*, III, 6–7
6 The Germans had also suspected that the *Metox* receiver itself, which emitted a small signal when operating, was the source of Allied detection, and on 14 August use of that receiver was banned. *Wanze* emitted only about one-fifth of *Metox*'s radiation. Ibid., 4
7 Ibid., 22. On page 24 Hessler notes the apparent success of *Wanze* but then

mentions that the general relaxation of Allied vigilance in the bay was the real reason for declining U-boat losses.
8 Notes on the 'German Acoustic Torpedo' attached to 'Notes on Homing Torpedoes from an Operational Standpoint,' Director of Operational Research Report, 20 Sept 1943, DHist, 81/520/1000-973
9 'German Naval Research Report G 6: German Navy U-Boat Torpedoes, 1939-45: Overview and Descriptive Catalogue,' DHist
10 Hessler, *U-Boat War in the Atlantic*, III, 23
11 Rohwer and Douglas, 'Canada and the wolf Packs,' 165
12 Ibid., 166-7
13 *Defeat of the Enemy Attack on Shipping*, v 1A, 119
14 'Christmas Festival: E.G. 9's First and Last Operation,' RCNMR 24 (Dec 1943) 50-62. 'Christmas' was the radio call sign for the commodore of ONS 18 and 'Festival' that for the commodore of ON 202.
15 As quoted in Roskill, *War at Sea*, III, pt 1, 39
16 As quoted in ibid.
17 As quoted in Rohwer and Douglas, 'Canada and the Wolf Packs,' 175
18 As quoted in ibid., 176
19 Fisher's account and the best detail of the fate of the crews of *St Croix* and *Itchen* are found in the RCNMR 31 (Aug 1944).
20 Rohwer and Douglas, 'Canada and the Wolf Packs,' 181
21 The account of ONS 18/ON 202 is compiled from ibid.; 'The End of H.M.C.S. "*St Croix*" ' RCNMR 20 (Aug 1944); 'Christmas Festival: E.G. 9's First and Last Operation,' RCNMR 24 (Dec 1943); the Admiralty analysis of 15 Nov 1943, found in DHist, 81/520/8280, ON 202; and Naval Historian's Narrative A, DHist.
22 Figures compiled from the contacts listed in Rohwer and Douglas, 'Canada and the wolf Packs,' which are based on a thorough study of both sides.
23 Information courtesy of Oscar Sandoz, who was on the development team for the PNM
24 'Notes on Homing Torpedoes from the Operational Standpoint,' DOR memo, 20 Sept 1943, DHist, 81/520/1000-973
25 This account is drawn from the first 'Interim Report' on CAT gear and differs significantly from that of John Longard, one of the scientists involved in the development of CAT, which appears in his history of the Naval Research Establishment. Longard claims that they settled on a 30-inch rod immediately, but the contemporary documentation indicates clearly that the CAT Mk I used 5-foot rods. The Mk II version, the type finally adopted for general service use at the *end* of 1943, used a 30-inch rod.
26 C-in-C, CNA to FONF, 221248Z/9/43, DHist, 81/520/1000-973
27 'Notes on Homing Torpedoes,' DHist, 81/520/1000-973

28 C-in-C, CNA to FONF 252110Z/9/43 as cited on *Waskesiu* Movement Card, DHist
29 Cdr R. Phillimore, RN, to the author, 20 Aug 1990
30 The evidence for the Canadian decision not to use a paravane is circumstantial but sound. The original drawing of the PNM as submitted to the Admiralty in 1940 shows a small kite just ahead of the yoke (NHB, M-files), and the first 'Interim Report' on CAT gear (see below) refers to an Admiralty signal warning the RCN not to stream PNMs without flotation. The issue of a paravane was also raised in a minute on G.H. Henderson's report on the 'Present State of CAT Gear,' 23 Nov 1943, and dismissed as unnecessary. See the reports in NAC, RG 24 83-84/167, 2050, S6060-3. I am grateful to Mike Whitby for passing them along.
31 As reported in his 'Interim Report' on the 'Development of Canadian Anti-Acoustic Torpedo Gear (C.A.T. Gear),' 7 Oct 1943, NAC, RG 24, 83-84/167, 2505, S6060-3
32 See 'Comments on Admiralty Signal 272125A of Nov. '43 RE "Appreciation of the present Trend of the U-Boat War,"' NAC, RG 24, 11463.
33 The discussion of USN trials is contained in Peers's 'Interim Report No.2,' while the recommendation to stick with the CAT – but cut down as per the USN trials – is contained in G.H. Henderson, 'Present Status of CAT Gear,' 23 Nov 1943, both in NAC, RG 24, 83-84/167, 2505, S6060-3. John R. Longard, one of the scientists who developed CAT, recognized the American gear as a direct copy, because his original model, put together in haste from materials available in the lab, contained oversized bolts. These bolts were copied into the original technical drawings of CAT Mk I, and the same oversized bolts appeared on the American gear. See Longard, *Knots, Volts and Decibels: An Informal History of the Naval Research Establishment, 1940–1967* (Dartmouth, NS: Defence Research Establishment Atlantic 1993).
34 Minute attached to Henderson's memo of 23 Nov 1943, NAC, RG 24, 83-84/167, 2505, S6060-3
35 Roskill, *War at Sea*, III, pt 1, 41
36 For a technical description of the FOXER Mk II, the most common gear in use by 1944, including drawings, see Confidential Admiralty Fleet Order 595, 16 March 1944, DHist. The Mk III gear is described in CAFO 1204, 25 May 1944, and major modifications are listed in CAFOs 1923 (1944), 441 (1945), 467 (1945), and 509 (1945).
37 *Monthly Anti-Submarine Report* (June 1944) 27, DHist
38 As quoted in Terraine, *U-boat Wars*, 640. Terraine considers the ratio of hits on ships streaming PNMs to be high, roughly 10.4 per cent of all GNAT hits, but he errs in assuming that PNMs were the sole defence against GNATs and that they were 100 per cent effective. The mistaken belief that FOXERs and CATs were

fully effective is widely held, an error repeated by Hinsley, who contended that 'No ship towing *Foxer* was ever hit by an acoustic torpedo.' (*British Intelligence*, III, pt 1, 223). More to the point, 387 of 464 GNATs *missed* entirely, and the percentage of GNATs fired that hit a ship streaming a PNM was tiny, 1.73 per cent.

39 They also very sensibly rejected the decision reached by the Admiralty, at the 14 Oct meeting of the U-Boat Warfare Committee, to remove ammunition from the after magazines of warships to prevent it being detonated by a GNAT hit. See NAC, RG 24, 8080, NSS 1271-35.

40 C-in-C, WA, to C-in-C, CNA, 212309Z/9/43, and NSHQ to C-in-C, CNA, 221346Z/9/43, in DHist, NHS 8440, EG 9

41 The account of the fall campaign and the role of support groups was compiled from Rohwer and Hummelchen, *Chronology of the War at Sea*, 'Ships' Movement Cards' held by DHist, and information supplied by John Burgess.

42 See Hadley, *U-Boats against Canada*, 170-5.

43 For an excellent account of this episode see ibid., 179-85.

44 See the memo on 'Operational Research Liaison' from RCAF Overseas Headquarters, dated 17 Nov 1943, which deals with hunts to exhaustion, NAC, RG 24, 11463.

45 The principles of the two types were different, and the American system was more efficient. Sonobuoys are expendable listening devices which deploy a hydrophone on a long wire into the sea. A small radio transmits the sounds heard to the circling aircraft. These first sonobuoys were entirely non-directional, but location and movement of a target could be estimated from the strength of signals received from a pattern of buoys. The American system permitted aircraft to track a submerged U-boat and be on the spot when next it surfaced. Moreover, the RAF's experience with coordinated hunts between aircraft and surface vessels in the Bay of Biscay during the summer was not entirely positive. Minutes of 'Anti-Submarine Warfare Meeting,' RCAF Headquarters, 26 Oct 1943, NAC, RG 24, 11463

46 See C-in-C, CNA, 082205Z/5/44, attached to Minutes of the RCN-RCAF Anti-Submarine Warfare Committee, 25 Apr 1944, NAC, RG 24, 11026, CNA 7-19-2.

47 W.A.B. Douglas, 'The Nazis' Weather Station in Labrador,' *Canadian Geographic*, 101 (Dec 1981 – Jan 1982) 42-7

48 The best summary of the first two Salmons is contained in 'Memorandum on Operation "Salmon,"' an RCN Operational Research report, 24 Dec 1943, NAC, RG 24, 11463. See also Douglas, *Creation of a National Air Force*, 570-71.

49 See King's speeches on the Italian capitulation in NAC, MG 26, J5, v 75.

50 Secretary of State, Dominion Affairs, to Secretary of State, External Affairs, Canada, 5 Nov 1943, NAC, RG 24, 8080, NSS 1271-35

51 Minister to CNS, 25 Nov 1943, Macdonald Papers, PANS, F276/28. For a more detailed discussion of the Macdonald-Nelles controversy see chap 9 of the author's *North Atlantic Run*.
52 As quoted in *North Atlantic Run*, 259
53 See 'Comments on Admiralty Signal 272125A,' NAC, RG 24, 11463.
54 The account is compiled from *RCN-RCAF Monthly Operational Review*, 2:1 (Jan 1944), and *Monthly Anti-Submarine Report*, Jan 1944, DHist. The latter claims that *Nene* delivered the attack that seriously crippled *U536*, but a reconciliation of the two accounts suggests it was *Snowberry*.
55 Figures compiled from Roskill, *War at Sea*, III, pt 1, app D
56 Naval Staff Minutes, 12 July 1943
57 Ibid., 4 Oct 1944
58 Naval Board Minutes, 30 Aug 1943
59 Naval Staff Minutes, 4 Oct 1944, and Naval Board Minutes, 4 Oct 1944
60 Naval Staff Minutes, 9 Nov 1943
61 'Report of Visit to U.K. of Captain W.B. Creery, R.C.N., 24th November - 6th December 1943,' NAC, RG 24, 11128 ('Visit to the U.K.')
62 Tucker, *Naval Service of Canada*, II, 83–4. Elliott, in *Allied Escort Ships*, claims that a total of ninty-seven frigate contracts were cut in Canada, but he is not specific on just when the cuts came, nor does he make it plain how many of them were on order for the RN.
63 Creery Report, 'Visit to the U.K.,' NAC, RG 24, 11128
64 NSHQ to ADM 111530Z/12/43, NAC, RG 24, 3979, NSS 1048-48-129
65 Elliott, *Allied Escort Ships*, 229
66 Creery Report, 'Visit to the U.K.,' NAC, RG 24, 11128
67 See *Particulars of Canadian Warships*, quarterly report for Jan 1944, DHist, which lists all frigates as having RX/C; this is a statement of policy rather than fact.
68 Naval Staff Minutes, 3 May 1943
69 Ibid., 6 Sept 1943
70 Ibid., 1 Nov 1943
71 Naval Board Minutes, 11 Nov 1943 and 11 Jan 1944
72 Naval Staff Minutes, 17 May 1943
73 The policy of separate SOs was not applied to WEF. Gordon A. Kennedy, a telegraphist aboard *Winnipeg*, the senior officer's ship of W 7 in 1944, recalls that her captain fulfilled both functions.
74 *Waskesiu* Movement Cards, DHist
75 C-in-C, CNA, War Diary, Nov 1943, DHist
76 C-in-C, CNA, to NSHQ, 201536Z/11/43, DHist, NHS 8440 EG 9
77 Ibid.

294   Notes to pages 93–107

78 DOP memorandum, 18 Aug 1943, DHist, NHS 1650-1, Naval Plans
79 Interview with Cdr A.F.C. Layard, RNR, 22 May 1987
80 See notes from *Western Approaches Monthly News Bulletin*, Dec 1943, DHist, NHS 8440 C 2.
81 Roskill, *War at Sea*, III, pt 1, 77
82 C-in-C, WA, to C-in-C, CNA, 021813A/12/43, DHist, NHS 8440 C 2
83 The incident is recorded in the 'Correspondence with Commodore Simpson' file in the Macdonald Papers, PANS.

CHAPTER 3: TRIUMPH WITHOUT CELEBRATION

1 The incident recorded is a discussion between Easton, then in command of *Matane*, and Cdr A.F.C. Layard, RNR, who had come aboard as senior officer of EG 9 in Feb 1944, 213.
2 Howard-Johnston's reply is in the First Sea Lord's Papers, PRO, ADM 199/1937.
3 NAC, MG 26, J 5, V 75
4 C-in-C, WA, to various, 142141A/12/43, NAC, RG 24, 11580, D 23-2-1
5 NSHQ to C-in-C, CNA, 301415Z/11/43, DHist, NHS 8440 EG 9
6 'Summary of Naval War Effort,' fourth quarter of 1943, 18
7 NSHQ to C-in-C, CNA, 112159Z/1/44. NAC, RG 24, 11575, D 01-18-7
8 For a discussion of the Murray-Jones animus see *North Atlantic Run*, 93–4.
9 The boundaries of the search areas are listed in the C-in-C, CNA, War Diary for Jan 1944, DHist.
10 Douglas, *Creation of a National Air Force*, 571–2
11 Hadley, *U-Boats against Canada*, 198
12 DHist, NHS 8440, W 10. The torpedo firing is confirmed in Rohwer, *Axis Submarine Successes*, 177.
13 Easton, *50 North*, 202
14 Macpherson and Burgess, *Ships of Canada's Naval Forces*, app 8
15 Captain (D), Halifax, to COAC, 2 Mar 1943, DHist, NHS 8000, HMCS *Somers Isles*
16 This is a revision of the information found in the author's article, 'HMCS *Somers Isles*: The Royal Canadian Navy's Base in the Sun,' *Canadian Defence Quarterly* 14:3 (Winter 1984–5) 41–7.
17 Macpherson and Burgess, *Ships of Canada's Naval Forces*, app 8
18 Roskill, *War at Sea*, III, pt 1, 74–5, and *RCNMR* (Mar 1944)
19 George Devonshire to the author, 26 Feb 1991
20 *RCN-RCAF Monthly Operational Review* (Mar 1944) 14–16
21 Ibid., 14
22 Report of Proceedings, C 2, 24 Jan 1944, PRO, ADM 199/354
23 *Monthly Anti-Submarine Report*, Nov 1943

24 Naval Staff Minutes, 27 Dec 1943
25 C-in-C, WA, 091102B/4/44, DHist, NHS 8440 EG 6
26 See Movement Cards for *Swansea, Frontenac,* and *North Bay,* DHist, signal cited variously as 282018, 282118, and '28218,' Jan 1944
27 *Western Approaches Monthly News Bulletin,* Jan 1944, DHist
28 C-in-C, WA, to C-in-C, CNA, 271216Z/1/44, DHist, NHS 8440 EG 9
29 C-in-C, CNA, to C-in-C, WA, 081754Z/2/44; C-in-C, WA, to C-in-C, CNA, 091745Z/2/44; and NSHQ to C-in-C, WA, 101947Z/2/44, DHist, NHS 8440 EG 9
30 'Summary of Naval War Effort,' first quarter of 1944
31 Douglas, *Creation of a National Air Force,* 574-5; C-in-C, CNA War Diary, Feb 1944, DHist. For a detailed account of *U845* operations off Newfoundland see Hadley, *U-Boats against Canada,* 199–208.
32 Poolman, *Escort Carrier,* 115–16
33 *Monthly Anti-Submarine Report,* Mar 1944. The British claim for Walker's accomplishment in the U-Boat war was justified, but the laurels in the A/S war went to the USS *England,* which alone sank six Japanese submarines between 19 and 31 May 1944.
34 DHist interview with Captain W. Willson, RCN, 47
35 The simplest evidence for this is the setting tables painted alongside the depth-charge throwers visible in a number of 1942 photos. See for example the photo of a depth charge being thrown from *Pictou* in March 1942 in *North Atlantic Run* (NAC PA-116838) and the starboard throwers of *Kitchener* as taken during the filming of 'Corvette K-225' in July 1942, DHist, *Kitchener* PRF.
36 Rössler, *The U-Boat,* 157
37 Ibid.
38 E.H. Chavasse, 'Business in Great Waters,' unpublished memoir, IWM, 128
39 'Summary of Naval War Effort,' last quarter 1943, 46
40 Admiralty to various, 111910Z/2/44, which refers to an earlier signal on the same subject dated 18 Aug 1943, DHist, 81/520/1000-973
41 *Monthly Anti-Submarine Report* (Feb 1944)
42 See the Report of Proceedings in PRO, ADM 199/313.
43 *RCN-RCAF Monthly Operational Review* (Mar 1944)
44 C-in-C, WA to Secretary of the Admiralty, 21 Mar 1943, PRO, ADM 199/468
45 PRO, ADM 199/313
46 'Excerpt from Minutes of 202nd Daily Informal Discussion of Naval Staff,' 25 Feb 1944, DHist, NHS 8440 EG 9
47 *RCN-RCAF Monthly Operational Review* (Apr 1944)
48 C-in-C, WA's minute sheet attached to C 2's Report of Proceedings, PRO, ADM 199/468
49 DAUD, Minute 11, Mar 1944, PRO, ADM 199/468

50 For an assessment of *Kenilworth Castle*'s attacks see PRO, ADM 217/100.
51 Interview with Rear Admiral P.W. Burnett, RN, 25 May 1987
52 Rössler, *The U-Boat*, 336, gives the surface speed for a type IXC/40 as 18.3 knots.
53 As quoted in Hadley, *U-Boats Against Canada*, 207. For a ringside account of the chase, including a recounting of the radio traffic, see Easton, *50 North*, 222-6.
54 'Remarks by Commodore (D), Western Approaches,' 1 Apr 1944, DHist, NHS 8000, *U-845*
55 Y'Blood makes no mention of special intelligence in his account of *Bogue*'s search that led to contact with *U575*. See *Hunter-Killer*, 151.
56 RCN-RCAF *Monthly Operational Review* (Apr 1944) Y'Blood, *Hunter-Killer*, 151-2
57 Hessler, *U-Boat War*, III, 58
58 See the correspondence in PRO, ADM 1/13698.
59 Naval Staff Minutes, 28 Feb 1944
60 'Deep and/or Fast U-Boats,' PRO, ADM 1/16495
61 Figures on hedgehog from Elliott, *Allied Escort Ships*, 531
62 Naval Staff Minutes, 7 Feb 1944
63 Secretary, Naval Board to C-in-C, CNA, COPC, FONF, etc., 12 Apr 1944, NAC, RG 24, 11575, D O1-18-7
64 NSHQ to SCNO(L) 181920Z/4/44, SCNO(L) to NSHQ 221103B/4/44, DHist, 81/520/1000-973
65 Secretary, Naval Board to RCN Depot and various schools, 23 Mar 1944, NAC, RG 24, 11575, D 1-18-6
66 Naval Staff Minutes, 28 Feb 1944. The Naval Staff was prepared to do what it could to make proven weapons systems work better.
67 C-in-C, WA, to NSHQ 241742B/5/44, NAC, RG 24, 11723, CS 148-1-1
68 Zimmerman, *Great Naval Battle*, chap 9
69 Figures compiled from the fighting equipment reports found in each ship file at DHist
70 Zimmerman, *Great Naval Battle*, 123
71 See comments on the Salmon of 16-26 Mar 1944 in NAC, RG 24, 11022, CNA 7-6.
72 RCN-RCAF Anti-Submarine Warfare Committee, 25 Apr 1944; see the attached signal, dated 082205Z/5/44, NAC, RG 24, 11026, CNA 7-19-2. Reports of the Tactical Unit are found in NAC, RG 24, 11053, 30-1-10.
73 Y'Blood, *Hunter-Killer*, 160-3, and Morison, *United States Naval Operations*, 318
74 RCN-RCAF Anti-Submarine Warfare Committee, 16 May 1944, NAC, RG 24, 11026, CNA 7-19-2
75 See C-in-C, CNA to various, 281557Z/3/44, NAC, RG 24, 11022, CNA 7-6-1.
76 DHist, NHS 1000-5-13, V 21
77 Operational status as given in Macpherson and Burgess, *Ships of Canada's Naval Forces*, app 8

78 Ibid.
79 The base was eventually established at St Georges and opened officially on 1 Aug 1944. See Milner, 'HMCS *Somers Isles.*'
80 For a detailed description of the sinking of *Valleyfield* see Hadley, *U-Boats against Canada*, 208–20.
81 Douglas, *Creation of a National Air Force*, 579–80
82 As quoted in Milner, 'Inshore ASW: The Canadian Experience,' 151
83 RCN-RCAF *Monthly Operational Review* (June 1944)
84 See Commodore (D), Western Approaches, Minute sheet, 2 May 1944, in PRO, ADM 199/469.
85 The British assessments are in PRO, ADM 199/469. Layard's harshest critic was Admiral Murray, who chided him for not delivering a deliberate attack when *Matane* had the chance. See the assessments by DWT in DHist, NHS 8440 EG 9
86 Rohwer and Hummelchen, *Chronology of the War at Sea*, 272–3; DWT assessment, DHist NHS 8440 EG 9
87 Minutes of the '55th Meeting in the First Lord's Room to Consider Trade Protection,' no date, but judging by surrounding evidence circa late March 1944, NAC, RG 24, 8080, NSS 1271-20

CHAPTER 4: THE SUMMER INSHORE

1 Dönitz, *Memoirs*, 395
2 Rössler, *The U-Boat*, 198–204
3 Hadley, *U-Boats against Canada*, 233–4
4 Hessler, *U-Boat War*, III, 67
5 Dönitz, *Memoirs*, 396
6 See 'Overinsuring versus Underinsuring Overlord from the U-Boat and W-Boat threat,' 30 Mar 1944, 'Notes on A/S Aspects of Overlord,' by L. Solomon, 23 Mar 1944, and 'The Case for Increasing the Number of A/S Escorts Employed in Overlord,' 30 Apr 1944, in Fawcett Papers, Churchill College.
7 Intelligence estimates drawn from Hinsley, *British Intelligence*, III, pt 2, 95–7
8 Naval Historian's Narrative B, 179, a figure drawn from contemporary Admiralty documents
9 Hinsley, *British Intelligence*, III, pt 1, 241–3
10 'Escorts Available for Operation Overlord,' DNOS Report 33/44, 12 Apr 1944, PRO, ADM 219/118
11 'U-Boat threat to "Overlord" Compared with Operation "Torch," ' DNOS Report 34/44, 24 Apr 1944, PRO, ADM 219/119
12 Roskill, *War at Sea*, III, pt 2, 17
13 C-in-C, CNA War Diary, Mar and Apr 1944, DHist. In addition, it was decided in

March that operations of support groups in the north Atlantic would be 'administered' by C-in-C, Western Approaches. This command extended to groups based in Halifax, so that the disposition of all support groups appears under Western Approaches in official lists. Strictly speaking, then, the groups latter formed in Canada and retained in the western Atlantic were 'on loan' to C-in-C, CNA, as those assigned to Rosyth, Plymouth, and Portsmouth were in the eastern Atlantic. Secretary, Naval Board, to C-in-C, WA, 23 Mar 1944, DHist, NHS 8440, Support Groups – General

14 C-in-C, CNA War Diary, Mar and Apr 1944, DHist
15 Tucker, *Naval Service of Canada*, II, 394 contains a plot and details of the movement of HXS 300.
16 *Monthly Anti-Submarine Report* (Feb 1944) 22–4
17 See CAFOs 367/40, 753/41, 1544/43. I am particularly grateful to Arnold Hague, an associate of the Naval Historical Branch, MOD, London, for drawing these sources to my attention.
18 See 'Wreck List and Amendments,' issued by the RCN 7 Sept 1942, NAC, RG 24, 11026, CNA 7-17-2.
19 Interviews with Hannington, Chance, and Willson, May 1990
20 'Wrecks and Other Permanent Non-Sub Contacts in British Inshore Waters,' DOR Report 86/44, Dec 1944, PRO, ADM 219/166
21 DECCA was used only for D-Day and then for 'sweeping the Scheldt estuary. Information courtesy of Sir Edward Fennessy
22 'Reports on Navigation Aids used in Operation Neptune,' Director of Navigation, Admiralty, ca 6 Sept 1944, PRO, ADM 1/16664. The value of GEE was confirmed by Chance and Hannington. For a summary of the development of various radio navigation aids as they pertained to air navigation, see W.F. Blanchard, 'Air Navigation Systems: Chapter 4: Hyperbolic Airborne Radio Navigation aids – A Navigator's View of their History and Development,' in *Journal of Navigation* 33:3, (Sept 1991) 285–315. I am grateful to John F. Kemp, editor of the journal, for clarifying some points about GEE and DECCA and drawing Blanchard's article to my attention. I am also particularly grateful to Sir Edward Fennessy and Mr Claude Powell for their lengthy and informative correspondence on GEE and DECCA.
23 Howard Quinn recalls that this was often a nuisance, especially when they sought a little peace and quiet by cruising over a deep minefield.
24 Nixon comments on draft and Willson diary, 6–9 May, DHist
25 Willson, diary, 20–25 May, DHist
26 Interview with Chadwick, 24 May 1990. See also Easton, *50 North*, 238.
27 Naval Historian's Narrative B, DHist, 179
28 Ibid., 180–1

29 *Monthly Anti-Submarine Report*, Oct 1944
30 See 'Operation Neptune, Naval Operations Orders, Allied Naval C-in-C, Expeditionary Force, 10.4.44,' DHist, Operation Neptune file, especially 'ON 5.- Instructions for Neptune Covering Forces.'
31 Prentice's S206 of 14 Apr 1944, Prentice personal file
32 The author got to know Hugh Pullen briefly before his death and recalls him noting with a sense of grace and irony that Prentice got his group.
33 Easton, *50 North*, 245
34 Details on asdic fittings drawn from tables compiled by naval historians, now found in DHist 81/520/1000-973
35 Naval Historian's Narrative B, DHist, 184
36 Easton, *50 North*, 251
37 Roskill, *War at Sea*, III, pt 2, 57
38 Easton, *50 North*, 252-3
39 The group reported four; Rohwer, *Axis Submarine Successes*, 181, reports that *U984* fired three.
40 As quoted in Naval Historian's Narrative B, DHist, 198
41 Ibid., 199
42 For a thorough account of the 10th DF action see M. Whitby, 'Masters of the Channel Night: the 10th Destroyer Flotilla's Victory off Ile de Batz, 9 June 1944,' *Canadian Military History* 2:1, (Spring 1993) 5-21.
43 Naval Historian's Narrative B, DHist, 199-201
44 Ibid., 186-8. See also Report of Proceedings, 5-22 June 1944, DHist, NHS 8440 EG 6.
45 See DHist, NHS 8000 HMCS *Haida* and *U971*.
46 Roskill, *War at Sea*, III, pt 2, 67
47 Prentice to Tucker, 15 Jan 1947, DHist, NHS 8000 HMCS *Chambly*
48 Rohwer and Hummelchen, *Chronology of the War at Sea*, 287
49 Naval Historian's Narrative B, DHist, 205-8
50 Ibid., 195
51 Roskill, *War at Sea*, III, pt 2, 68-9; figures mentioned above are drawn from the same passage.
52 Admiralty to C-in-C, (Home Ports) 162328B/1/44, DHist 81/520/1000-973
53 Commodore (D), Western Approaches to DAUD, 17 July 1944, PRO, ADM 217/90
54 For an easily comprehensible discussion of the problems of A/S searches, both inshore and in deep water, see *Asdic Operating and Control: Classification and Interpretation of Asdic Contacts*, TASW, 481/46, Admiralty, May 1946, PRO, ADM 303/239. I am also particularly grateful to Commander Mark Tunneycliffe, command oceanographer for Maritime Command, for taking the time and effort to explain the complexities of sound propagation in inshore waters.

55 *Monthly Anti-Submarine Report*, July 1944
56 See SO EG 11 to C-in-C, Portsmouth, 14 July 1944, DHist NHS 8440 EG 11
57 Willson diary, 15 Apr 1944, DHist
58 R.W. Timbrell, 'There but for the flip of a coin,' *Salty Dips*, III, 33
59 Naval Historian's Narrative B, DHist 273
60 The account of the sinking of *U678* is derived from Naval Historian's Narrative B, the Reports of Proceedings in DHist, NHS 8440 EG 11, and interview with W.H. Willson, May 1990.
61 Commodore (D), Western Approaches, to DAUD, 16 July 1944, PRO, ADM 217/90 and DHist, NHS 8440 EG 11. Willson's preference was expressed in his interview of May 1990.
62 SO EG 11 to C-in-C, Portsmouth, 16 July 1944, DHist, NHS 8440 EG 11
63 Commodore (D), Western Approaches, Minute attached to Senior Officer EG 11 to Commodore (D), Western Approaches, 'Submarine Warfare in the Channel,' 17 July 1944, NAC, RG 24, 11938, NS 8100-1, v I
64 See the 'Brief Narrative' attached to SO EG 11's report of 17 July to C-in-C, Portsmouth, NAC, RG 24, 11938, NS 8100-1, v I
65 Details of the 'Battle of Pierres Noires' drawn from Naval Historian's Narrative B, DHist 262-4, DHist EG 12 files and interview with Peter Chance, 25 May 1990
66 Easton, *50 North*, 285
67 Insights and story courtesy of Dan Hannington, interview 23 May 1990
68 W.E.S. Briggs, S206 from Baker Creswell, CO HMS *Philante*, 25 June 1943. Briggs personal file
69 Comments on EG 9's operations 15 June to 11 July 1944 in PRO, ADM 199/472
70 As quoted in *Matane*'s Report of Proceedings, 23 Aug 1944, DHist, NHS 8440 EG 9
71 For a graphic representation of these patterns see the maps in the respective editions of the *Monthly Anti-Submarine Report*.
72 Prentice explained in his report that gunfire from the ships was controlled by a circling Liberator. The presence of the aircraft was not confirmed by Nixon, captain of *Chaudière* and may have been inserted in the original report to cover for the collateral damage to the village. EG 11 Report of Proceedings, 28 July to 21 Aug 44, and comments on the draft chapter by Nixon
73 Naval Historian's Narrative B, DHist 341
74 Ibid., 347-8, and interview with Chance, May 1990
75 Account of sinking of *U621* compiled from Naval Historian's Narrative B, 344-5, and Reports in DHist, NHS 8440 EG 11
76 The staff of the DAUD noted, on 2 Sept 1944, that the attack on the 20th was assessed as a 'C,' indicating probable damage. The RCN's Naval Historian's

Narrative B shows the assessment as 'insufficient evidence of a presence of a U-Boat,' an 'H' assessment. The USN's *Anti-Submarine Bulletin* for Nov 1944, which presents the assessments for attacks over this period and lists *all* A to E assessments, does not include EG 11's attack on the 20th.
77 Nixon comments on the draft
78 Naval Historian's Narrative B, DHist 345
79 See DAUD's Minute sheet attached to EG 11's Report for the period 28 July to 21 Aug, PRO, ADM 199/1460.
80 Prentice's S206 from Simpson, ca 25 Sept 1944, Prentice personal file
81 As quoted in DWT/Tactics memo of 16 Oct 1944, DHist, NHS 8440 EG 9
82 C-in-C, Plymouth, on EG 9's Report of Proceedings, 31 Aug – 14 Sept 1944, PRO, ADM 199/1462
83 Comments on file in PRO, ADM 199/1462. The article on the sinking of *U247* appeared in Oct 1944.
84 Hessler, *U-Boat War*, III, 74
85 Ibid., 79
86 Compilation of U-boat kills derived from Roskill, *War at Sea*, III, pt 2, app Y; *Defeat of the Enemy Attack on Shipping*, 1A, app 2; and Western Approaches, 'Dispositions at 1200 ...,' July and Aug 1944, DHist, NHS 1650-DS
87 Nixon comment on the draft
88 Rohwer and Hummelchen, *Chronology of the Ware at Sea*, 291

CHAPTER 5: A SEA OF TROUBLES

1 Roskill, *War at Sea*, III, pt 2, 180; correspondence in the Roskill Papers, file 6/47; and Rohwer, *Axis Submarine Successes*, 185
2 See 'Life and Letters of Gilbert Howland Roberts,' IWM, II, pt VII, 'Schnorkel.'
3 Captain (D), Halifax, minute on SO (A/S) Gaspé's 'Report on Asdic Conditions in the Gulf,' 17 Dec 1944, NAC, RG 24, 11026, CNA 7-16
4 'Review of the Asdic Oceanographic Conferences,' Jan 1944, NAC, RG 77, accession 85/86, 044, PN-2
5 'Asdic Ranging Conditions in the Halifax Approaches,' NRC, National Research Laboratories Report, PSA-1, 18 Aug 1944, NAC, RG 24, 11463, Bathythermograph – General
6 Secretary of the Naval Board to C-in-C, CNA, 12 Apr 1944, DHist, 81/520/1000-973, anticipated supplies becoming available the next month; and the SNWE from the second quarter of the year mentions that Q supplies were to hand.
7 The RCN ordered seventy-three sets from the Admiralty in the fall of 1943, but

by March 1944 only one complete set had arrived; it was installed at Cornwallis for training. By the end of June 1944 only one further partial set had been obtained. As the 'Summary of Naval War Effort' observed for the second quarter of 1944, 'Strenuous efforts have been made to expedite U.K. deliveries of 147B, without avail.' The RCN was told by the Admiralty in November 1943 that it would provide only enough sets for training purposes, and despite much prodding by the RCN, even those had not been acquired. An RCN report on type 147B supply noted in April 1944, that U.K. production was reserved for the RN and that 'the Admiralty satisfy their own demands, before considering the demands of others.' In fact, RN supplies were also slow to materialize, and the British expected that the RCN's needs would be met from Canadian production, scheduled to begin in March 1944. This production was controlled by the British Admiralty Technical Mission. The British were kind – or astute – enough to grant the RCN access to Canadian production in proportion to its needs. Owing to problems in technical liaison, however, Canadian production of type 147B did not start until August, and the RCN was forced to wait. See correspondence in DHist, 81/520/1000-973.
8 C-in-C, WA, to NSHQ 241742B/5/44, DHist, NHS 8440, Support Groups – General
9 'RCN Operational Research Report No. 21: Probable Operational Returns of an R.C.N. [Squid] Installation Program,' June 1944, DHist, 81/520/1000-973
10 FONF to C-in-C, CNA, 091701Z/3/44, DHist, NHS 8440, Support Groups – General
11 See Y'Blood, *Hunter-Killer*, the endpapers and app II for USN CVE kills; Hadley, *U-Boats against Canada*, passim, explains their relationship to operations in Canadian waters, as does Douglas, *Creation of a National Air Force*, IV.
12 C-in-C, CNA, to NSHQ 141453Z/5/44 DHist, NHS 8440 EG 16
13 NSHQ to Admiralty 101538Z/7/44, DHist, NHS 8440 EG 16
14 Admiralty to NSHQ 220606B/7/44, DHist, NHS 8440 EG 16
15 CNMO to NSHQ 02105A/8/44 and NSHQ to Admiralty 031935Z/8/44, DHist, NHS 8440 EG 16
16 Macpherson and Burgess, *Ships of Canada's Naval Forces*, app 8
17 DOD(H), memo 25 July 1944, NAC, RG 24, 11752, CS 346-1
18 NA(PP) to CNMO, 28 July 1944, NAC, RG 24, 11752, CS 346-1
19 CNMO to NSHQ 61531B/8/44, and CNMO to CNMO 102114Z/8/44, NAC, RG 24, 11752, CS 346-1
20 C-in-C, WA, to ADM 131917B/8/44, ADM to NSHQ 171452B/8/44, and NSHQ to ADM 242203/8/44, NAC, RG 24, 11752, CS 346-1
21 Hessler, *U-Boat War*, III, 83
22 For a detailed account of the attack on *U1229* see Y'Blood, *Hunter-Killer*, 241–6. Comments on the implications of RX/C failure are drawn from DWT/Tactics

assessment, 4 October 1944, DHist, NHS 8440 EG 16. No report for EG 16 covering this period has been found.
23 C-in-C, CNA War Diary, Aug 1944, DHist
24 Douglas, *Creation of a National Air Force*, 587–99, covers the initial operations of *U802* and *U541*.
25 ADM to C-in-C, WA, 21605B/9/44, ADM to NSHQ 21713B/9/44, and NSHQ to ADM 52013Z/9/44, NAC, RG 24, 11752, CS 346-1
26 Details taken from Prentice's personal records
27 See DHist, NHS 8440, EG 11 files
28 Captain Denis Jermain, RN, to the author, 29 Jan 1991
29 Ibid., Alastair Carrick Smith, who served in the frigate *Inman*, to the author 23 Jan 1991; and J.S. Filleul, who served with EG 2, to the author 25 Feb 1991
30 Interview with A.B. Sanderson, 29 May 1990
31 E. Doctor to author, 11 September 1990
32 Details from the *RCN-RCAF Monthly Operational Review* (Dec 1944). The kill is confirmed in Roskill's account but does not appear in the 1951 edition of Schull, *Far Distant Ships*.
33 Roskill, *War at Sea*, II, pt 2, app Y. Axel Niestle to Doug McLean, 14 Jan 1994; My thanks to Doug McLean for providing this information.
34 *Western Approaches Monthly News Bulletin*, Sept 1944, DHist, NHS 1650-DS
35 'Wrecks and Other Permanent Non-Sub Contacts in British Inshore waters,' DOR Report 86/44, Dec 1944, PRO, ADM 219/166
36 See 'GEE System (radar) – North and South Western Approaches,' PRO, ADM 1/16172. I am also grateful to Sir Edward Fennessey for providing a map of GEE coverage as of 1 Nov 1945.
37 Interview with Howard Quinn, 25 May 1990
38 Hinsley, *British Intelligence*, III, pt 2, 466–7
39 'Operation "S.J.," ' C-in-C, Rosyth, 26 Sept 1944 and 'Operation "S.K." [in support of Operation SJ],' 28 Sept 1944, PRO, ADM 199/500
40 *RCN-RCAF Monthly Operational Review* (Oct 1944) and DHist, NHS 1650 U-1006
41 'Operation CW, Analysis for N.W. Approaches, August 25 – October 17, 1944,' DOR 85/44, PRO, ADM 219/165
42 C-in-C, WA, AIG 321319O3Z/9/44, DHist, 8440, Support Groups – General
43 *Orders for Anti-U-Boat Operations in Coastal Waters of the Western Approaches Command (Short Title:- 'Operation CE')*, Western Approaches Command, 11 Oct 1944, PRO, ADM 199/501
44 As quoted in WAGM 185N amendment to Operation CE orders, 092249A/2/45, PRO, ADM 199/501
45 RCN-RCAF Joint Anti-Submarine Warfare Committee, minutes of 2 Nov 1944, NAC, RG 24, 11026, CNA 7-19-2

46 It was also a departure from earlier orders for support group operations that were concerned primarily with offensive measures. See, for example, Operation CW orders, PRO, ADM 199/468.
47 *Defeat of the Enemy Attack on Shipping*, v IA, 129
48 Roskill, *War at Sea*, III, pt 2, 290
49 Capt (D), Halifax, to C-in-C, CNA, 312120Z/7/44, NAC, RG 24, 11575, DO18-18-7
50 *Western Approaches Monthly News Bulletin* (Oct 1944), DHist, NHS 1650-DS
51 See, for example, the comments on EG 26's operations in Nov 1944, PRO ADM 217/731.
52 NSHQ to C-in-C, CNA, 182202Z/9/44, Halliday personal records
53 The precedence of officers was of sufficient importance for WAC to issue a periodic publication called *Western Approaches Escort Force Seniority List*, in which the various grades and seniority of officers were listed. I am indebted to W.H. Willson for the loan of his copies. The problem of being a VR senior officer in a sea full of anxious professionals was mentioned often in interviews with Canadian veterans, to the point where some felt that U-boat kills were stolen by RN groups.
54 CNA War Diary, Sept 1944
55 For a complete account of the incident see Hadley, *U-Boats against Canada*, 230-1.
56 As quoted in ibid., 229-30
57 CNA War Dairy, Sept 1944, and ibid., 232
58 CNA War Diary, Sept 1944, and Hadley, *U-Boats against Canada*, 234
59 See 'Radar Trials on Dummy Snort,' Flag Officer, Gibraltar and Mediterranean Approaches, 16 Nov 1944, PRO, ADM 1/16198
60 They presented the information on range and bearing on separate screens in the form of a horizontal line, with the contact indicated by a sharp spike (or 'A'). Comparable British and American sets employed the new – and now familiar – Plan Position Indicator, in which the ship was at the centre of the screen and the radar displayed a 'map' of the surrounding area while a bright line radiating from the centre, representing the radar beam itself, swept around in a continuous search.
61 See PRO, ADM1/16121, for minutes of a special meeting
convened by Horton on 3 Oct 1944, to discuss the radar vs snort problem.
62 Zimmerman, *Great Naval Battle*, 160
63 Elliott, *Allied Escort Ships*, 524
64 Zimmerman, *Great Naval Battle*, 158-9
65 CNMO and NSHQ exchange of signals in DHist, 81/520/1000-198, v I
66 See the report for Naval Officer in Charge, Digby, for the period, NAC, RG 24, 11053, 30-1-10, v 37-39.

67 Rear Admiral Robert Murdoch, RCN (Ret'd), recalls being ordered by Worth to report on which communications system – RN or USN – the RCN should adopt after the war. Murdoch recommended the RN system, whereupon he was instructed to rethink his proposal. Assuming that Worth wanted a stronger case, Murdoch returned with a more enthusiastic proposal in favour of the British system. Worth flew into a rage and told Murdoch that if 'he loved the British so much he could go live with them for two years.' Murdoch was posted to the RN and, despite his attempts to come home early, did not receive his orders to return until the second anniversary of his banishment. Interview with R.Adm. R. Murdoch, 29 May 1990.

Jones's anti-British sentiment is recorded in the Cunningham papers at the British Museum in the First Sea Lord's correspondence with Sir James Somerville, head of the British Admiralty Delegation in Washington; see file 52563, 108–10. A similar report on Jones's anti-British feelings was filed by the USN attaché in Ottawa in November 1945. See J. Sokolsky, 'Canada and the Cold War at Sea,' in *The RCN in Transition*, ed W.A.B. Douglas (Vancouver: University of British Columbia Press 1988), 211–12.

68 Captain (D), Halifax, to WEF and Halifax Force ships, 17 Nov 1944, NAC, RG 24, 11575, D O1-18-1

69 Douglas, *Creation of a National Air Force*, 398

70 All these ideas are discussed in the minutes of the RCN-RCAF Joint Anti-Submarine Warfare Committee, NAC, RG 24, 11026, CNA 7-19-2.

71 National Research Laboratories, Division of Physics and Electrical Engineering, Report No. PSA 1, 18 Aug 1944, 'Asdic Ranging Conditions in the Halifax Approaches,' NAC, RG 24, 11463, Bathythermography – General

72 It is Interesting that no report of proceedings for EG 16 during the period of the BT trials has yet been unearthed.

73 National Research Laboratories, Division of Physics and Electrical Engineering, Report No. PSA 2, 17 Nov 1944, 'Asdic Ranging Conditions in the River and Gulf of St Lawrence in the Late Summer,' NAC, RG 24, 11463, Bathythermography – General

74 Secretary of the Naval Board to C-in-C, CNA, and various, 15 Nov 1944, 'Introduction of Bathythermographic Equipment in Escort Vessels,' NAC, RG 24, 11026, CNA 7-16

75 *Culver* was in poor shape when the RCN took her over, and the Royal Society refused to take her back! The RCN had to pay the society a settlement in order to dispose of the vessel. See DHist, Naval Historian's Auxiliary Ship files, *Culver*, for the details of her brief RCN career.

76 National Research Laboratories, Division of Physics and Electrical Engineering, Report no. PSA 3, 'Bottom Sediments and Their Effect on Shallow Water

306 Notes to pages 208–15

Echo Ranging in Canadian Atlantic Coastal Waters,' NAC, RG 24, 11463, Bathythermography – General
77 The preference for active over passive was first emphasized in the Admiralty's 092030Z/9/44, NAC, RG 24, 11580, D 23-2-1
78 'Investigation of the Conditions affecting Asdic results during Summer,' CB 1835[2], HMS *Osprey*, 1931, PRO, ADM 186/479
79 See correspondence in PRO, ADM 1/15410.
80 'Notes on Asdic Conditions in Northern Waters,' Cazalet Papers, IWM
81 Cazalet to C.D. Howard-Johnston, the DAUD, no date but clearly late 1944, Cazalet Papers, IWM
82 British estimates are drawn from notes from Cdr Gilbert Roberts of the Western Approaches Tactical Unit as passed to the RCN in the fall of 1944. NAC RG 24, 11022, CNA 7-6-1.
83 Hinsley, *British Intelligence*, III, pt 2, 477
84 Naval Intelligence Division estimates, PRO, ADM 1/16848
85 For technical details, see Rössler, *The U-Boat*, passim and Hessler, *U-Boat War*, III, 86–7. The postwar American evaluation of the type XXI is gleaned from 'Proceedings of Anti-Submarine Warfare Conference, 17 June 1946,' Naval Historical Center, Washington, DC, Command File Post-1 Jan 1946, Conferences ASW 1946. I am grateful to Dr Michael Palmer for passing along a copy of this key document.
86 See the various reports on the subject in PRO, ADM 219, records of the Director of Naval Operational Studies.
87 Report on the Rockabill Trials, 10–30 Oct 1944, PRO, ADM 219/154 and 158
88 'Proceedings of Anti-Submarine Warfare Conference, 17 June 1946,' 6, 15–16
89 Admiralty to various, 132122A/12/44, DHist, 81/520/1000-973
90 'Tactics against Fast Submerged U-Boat,' Director of the Tactical Unit to Captain (D), Halifax, 25 Nov 1944, NAC, RG 24, 11022, CNA 7-6-1
91 See the various reports of training establishments under Flag Officer, Atlantic Coast, in NAC, RG 24, 11053, 30-1-10
92 See Murray's Minutes on AOC, Eastern Air Command's report to G.C. Power of 28 Oct 1944; NAC, RG 24, 11022, CNA 7-6.
93 As quoted in Hinsley, *British Intelligence*, III, pt 2, 477
94 'Prospects in the U-Boat War,' DNOS Report No. 81/44, 17 Nov 1944, PRO, ADM 219/161
95 Roskill, *War at Sea*, III, pt 2, app Y
96 Hessler, *U-Boat War*, III, 85
97 See EG 26's Report of Proceedings for 8–18 November 1944, PRO, ADM 217/731.
98 See the assessment by DAUD staff in PRO, ADM 199/503.

99 Cunningham Diary, 19 Dec 1944, British Museum

CHAPTER 6: VICTORY ... OF A SORT

1 War Diary, Commander, Eastern Sea Frontier, Aug 1944, Scholarly Resources microfilm
2 Naval Assistant (Policy and Plans) note on telephone conversation, DHist, NHS 8440, Support Groups – General
3 CNMO to NSHQ 191115A/12/44, NAC, RG 24, 8080, NSS 1271-35
4 Figures drawn from Western Approaches Command 'Disposition of Forces at 1200' (Hereafter 'WAC Dispositions'), DHist, NHS 1650-DS; details of RCN group redeployment from their Reports of Proceedings, DHist, NHS 8440
5 The story is recounted in Howard-Johnston to Roskill, 31 May 1979, Roskill Papers, ROSK 5/25. Howard-Johnston gives no specific date, referring simply to Christmas 1944. Cunningham's Diary in the British Museum refers to a stormy meeting of the Anti-U-Boat Committee on the 19th. On the basis of internal evidence the incidents appear to be the same.
6 Cunningham Diary, 31 Oct 1944, British Museum
7 FSL to PM, 30 Dec 1944, 'U-Boat Activities in the Channel,' PRO, ADM 199/1937
8 Roskill, *War at Sea*, III, pt 2, 183
9 See survivors comments in DHist NHS 1650-U-877.
10 *RCN-RCAF Monthly Operational Review* (Jan 1945); interview with L.P. Denny, Chester, 30 Oct 1990; G.A. Elsey to author, 3 Aug 1990; and C.K. Hurst to author 9 Mar 1990
11 CNMO to NSHQ, 271726A/12/44, and CNMO to NSHQ 41532A/1/45, PRO, ADM 199/1937
12 At the end of October the air officer commanding of Eastern Air Command, G.O. Johnson, reported upon his return from Britain that the Germans were organizing in Norway for a renewed offensive. In addition, there was a threat from new U-boat types, whose construction, Johnson warned, was going ahead. 'We must, therefore, expect and be prepared for a renewed submarine offensive in our area,' Johnson wrote to his minister in late October, 'which may be difficult to combat and will strain all available resources to the utmost.' Murray's reaction to the prospect of high-speed submarines in his zone was to observe that 'If we are blessed with these brutes,' he would have to recall EG 11 – clearly his best option in October. See Johnson to G.C. Power, 28 Oct 1944, and Murray's Minute of 2 Nov attached, NAC, RG 24, 11022, CNA 7-6
13 ADM to NSHQ 81450A/12/44, NAC, RG 24, 11752, CS 389-1
14 NSHQ to ADM 122201Z/1/45, NAC, RG 24, 11752, CS 389-1
15 WAC Dispositions, DHist, NHS 1650-DS. EG 22 was composed of sloops, EG 23 of

two Lochs and four Colony class frigates, and EG 24 of two Lochs and three Canadian-built River class frigates.
16 Cunningham to Fraser, 19 Jan 1945, Roskill Papers, ROSK 5/25
17 See signal listed in *Kirkland Lake*'s Movement Card, DHist.
18 For a detailed account of the incident see Hadley, *U-Boats against Canada*, 269. Hadley is too harsh on both Aubrey and his RN colleague, Puxley. Aubrey was the logical, and probably only, choice to lead the search, given the haste with which it was organized.
19 As quoted in ibid., 268
20 C-in-C, CNA to C-in-C, Lant, 011520Z/12/44, and C-in-C, Lant, to C-in-C, CNA, 011933Z/12/44, NAC, RG 24, 6901, NS 8910-166/10. I am grateful to Roger Sarty for bringing this point to my attention.
21 C-in-C, U.S. Fleet, to Commander, Task Units 27.1.1, 27.1.2, and 27.1.3, 13 Dec 1944, NAC, RG 24, 11022, CNA 7-6; the quote is from the signal to the task units.
22 Hadley, *U-Boats against Canada*, 276–7
23 As quoted in ibid., 277
24 As quoted in ibid., 280
25 Ibid., 280–1
26 McLean, 'The Last Cruel Winter,' 68
27 Report of Proceedings, 20–23 Jan 1945, DHist, NHS 8440 EG 27
28 C-in-C, CNA, to EG 16, EG 27, and TG 22.9, 19 Jan 1945, NAC, RG 24, 11022, CNA 7-6. For the pattern of operations see the relevant reports of proceedings for EGs 16, 27, and 28, DHist, NHS 8440.
29 Training Report, CNA, Dec 1944, DHist, Naval Staff 1000-5-13
30 See Report of Captain (D), Halifax, for the period, DHist, Naval Staff 1000-5-13. Support groups often underwent gunnery exercises off Halifax just after sailing, but there was no operational training *as a group with tame submarines.* 'Operational' training concentrated on shore-based simulators, where refresher training of weapons and sensor specialists was offered. A section in Tucker's *Naval Service of Canada* is suggestive of this approach: he describes individual and team refresher training almost exclusively when discussing 'operational training'; see II, 261–7.

When EG 16 arrived in Britain in mid-March, its first six days were taken up by harbour training, followed by two days at sea off Londonderry, and then by four days of group exercises under the direction of HMS *Philante*. The assessment of this training period – one of a very few to survive – described EG 16 as a group 'of average efficiency.' Individual skills, such as radar, communications, gunnery, and A/S were generally satisfactory. But the group as a whole 'suffered from lack of sea training which they have not obtained since last July.' See training reports for the spring of 1945 in NAC, RG 24, 11744, CS 165-1-1.

31 'Western Approaches Tactical Unit, Annual Report 1944,' PRO, ADM 1/17557
32 'A/S Training for Operations in Inshore Waters,' DNOS Report 83/45, 8 Jan 1945, PRO, ADM 219/283
33 See CNMO to Secretary, Naval Board, 17 Jan 1945, NAC, RG 24, 8080, NSS 1271-35.
34 C-in-C, CNA, to various 152145Z/1/45, NAC, RG 24 11022, CNA 7-6
35 Joint RCN-RCAF Joint Anti-Submarine Warfare Committee minutes 25 Jan 1945, NAC, RG 24, 11026, CNA 7-19-2
36 See the memo from the Secretary of the Naval Board on HE ranges, 29 Mar 1945, NAC, RG 24, 83-84/167, NSS 8100-166/10.
37 See Reports for EGs 16, 27, and 28. The quote is from Report of Proceedings, 10-17 Mar 1945, DHist, NHS 8440 EG 27.
38 Commander, Task Group 22.9, to C-in-C, U.S. Fleet, 21 Jan 1945, NAC, RG 24, 11022, CNA 7-6
39 CTG 22.9 to C-in-C, Lant, 21 Jan 1945, and C-in-C, Atlantic Fleet, to C-in-C, CNA, 20 Jan 1945 NAC, RG 24, 11022, CNA 7-6
40 Patrol of Halifax Harbor Approaches, 6 Feb – 1 Mar 1945, CTG 22.10 and attached Report of Exercises with Bottomed Submarine, 13–14 Mar 1945, copies provided by DHist
41 NSHQ to CNMO 311650Z/1/45, and CNMO to NSHQ 31320A/2/45, NAC, RG 24, 11752, CS 389-1 See also the discussion on CVEs at the 39th meeting of the RCN-RCAF Joint Anti-Submarine Warfare Committee, 25 Jan 1945, NAC, RG 24, 11026, CNA 7-19-2
42 Report of Proceedings, 9–17 Feb 1945, DHist, NHS 8440 EG 16
43 Equipment details from Tucker's notes in DHist, 81/520/1000-973, v 5, and the Fighting Equipment Co-ordinating Authority (FECA) reports in the same collection. Tucker lists *Ste Thérèse* with type 271Q radar, but the group's Report of Proceedings refers directly to her RX/C set.
44 Report of Proceedings, 26 Feb – 8 Mar 1945, DHist, NHS 8440 EG 28
45 See the amendments attached to the front of the orders for Operation CE in PRO, ADM 199/501
46 WAC Dispositions, DHist, NHS 1650-DS, and Reports of Proceedings, DHist, NHS 8440 EG 6
47 Report of Proceedings, 20 Feb 1945, and the comments of WAC staff, PRO, ADM 199/198
48 *Monthly Anti-Submarine Report* (Mar 1945) 10–11
49 Report of Proceedings, 4–18 Feb 1945, DHist, NHS 8440 EG 9
50 See WAC Staff Comments in PRO, ADM 199/198.
51 Report of Proceedings, 4–18 Feb 1945, DHist, NHS 8440 EG 9
52 Hinsley, *British Intelligence*, III, pt 2, 627–8
53 Ibid., 630–1

54 Cunningham Diary entry for 1 March indicates the three-month period needed to end the war; British Museum. The desire to drive the U-boat back into deep water was an old one, but it remained current in 1945. See Fawcett's 'Second Thoughts on Anti-SNORT Obstructions,' a pencilled draft memo dated 30 Dec 1944 in the Fawcett Papers, FWCT 2/4/5, Churchill College, Cambridge
55 Cunningham, *A Sailor's Odyssey*, 631
56 The regular senior officer, A/Cdr V. Browne, RCNVR, was ashore for a court of inquiry.
57 See Commanding Officer's Remarks appended to the Report of Proceedings for the action, DHist, NHS 8440 EG 25.
58 PRO, ADM 199/199
59 As quoted in C-in-C, WA, to the Secretary of the Admiralty, 21 Apr 1945; EG 25's Report of Proceedings file, DHist, NHS 8440 EG 25
60 PRO, ADM 199/199
61 The account of the *U1003* was drawn from EGs 26 and 25, Reports of Proceedings, DHist, NHS 8440, the *RCN-RCAF Monthly Operational Review* (Feb–May 1945), and the *Monthly Anti-Submarine Report* (Mar 1945)
62 See staff comments in PRO, ADM 199/203.
63 The RCN official history described the incident in a way that implied that *New Glasgow* had rammed the U-boat. See Schull, *Far Distant Ships*, 194.
64 Hessler, *U-Boat War*, III, 97
65 Report of Proceedings, 1–26 Apr 1945, DHist, NHS 8440 EG 9; interview with A.B. Sanderson, 29 May 1990
66 A detailed account of the saga of *U1024* was printed in *Monthly Anti-Submarine Report* (Sept–Dec 1945). Other details are from EG 25's Report of Proceedings, DHist, NHS 8440, and an interview with Howard Quinn, 25 May 1990.
67 *United States Fleet Anti-Submarine Bulletin* (Apr 1945) 31–2. Details of Canadian operations in the area are drawn from EGs 16 and 27, Reports of Proceedings, and from the Anti-Submarine Action Report of TG 22.14, DHist, NHS 8440.
68 *St Croix*'s destruction of *U90* in July 1942 was west of the CHOP line established in March 1943 and therefore technically within the Canadian zone as constituted. That kill was part of a convoy battle which carried over into the Grand Banks, however, and occurred before the Canadian northwest Atlantic was established.
69 CNA War Diary, Mar 1945, DHist
70 This is the zone referred to in CNA's U-boat estimates and roughly conforms to the zone operated by Canada under NATO.
71 C-in-C, Lant, A16-3/(0500), 26 Feb 1945, NAC, RG 24, 11022, CNA 7-6-1
72 Articles 1120-1161, NSHQ to various 101630Z/3/45, NAC, RG 24, 11022, CNA 7-6

Notes to pages 252–62   311

73 DTU to Capt (D), Halifax, 2 Mar 1945, Secretary, Naval Board, to C-in-C, CNA, 10 Apr 1945, NAC, RG 24, 11022, CNA 7-6-1
74 NSHQ to C-in-C, CNA, 152154Z/3/45, NAC, RG 24, 11022, CNA 7-6-1
75 Captain (D), Halifax, to C-in-C, CNA, 25 Mar 1945 and SO EG 28 to Captain (D), Halifax, 11 Apr 1945, NAC, RG 24, 11575, D-1-18-1
76 For a description of Operation Teardrop see Y'Blood, *Hunter-Killer*, 260–72; details of the sinking of *U879* can be found in the *United States Fleet Anti-Submarine Bulletin*, May 1945
77 The U-boat was probably *U548*, thought to have been sunk off Cape Henry in the United States at about the same time. The number recovered from the wreckage of the U.S. kill off Sable Island was '*U369*,' which corresponds to the contractor's code number for *U548* during construction. Information from Axel Niestle, via Doug McLean 31 March 1994
78 Douglas, *Creation of a National Air Force*, 608
79 The cabinet objection to Roskill's draft is found in the Roskill Papers, ROSK 6/47. The final assessment of the U-boat campaign, as modified, is found on in *War at Sea*, III, pt 2, 305.

EPILOGUE: LOCH ERIBOLL

1 See Milner, 'HMCS *Somers Isles*.'
2 See *Western Approaches Seniority Lists* for early 1945.
3 'Summary of Information Regarding Departmental Activities,' Jan 1945, NAC, King Papers, MG 26, J4, v 264, f. C180727
4 Tucker, *Naval Service of Canada*, II, 253, lists thirty-six officers in the executive branch as 'being' qualified by Jan 1945, but does not state how many of them qualified before the war.
5 Figures compiled from the May 1945 RCN *Navy List*, DHist
6 See 'A/S Warfare' report submitted to the VCNS on 1 Sept 1949 by the director of Scientific Services, and the attached Minutes, NAC, RG 24, 83-84/167 NSS 1670-1-4, v 1.
7 From C.P. Nixon's comments on the draft
8 As quoted in *North Atlantic Run*, 64
9 See, for example, his memoir, *U-Boat Killer*.
10 This issue is discussed briefly in the author's 'A Canadian Perspective on Canadian and American Naval Relations Since 1945,' in *Canada-United States Defense Cooperation: The Road From Ogdensburg*, ed. Joel J. Sokolsky and Joseph T. Jockel, (Edwin Mellen Press 1992), 145–74.
11 See 'Western Approaches Command, Disposition of Forces at 1200, 1-4-45,' DHist, NHS 1650-DS

12 See Adm K.G.B. Dewar to Roskill, 3 Nov 1957, Roskill Papers, ROSK 6/54.
13 'A Comparison of A/S Hunts in Canadian and British Coastal Waters,' RCN Operational Research Report No 35, 27 Aug 1945, NAC, RG 24, 11463
14 Mentioned by veterans and also in Coates's report of his time spent aboard the German depot ship, see DHist, NHS 8440 EG 9.
15 The story was recounted by A.B. Sanderson, one of *Monnow*'s officers.
16 Report of Proceedings, 10–22 May 1945, DHist, NHS 8440 EG 9

APPENDIX I: ESCORT BUILDING AND MODERNIZATION, 1941–3

1 Tucker, *Naval Service of Canada*, II, 66 n2
2 Ibid., 66–7
3 Naval Staff Minutes, 25 Sept 1941
4 Tucker, *Naval Service of Canada*, 68–9
5 Naval Council Minutes, 3 Nov 1941
6 Tucker, *Naval Service of Canada*, II, 69
7 Ibid., 70
8 CNS to Macdonald, 23 June 1942, as quoted in ibid., 70
9 Interview with Rear Admiral H.G. DeWolf, RCN, who was director of Plans at the time
10 I am grateful to Michael Hennessey for drawing my attention to the restriction on the use of American yards and on the pernicious influence of the Department of Munitions and Supply.
11 Four Tribal class destroyers were begun in Halifax in 1942 but were not completed in time to see action.
12 For a discussion of the problems of east coast yards see *North Atlantic Run*, 216–18.
13 See Admiralty to British Advisory Repair Mission, 6 March 1942, NAC, RG 24, ACC 83-84/167, v 2585, NSS 6101-1, pt 1.
14 See the Director of A/S, 'Record of Progress on A/S Projects in Connection with Production, Supply and Conversion [of asdic type 123 to type 127],' 24 Aug 1942, NAC, RG 24, 83-84/167, v 2571, NSS 6100-330, pt 1.
15 See 'Modernization of Armament and Equipment,' 13, DHist, NHS 8060, Repairs and Alterations.
16 The types 144 and 145 were basically the same set, except that the 145 had a portable dome. The RCN's ultimate objective was the type 145 set, but many corvettes only got as far as fitting the type 144
17 As cited in the 'Record of Progress on A/S Projects ....,' DA/S memo of 24 Aug 1942

Bibliography

PRIMARY SOURCES, CANADA

National Archives of Canada

MG 26	William Lyon Mackenzie King Papers
MG 30	Rear Admiral L.W. Murray, RCN, Papers
RG 2	War Cabinet Committee Papers and Minutes
RG 24	NSHQ Central Registry Files
	Atlantic Command Files
	Captain (D), Halifax, Files
	Senior Canadian Naval Officer (London) Files
	Naval Member, Canadian Staff (Washington) Files
	Flag Officer, Newfoundland, Files
	Accession 83-84/167, various Files

Directorate of History, National Defence Headquarters (NDHQ)

Naval Board Minutes
Naval Staff Minutes
'Summaries of Naval War Effort'
C-in-C, CNA War Diary
Naval Historian's Files on EG 5, EG 6, EG 11, EG 12, EG 16, EG 25, EG 26, EG 27, EG 28, W 10, W 13
Ships' Movement Cards
Naval Historian's Narrative A: 'Canadian Participation in North Atlantic Convoy Operations, June 1941 to December 1943'
Naval Historian's Narrative B: 'The Royal Canadian Navy's Part in the Invasion of Northern France – Operation Overlord'

314 Bibliography

Naval Historian's Narrative: 'Modernization of Armament and Equipment'
Monthly Anti-Submarine Reports
US Fleet Anti-Submarine Bulletin
RCN-RCAF Monthly Operational Review
RCN Monthly Review
P.W. Nelles Papers
Interviews by H. Lawrence
Canadian Confidential Naval Orders
Confidential Admiralty Fleet Orders
Defeat of the Enemy Attack on Shipping
Handbooks for the Type 147B Asdic

Public Archives of Nova Scotia (PANS)

Angus L. Macdonald Papers

PRIMARY SOURCES, BRITAIN

Public Record Office, Kew

ADM 1	General RN Subject Files
ADM 199	Naval Historian's Files
ADM 205	First Sea Lord's Files
ADM 217	Western Approaches Command Files
ADM 219	Director, Naval Operational Studies Files
ADM 239	Technical Staff Monographs
PREM 3	Prime Minister's Papers
PREM 4	Prime Minister's Papers

Churchill College, Cambridge

Captain S.W. Roskill Papers
Commander H.J. Fawcett Papers
Cunningham of Hyndhope Papers
Sir Charles Goodeve Papers

Imperial War Museum, London (IWM)

G.V. Ball Papers

Bibliography 315

P.G.L. Cazalat Papers
E.H. Chavasse Memoir
S. France Memoir
W.J. Moore Papers
J. Moose Papers

British Museum, London

A.B. Cunningham Papers

Naval Historical Branch, London

Convoy and Anti-Submarine Warfare Reports
Annual Report of the Torpedo School
M Files
'The War At Sea,' NHB Narrative, CA 1648(d)

SECONDARY SOURCES: SELECT BIBLIOGRAPHY

Chalmers, R/Adm W.S. *Max Horton and the Western Approaches.* London: Hodder and Stoughton 1954
Cunningham, Viscount of Hyndhope [A.B.]. *A Sailor's Odyssey.* London: Hutchinson & Co. 1951
Dönitz, Grand Admiral Karl. *Memoirs: Ten Years and Twenty Days.* Annapolis, Md: U.S. Naval Institute Press 1990
Douglas, W.A.B. *The Creation of a National Air Force: The Official History of the Royal Canadian Air Force,* vol II. Toronto: University of Toronto Press 1986
Easton, Alan. *50 North.* Toronto: Ryerson Press 1964
Elliott, Peter, *Allied Escort Ships of World War II.* London: Macdonald and Janes 1977
Graves, Donald E. 'The Royal Canadian Navy and Naval Aviation, 1942–1944.' Unpublished narrative, DHist, NDHQ, May 1989
Gretton, V/Adm Sir Peter. *Convoy Escort Commander.* London: Cassell 1964
– *Crisis Convoy.* London: P. Davis 1974
Hackmann, Willem. *Seek and Strike: Sonar, Anti-Submarine Warfare and the Royal Navy 1914–54.* London: HMSO 1984
Hadley, Michael. *U-Boats against Canada.* Montreal and Kingston: McGill-Queen's University Press 1985
Hinsley, F.H. *British Intelligence in the Second World War,* vol III, pts 1 and 2. Cambridge: Cambridge University Press 1984, 1988

[Hessler, Gunther] *The U-Boat War in the Atlantic 1939-1945*, 3 vols. London: HMSO Books 1989

Macintyre, Donald. *U-Boat Killer.* London: Weidenfeld and Nicholson 1956

McLean, D.M. 'The Last Cruel Winter: RCN Support Groups and the U-Boat Schnorkel Offensive.' Unpublished MA thesis, Royal Military College, Kingston 1992

Macpherson, Ken. *Frigates of the Royal Canadian Navy 1943-1974.* St Catharines, Ont: Vanwell 1989

Macpherson, Ken, and J. Burgess. *The Ships of Canada's Naval Forces 1910-1981.* Toronto: Collins 1981

Macpherson, Ken, and Marc Milner. *Corvettes of the Royal Canadian Navy, 1939-1945.* St Catharine's, Ont: Vanwell 1993

Milner, Marc. *North Atlantic Run: The Royal Canadian Navy and the Battle for the Convoys.* Toronto: University of Toronto Press 1985

- 'The Dawn of Modern ASW: Allied Responses to the U-Boat 1943-1945,' *RUSI Journal* 134:1 (Spring 1989) 61-8
- 'Inshore ASW: The Canadian Experience in Home Waters,' in *The RCN in Transition*, ed. W.A.B. Douglas. Vancouver: University of British Columbia Press 1988
- 'The RCN and the Offensive against the U-Boats, 1943-1945.' Unpublished narrative, DHist, NDHQ, June 1986
- 'HMCS *Somers Isles*: The Royal Canadian Navy's Base in the Sun.' *Canadian Defence Quarterly* 14:3 (Winter 1984) 41-7

Morison, S.E. *History of United States Naval Operations in World War II: volume X: The Atlantic Battle Won, May 1943 – May 1945.* Boston: Little, Brown and Company 1956

Poolman, Kenneth. *Escort Carrier 1941-1945.* London: Ian Allan 1972

Rohwer, J. *Axis Submarine Successes 1939-1945.* Annapolis, Md: U.S. Naval Institute Press 1983

Rohwer, J. and W.A.B. Douglas. 'Canada and the Wolf Packs, September 1943.' In *The RCN in Transition*, ed. W.A.B. Douglas. Vancouver: University of British Columbia Press 1988

Rohwer, J., and G. Hummelchen, *Chronology of the War at Sea 1939-1945.* Annapolis, Md: U.S. Naval Institute Press, rev. ed. 1992

Roskill, S.W. *The War at Sea*, 3 vols. London: HMSO 1954-61

Rössler, Eberhard. *The U-boat.* London: Arms and Armour Press 1981

- *Die Torpedos der Deutschen U-Boote.* Herford, Germany: Koehlers Verlagsgesellschaft mbH 1984

Salty Dips, 3 vols. Ottawa: Naval Officers Association of Canada, Ottawa Branch 1983-8

Schull, Joseph. *The Far Distant Ships.* Ottawa: King's Printer 1950
Terraine, John. *The U-boat Wars, 1916–1945.* New York: Putnam's 1989
Thomas, Captain (N) Robert H. 'The War in the Gulf of St Lawrence: Its Impact on Canadian Trade.' *Canadian Defence Quarterly* 21:5 (Spring 1992) 12–17
Tucker, Gilbert N. *The Naval Service of Canada: Its Official History,* vol II. Ottawa: King's Printer 1952
Wemyss, Cdr D.E.G. *Walker's Groups in the Western Approaches.* Liverpool: Liverpool Daily Post & Echo Ltd 1948
Y'Blood, William T. *Hunter-Killer.* Annapolis, Md: U.S. Naval Institute Press 1983
Zimmerman, David. *The Great Naval Battle of Ottawa.* Toronto: University of Toronto Press 1988

Index

Adams, Cdr K.F., RCN 49, 105, 124–5
aircraft carriers: A/S role 35; MAC ships 64; RCN plans for 52, 128, 182–3, 231
Algerine class minesweepers 273
Allied Anti-Submarine Survey Board 26, 42, 51
anti-submarine warfare 5–6, 6–8, 98–9, 110–11, 113–14, 122; barrage attacks 112, 122; creeping attacks 99, 112; impact on personnel 188–90; impact of schnorkel 202–3; in arctic waters 208–9; in deep water 182–3; inshore problems 140–2, 152–3, 154–6, 177–82, 190–1, 199–200, 206–8, 229, 251, 264; need for accurate inshore navigation 141–2, 190–1, 205; problems of fast U-boats 211–13; use of minefields 191; wreck charts 140–1
asdic: final RCN policy 125; limits inshore 17–18; 'Q' attachment 89, 112–13, 122, 124, 125, 146, 168, 179, 200, 229; seabed 205; type 123A 13, 276–7; type 123D 44, 276–7; type 127DV 276–7; type 128, 34, 277;
type 144, 89, 112–13, 122, 125, 146, 232, 276–7; type 145, 125, 276–7; type 147B 87–90, 118, 122, 124, 125, 146, 168, 179–81, 200, 209, 232, 235, 263, 301–2 (n 7); *see also* anti-submarine warfare
Ashe, Cdr P.B., RCN 127
Atlantic Convoy Conference: *see* Washington Convoy Conference
Atlantic Oceanographic Research Group 178, 206
Aubrey, Cdr R., RN 224, 241, 258
Audette, L.C. 226

Balfour, Cdr St Clair, RCNVR 197, 226
bathythermography 18, 177–9, 206, 208–9, 229–30, 251
Bermuda 103–4, 129, 213, 256–7, 260
Bidwell, Capt R.E.S., RCN 43, 53, 94
Birch, Cdr J.D., RNR 45, 85, 93, 115, 145, 162
Biscay Offensive 46, 55–8
Bliss, Cdr P.M., RN 127–8, 178
Bridgeman, Cdr C.E., RNR 54, 65, 67, 93
Briggs, Lt Cdr W.E.S., RCNVR 162–3
Brock, Lt Cdr John, RCNVR 239–40

Brock, Cdr J.V., RCN 258
Brodeur, RAdm V.G., RCN 26, 33
Browne, Cdr Victor, RCNVR 195, 240
Burnett, Cdr P.W., RN 65, 66, 93, 95, 116–19, 233

CAT gear 72–6, 148, 175, 211
Cazalet, Cdr P.G.L., RN 209
Chadwick, Cdr E.M., RCN 142, 143
Chavasse, Cdr E.M., RN 106, 111, 179
CHOP line 22, 23, 26, 27
Churchill, Rt Hon W.S. 18, 19, 22, 50, 216, 219–20, 223
Coates, Lt J.J., RCNVR 265–6
Connolly, J.J. 37, 80
convoys: BB 80, 248; BTC 76, 234; BX 141, 226–8; BX 142, 228; CU 36, 176; HX 237, 39; HX 256, 63; HX 265, 77, 82; HX 280, 116; HXF 305, 176; HX 306, 176; HX 346, 245–6; JW 67, 265; ON 113, 12; ON 115, 11, 14; ON 207, 77; ONS 5, 37–8, 71; ONS 18 / ON 202, 63–71, 85, 108, 261; ONS 21, 77: ONS 27, 107; ONS 33, 201; ONS 154, 12; ONS 251, 176; OS 64 / KMS 38, 106; RA 59, 132; SC 42, 9; SC 97, 12; SC 107, 16; SC 118, 22; SC 153, 114; SC 154, 119; SH 194, 225; SL 139 / MKS 30, 82; VWP 17, 244; WN 74, 234
convoys: routing of 39–40, changes for D-Day 139–40
corvettes: Castle class 88, 181, 184, 185, 209, 263; employment of 34; equipment of 13–14, appendix I passim; increased endurance class 272–3; modernization of 40–4, 50, 100, appendix I passim; pre-1941 use of 6; Prentice on 9; unsuitability for support groups 107–8

Creery, Capt W.B., RCN 86, 89
Cunningham, Adm Sir Andrew, RN 60, 219–20, 223, 237, 238

Dawson, Lloyd 126, 241
D-Day 137–46
DECCA 142, 205
Denny, Lt Cdr L.P., RCNR 221
depth charges 44, 83, 110–12, 123, 168, 235–6, 250
destroyers 15–16, 86, 99–100; *see also* support groups EG 11 and EG 12
Devonshire, George 106
Doctor, Ernie 189
Dönitz, Grand Admiral Karl 21, 26, 39, 46, 47, 59, 62, 63, 70, 82, 85, 137, 177, 210, 218, 222, 254

Easton, Cdr Alan, RCNR 102, 145, 147, 148, 160–2
E-boats 142
Edelsten, RAdm J.H., RN 60
Elsey, G.A. 221
escort groups: *see* support groups
Evans, Cdr M.B., RN 64, 66, 68, 71

Fawcett, Cdr H.J., RN 137–8
FOXER 72–6, 175, 211–12
Fraser, Lt Cdr J.P., RCNVR 114–15
frigates: acquisition of 86–8, 105; Captain class 86–7, 263; delay in arrival 102; employment 92, 184–5, 218–19; equipment 88–92; Loch class 87, 125; numbers in early 1944 130; River class 15–16, 270–2

GEE: *see* QH
George, Lt J., RCNVR 48–9
glider bombs 56, 84–5, 164, 166
Gretton, Cdr Peter, RN 38, 49, 116

Groos, Lt Cdr D., RCN 168

Hadley, Michael 200
Halliday, Lt Cdr W.C., RCNR 186
Hannington, R/Adm Dan, RCN 141
Headland, Cdr E.H., USN 231
hedgehog 44, 89, 275–6; depth limitations 83, 123–4
HF/DF 15, 156
Horton, Adm Sir Max, RN 11, 23, 31, 73, 76, 94, 98–9, 100, 108, 109, 125, 131, 176, 185, 190, 194, 219, 240, 244
Houghton, Capt F., RCN 53–4, 258
Howard-Johnston, Capt C.D., RN 97, 219–20, 229, 255

Isaacs, Cdr V.A., USN 230

Jermain, Capt D., RN 188–9
Jones, V/Adm G.C., RCN 52, 81–2, 100, 204, 222, 258, 261

King, Lt Cdr C.A., RCNR 164, 258
King, Adm E.J., USN 29–30
King, Rt Hon W.L.M. 3–5, 13, 18, 19, 20, 30, 37, 39, 50, 52, 80, 97, 133

Layard, Cdr A.F.C., RNR 94, 108, 145, 170, 171, 235, 245, 265–6
LORAN 142, 205

Macdonald, A.L. 19, 50, 51–2, 80–2, 272
Macintyre, Capt Donald, RN 259
MacKillop, Cdr A.M., RN 145, 147–8, 160–2
MacLean, Doug 227
McMaster, Cdr H., RCN 89
Mainguy, Capt E.R., RCN 11

Medland, Lt Cdr M.A., RCN 104
merchant ships: *Alsterufer* 106; *Athelviking* 226; *British Freedom* 226–7; *Cuba* 244–5; *Empire Heritage* 176; *Empire MacAlpine* 64, 68–9; *Fjordheim* 176; *Fort Thompson* 202; *Frederick Douglas* 65–6; *Jacksonville* 176; *James W. Nesmith* 245–6; *Kelmscott* 109; *Leopoldville* 216; *Livingston* 187, 199; *Martin van Buren* 226; *Nipiwan Park* 225; *Pinto* 176; *Polarland* 225; *Samtucky* 216; *San Francisco* 119–20; *Silverlaurel* 216; *Waleha* 70; *Watuka* 127; *Will Rogers* 248;
MF/DF 15
Mid-Ocean Escort Force 8, 12, 15, 18, 19, 20, 22, 27, 29–31, 38, 40, 45, 46, 48, 53, 60, 92, 98, 99, 103, 106, 108, 139, 185, 187, 189, 222, 259, 260, 263, 278
Mitchell, Lt Cdr J.E., RCNVR 199, 232, 252, 253
modernization of escorts 13–16, 40–3, 48–53, 80–2, 93, appendix I passim
Moore, V/Adm Sir Henry R., RN 29, 32
Moore, Flying Officer K.O., RCAF 147
Murray, R/Adm L.W., RCN 8, 9, 26, 28, 31, 32, 73, 78, 92, 93, 100–1, 102, 103, 105, 108, 109, 128–9, 130, 144–5, 163, 199, 213, 224, 229, 232, 252, 259

Nairn, Lt R.A., RCN 206–7
National Research Council 18, 178
Naval Service headquarters 24, 31, 32–3
Nelles, V/Adm P.W., RCN 6, 16, 19, 40, 48, 51–2, 80–2, 94, 96, 125, 184, 222

Newfoundland Escort Force 7–8, 10, 259, 269
Niestle, Axel 190
Nixon, Capt C.P., RCN 142

Operation CE 194–5, 228, 232–3
Operation Dredger 159–62
Operation SJ 192–3

Pavillard, Lt Cdr L.R., RCNR 106–7
Peers, Cdr A.F., RCN 74
Phillimore, Cdr Richard, RN 73
Pickard, Lt Cdr A.F., RCNR 65
Piers, Cdr D., RCN 49
Pizey, Capt C.T.M., RN 184–5
Plomer, Cdr J., RCN 212–13, 251–2
Pound, Adm Sir Dudley, RN 19, 40, 51, 60
Prentice, Capt J.D., RCNR 8–11, 35–6, 101, 103, 105, 129, 144–5, 152, 156–9, 166, 168, 170, 173, 188, 197–8, 256, 258
Pressy, Cdr A.R., RCN 18, 104, 178
Pullen, Cdr H.F., RCN 145
Puxley, Capt W.L., RN 195–6, 204–5, 252, 253

QH 142, 159, 164, 191, 205, 235
QM 142
Quinn, Lt Cdr Howard, RCNVR 238–40, 247–9, 262

radar: ability to detect schnorkel 202–3; German countermeasures 61–2, 174; RX/C 89, 92, 125–6, 186, 198, 202, 232, 241; RX/U 92, 3-cm 203–4, 232, 263; shore based 142, 205; SW1/2C 14; type 271, 14, 88, 89, 125, 200, 202, 232, 274; type 272, 88, 202; type 277, 87, 202; U-boat 175–6

Ralston, Col the Hon J.L. 19
rangefinders 122–3
Rayner, Lt Cdr H.S., RCN 10
R-boats 142
Reid, Cmdr H.E., RCN 130, 261
Roosevelt, F.D. 22, 50
Roskill, Capt S.W., RN 154, 176, 216, 220, 255
Rowland, Cdr J.M., RN 249, 258
Royal Canadian Air Force 17, 24–5, 34–5, 38, 69, 78–9 85, 102, 127–8, 182, 205, 220
Russell, Lt Cdr P.F.X., RCN 161
Rutherford, Cdr C.A., RCNR 67

St Lawrence, gulf and river 16–8; attacks in 1944 198–202; in 1943 33–4, 47–8; layering of water mass 206–7
Salmon operations 35; for U_{537} 78–80; for U_{543} 101–2; for U_{845} 109; for U_{541} and U_{802} 199–200; operational orders for 128
Sanderson, Lt A.B., RCNVR 189, 236, 245
seniority of officers 198, 262
Simmons, Cdr E.T., RCNVR 197, 241
Simpson, Cmdr G.W., RN 49, 170; see also Western Approaches Command
SONAR: see asdic
SOSSUS 205
squid 87–8, 116–17, 123, 233, 263; first kill 165; RCN policy 124–5, 181–2
Stacey, A/Lt Cdr W.R., RCNR 171–2, 235
Stephen, Cdr G.H., RCNR 119, 120, 121, 245
Storrs, Lt Cdr A.H.G., RCNVR 67
Strange, Capt W., RCNR 49, 51, 104

support groups
- American: EG 6, 45; TG 22.8, 254; TG 22.9, 226, 228, 230; TG 22.10, 226, 231, 232, 251; TG 22.14, 249–50, 253; TU 27.1.1, 251; TU 27.1.2, 225; TU 27.1.3, 225
- British: EG 1, 56, 143, 152, 165, 173, 190; EG 2, 45, 76, 77, 88, 92, 94, 108, 109–10, 112, 122, 132, 143, 165, 166, 173, 233, 248; EG 3, 45, 77, 143, 165, 173, 233, 234; EG 4, 45, 77, 152; EG 5, 45, 143, 152, 173, 238, 246; EG 7, 45, 77, 82, 94, 131; EG 8, 45, 77, 248, 262; EG 10, 112–13, 118, 125, 233, 234, 235; EG 14, 143, 147, 151, 159, 162, 173, 188; EG 15, 143, 146, 234, 238; EG 16, 109; EG 40, 55, 82; EG 21, 238; EG 22, 222; EG 23, 222; EG 24, 222, 223
- Canadian: 'Canadian support group' 31, 40; C 2, 94, 107–8, 114, 116–19, 130; EG 5, 45, 53, 54–8, 77–8, 82–4, 85, 93, redesignated EG 6, 105–6; EG 6, 106–7, 114–15, 125, 130, 131–2, 141, 143, 145, 146, 150, 151, 162–3, 172, 173, 181, 190, 192–4, 198, 215, 219, 233, 234, 238, 262, 264–5; EG 9, 53, 54, 63–71, 73, 85, 93–4, 102, re-established 108–9, 116, 119–21, 125, 130–1, 141, 143, 145, 146, 150, 151, 153, 163–4, 170–2, 173, 181, 189, 190, 192, 198, 215, 233, 235–7, 244–5, 262, 265–7; EG 11, 139, 141–6, 152–3, 156–9, 162–3, 165–6, 167–9, 169–70, 173, 188, 190, 192, 198, 222, 232, 236, 261, 262, 265; EG 12, 139, 141–6, 147–9, 150, 159–62, 166–7, 173, 188, 198, 236; EG 16, 108, 183, 185–8, 198, 199–201, 202, 206, 224, 225, 227–8, 231–2, 246, 249, 251, 262; EG 18, 108; EG 25, 188, 191, 195–6, 198, 199, 215, 234, 238–40, 242, 243, 245–9, 258, 262, 264; EG 26, 188, 189, 197, 198, 215–16, 234, 238, 241–4, 258, 262, 265; EG 27, 188, 196, 197–8, 224, 226–8, 231, 249, 250, 251, 264; EG 28, 222, 232, 252, 253, 254; Western Support Force 44; W 10, 31, 100, 102–3, 109, 127; W 12, 224–5; W 13, 199–201

support groups: 10, 31, 37; command of 31–2; concept of operations 1943–4 98; need for in CNA 128–9, 183–4; need for more 100–1, 218–20; reorganization for D-Day 139

Timbrell, Lt R., RCNVR 156, 157–8
torpedoes: GNAT 62–3, first use 65, countermeasures 72–6, 211; seabed 205–6; Taffy 62, 72, 73; T-11, 211
training 103–5, 126–7, 128–9, 213, 228–9, 252, 263, 308 (n 30)
Tucker, Dr G.N. 9
Tully, Dr J.P. 178

U-boats: *Avorio* 30; *Tritone* 30, 197; *U90* 12, 15; *U94* 30; *U163* 30 *U190* 252–4; *U210* 12; *U218* 172; *U224* 30; *U226* 112; *U229* 69; *U233* 182; *U234* 254; *U238* 65, 66, 70; *U241* 264; *U247* 171–2, 281; *U257* 133, 280; *U260* 70; *U262* 78; *U264* 110, 122, 135; *U270* 65, 166; *U278* 265; *U305* 66–7, 106; *U309* 235–6, 281; *U311* 131–2, 280; *U333* 165; *U338* 66, 67; *U356* 12; *U358* 122; *U369* 250; *U378* 68; *U385* 173; *U386* 66, 113; *U406* 113; *U448* 130–1, 132, 280;

324   Index

U480 174, 234; *U482* 175–6; *U484* 189–90, 281; *U486* 216; *U490* 182; *U501* 9, 10, 197; *U516* 265; *U536* 78, 83–4, 85, 88, 93, 132, 280; *U537* 78–80; *U539* 109, 179; *U541* 186–7, 199–201, 204; *U543* 101–2; *U548* 129–30; *U558* 11, 12, 15, 88, 112; *U575* 121–2, 132, 280; *U608* 173; *U618* 173; *U621* 168–9, 281; *U648* 83; *U666* 70; *U678* 156–9, 281; *U731* 66, 67, 70; *U736* 166; *U743* 190; *U744* 116–19, 122, 280; *U753* 280; *U756* 12; *U757* 280; *U763* 173; *U764* 150; *U767* 150, 151; *U772* 216, 220; *U802* 127–8, 186–7, 199–201, 204; *U805* 254; *U806* 224; *U842* 112; *U845* 109, 119–21, 131, 280; *U853* 253; *U854* 179; *U856* 128, 182; *U858* 254; *U866* 249–52; *U877* 220–1, 281; *U879* 253; *U881* 253; *U889* 254; *U952* 70; *U953* 148–9; *U971* 151–2, 154, 173, 280; *U984* 147–8, 281; *U989* 174, 233; *U1003* 241–4, 281; *U1004* 234; *U1006* 192–3, 214, 215, 281; *U1010* 265; *U1024* 245–8; *U1195* 244–5; *U1209* 215, 243–4; *U1223* 201–2; *U1227* 201; *U1228* 202, 254; *U1229* 186–7; *U1232* 225–8, 229, 264; *U1278* 233; *U1279* 233; *U1302* 239–40, 243–4, 281; *U2326* 254; *U2511* 237, 254; *U2513* 212
U-boats: 'Alberich' skin 174; diving depths 111; equipment 61; range of old-type late 1944 185–6; schnorkel 135–7; type XXI 61, 137, 138, 176, 209–10, 212–13, 214, 216, 221–2, 232, 237, 254–5, 262; type XXIII 61, 137, 138, 210, 221–2; Walter boat 138, 142, 143, 211

United States Navy 7–8, 24–5, 85, 178, 208, 224–5, 230–1, 250–1; *see also* support groups

Vigder, Dr J.S. 258

Walker, Capt J., RN 76, 108, 109–10, 113–14, 145
Walter, Prof Helmuth 60, 135
warships
– Allied: *Activity* 109; *Albrighton* 167; *Balfour* 102; *Bayntun* 106–7, 132, 233, 280; *Blackwood* 150, 163; *Bogue* 31, 38, 45, 64, 121, 186–7, 280; *Card* 64; *Chanticleer* 82; *Coatan* 128; *Egret* 56, 57; *Eskimo* 151–2, 154, 280; *Fencer* 77, 109, 110, 132; *Forester* 120, 156, 280; *Galatea* (Italian) 202; *Goodson* 102; *Grenville* 55–7; *Hadleigh Castle* 123; *Haverfield* 121–2, 280; *Helmsdale* 190; *Hobson* 122, 280; *Hurst Castle* 176; *Icarus* 65–7, 107, 116–17, 280; *Itchen* 54, 65, 66, 67–70, 71, 73, 93; *Kenilworth Castle* 116–18, 280; *Keppel* 64, 66, 69, 71; *King Haakon* 200; *Lagan* 39, 65, 71, 76, 280; *L'Escarmouche* 244; *Lobelia* 64; *Loch Dunvegan* 233; *Loch Eck* 233; *Loch Fada* 88; *Loch Glendhu* 248; *Loch Killin* 165, 166; *Lowe* 250; *Loyalty* 174; *Menges* 250; *Montgomery* 79; *Mourne* 150; *Nairana* 109; *Narcissus* 64, 66, 68, 69; *Natal* 238; *Nene*: *see* Canadian

Index 325

warships; *Orchis* 64; *Pelican* 130–1, 280; *Philante* 49, 54; *Polyanthus* 65–9, 76; *Portchester Castle* 190; *Pursuer* 146; *Rathlin* 67; *Renoncule* 64; *Roselys* 64; *Rowley* 158; *Santee* 64; *Saumarez* 209; *Seraph* 211–12, 215; *Spey* 113–14; *Starling* 165; *Statice* 157–9, 281; *Striker* 109, 165; *Tartar* 150; *Tavy* 165; *Towy* 64; *Tracker* 77, 146, 150; *Tweed* 40, 45, 56, 77, 87, 105, 106; *Unseen* 204; *Vindex* 122, 131; *Wanderer* 165; *Watchman* 244–5; *Woodpecker* 110
- Canadian: *Alberni* 127, 174; *Algoma* 44; *Amherst* 274; *Annan* 192–3, 280; *Assiniboine* 12, 49, 50, 145–6, 167, 188; *Athabaskan* 55–7; *Atholl* 108; *Battleford* 127; *Beacon Hill* 197, 241; *Belleville* 245–6; *Buckingham* 232; *Calgary* 41, 43, 44, 50, 54, 77, 82–4, 279, 280; *Camrose* 44, 106–7, 132, 280; *Cape Breton* 132, 145; *Chambly* 9, 10, 53, 65, 197; *Charlottetown II* 183; *Chaudière* 116–18, 142, 145, 146, 153, 166, 167–9, 173, 280–1; *Chebogue* 201–2, 215; *Chilliwack* 107, 116–18, 280; *Clayoquot* 129, 216, 224; *Coaticook* 197, 226; *Columbia* 102; *Culver* 178, 207; *Drumheller* 38, 65–7, 95, 280; *Dunvegan* 127; *Dunver* 189–90, 281; *Edmundston* 41, 43, 50, 54, 77, 84, 106–7, 221, 278, 281; *Ekholi* 207; *Esquimalt* 252–3; *Ettrick* 105, 197, 227, 264; *Fennel* 107, 116–18, 280; *Fort Erie* 232; *Frontenac* 108–9; *Gatineau* 65–6, 95, 107, 116, 142, 153, 280; *Grou* 132, 141, 145; *Haida* 149, 151–2, 154, 173, 280; *Hespeler* 189–90, 281; *Huntsville* 245–6; *Huron* 149; *Inch Arran* 232; *Ingonish* 129; *Iroquois* 55; *Joliette* 195; *Jonquière* 197; *Kamloops* 65, 95; *Kamsack* 186; *Kenogami* 127; *Kenora* 79; *Kentville* 225; *Kirkland Lake* 224; *Kitchener* 44; *Kootenay* 40, 110, 143, 146, 156, 157–8, 166, 167–9, 281; *La Hulloise* 126, 195, 239–41, 281; *LaSalle* 197, 226; *Lethbridge* 197; *Lévis II* 197, 226; *Lunenburg* 44, 77, 84; *Loch Achanalt* 125, 193; *Loch Alvie* 125, 235, 236, 244, 265; *Loch Morlich* 125; *Lockeport* 79; *Magnificent* 256; *Magog* 183, 199, 201–2; *Matane* 102, 108–9, 119–20, 130, 131, 145, 164, 265–6, 280; *Matapedia* 129; *Mayflower* 195; *Meon* 105, 145, 146, 164, 171, 197, 226; *Moncton* 274; *Monnow* 171, 188, 236, 244, 245, 265–6; *Montreal* 102, 108–9, 197, 215; *Moose Jaw* 9; *Morden* 53, 65, 70, 76, 95, 106; *Nabob* 183; *Nene* 40, 45, 55, 77, 82–4, 87, 105, 106, 115, 145, 146, 236, 244, 265; *New Glasgow* 128, 197, 241–4, 281; *New Waterford* 141, 145, 162; *Niagara* 102; *Nipigon* 226; *Norsyd* 199, 264; *North Bay* 108; *Oakville* 30; *Ontario* 256; *Orangeville* 181; *Orkney* 183, 186, 195; *Ottawa I* 15; *Ottawa II* 38, 40, 145, 146, 157–9, 166, 167–9, 281; *Outremont* 132, 150; *Owen Sound* 108, 119, 130, 280; *Port Arthur* 30, 44, 197; *Port Colborne* 128, 145, 163, 171, 215; *Prescott* 44; *Prestonian* 232; *Prince Robert* 84; *Prince Rupert* 121–2, 132,

326 Index

280; *Puncher* 183; *Qu'Appelle* 145, 146, 147–9, 159–61, 162, 167, 188; *Red Deer* 127; *Regina* 30; *Restigouche* 145, 146, 147–8, 160, 162, 167–8; *Ribble* 197, 215, 241; *Runnymede* 245; *Sackville* 14, 42, 53, 65, 68, 76, 95, 274, 278; *Saguenay* 195; *St Catharines* 95, 103, 107, 116–18, 280; *St Clair* 186; *St Croix* 12, 53, 65, 66, 67–9, 70–3, 76; *St Francis* 53; *Saint John* 145, 171–2, 190, 235, 236, 281; *St Laurent* 10, 50, 102, 119–20, 166, 197, 280; *St Pierre* 265–6; *Ste Thérèse* 195, 199, 200, 232; *St Thomas* 220–1, 281; *Sans Peur* 186; *Sarnia* 253; *Saskatchewan* 145, 147–9, 160–1, 166; *Saskatoon* 278; *Sault Ste Marie* 104; *Sea Cliff* 221; *Shawinigan* 202; *Shediac* 30, 127; *Skeena* 11, 12, 88, 145, 147–8, 160, 162, 167, 188; *Snowberry* 44, 54, 56, 77, 82–4, 88, 106, 280; *Springhill* 183, 186, 200, 206–7; *Stettler* 183, 199–201; *Stormont* 102, 130, 145, 170, 171; *Strathadam* 238–40, 245–9, 281;

Sussexvale 241; *Swansea* 93, 102, 108–9, 119–20, 130–1, 145, 164, 171, 280–1; *Swift Current* 127; *Teme* 145, 150; *The Pas* 127; *Thetford Mines* 195, 199, 239–40, 243, 245–8, 281; *Toronto* 183; *Truro* 127; *Uganda* 145, 256; *Valleyfield* 129–30; *Ville de Québec* 4, 30; *Wallaceburg* 127; *Warrior* 256; *Waskesiu* 92, 93, 103, 104, 106, 107, 114–15, 132, 133, 145, 146, 280; *Wentworth* 103, 127; *Westmount* 226–7; *Wetaskiwin* 11, 12, 88; *Woodstock* 44

Washington Convoy Conference 25–32, 260

Western Approaches Command 6–8, 11, 12, 16, 23, 43, 48, 53, 73, 94, 103, 104, 110, 115–16, 131, 177, 218, 222

Willson, Capt W.H., RCN 110, 142–3, 156, 168

Winn, Capt R., RNR 59, 237

Worth, Capt S., RCNR 204, 261

Zimmerman, David 126

Picture Credits

Bibliothek fur Zeitgeschichte: Sleek type XXI U-boats, 1945, 115/11; schnorkel head of *U2548*, 215/13

Defence Research Establishment Atlantic: CAT Mk II, Canadian solution to the acoustic torpedo

Department of National Defence: *Saskatoon*, 1942, CN-3836; *Camrose*, 1944, GM-1159; *Athabaskan*, 1944, R-1041; HMS *Nabob*, 1944, O-204-4; *Swansea*, 1943–4, GM-1441R; *Loch Alvie*, 1944, CN-6413; swastikas on *Chaudière* funnel, A-1015; *Chebogue*, 1944, PMR 94-095; It helped to be young, A-1017; Naval Staff in the last winter, GM 2992; photographer on *Unseen*'s dummy schnorkel mast, PMR 94-094; *Unseen*'s dummy schnorkel and search periscope, PMR 94-096; *U889* surrenders, A-1424

Ken Macpherson: *Moncton*, 1943; *St Croix*, 1943; *Long Branch*, 1944; *Hespeler*, 1944; *Redmill* and *Rupert*, 1945

National Archives of Canada: Nelles and Macdonald with *Halifax* model, PA-134335; type 123A asdic control equipment, PA-136247; James Douglas Prentice, PA-191076; 'data processing,' PA-184187; *Ottawa II*, PA-191034; type 127DV asdic control equipment, PA-139916; anti-submarine weapon of choice, PA-112918; British FOXER gear, PA-191072; type 144 asdic control equipment, PA-134330; *Coaticook*, 1945, PA-114638; *Niagara* and *Columbia*, 1943, PA-191070; shell-torn *U744*, PA-112996; Cdr Clarence King, RCNR, PA-191029; Nelles explaining depth-charge attacks, PA-146005; hedgehog bombs striking water, PA-190084; depth-charge explosions in shallow water, PA-133246; *Chaudière* and unidentified ship, 1944, PA-143939; *Saint John*, 1944, PA-134503; *Magog* adrift in the St Lawrence, PA-137797; Lt Cdr Craig Campbell, RCNVR, PA-190231; survivors of *U877*, PA-191069; surveying damage to *Strathadam*, PA-191028; *Esquimalt*, 1944, PA-116954; Ted Simmons, PA-191074; The enemy too was young, PA-191077; conning tower of *U190*, PA-134173; EG 9 in a British port, PA-191026; remnant of the German arctic fleet, PA-191027; Coates, Parrish, and Massey, PA-191030

Map: Operation Salmon for U543, Dec 1943 – Jan 1944

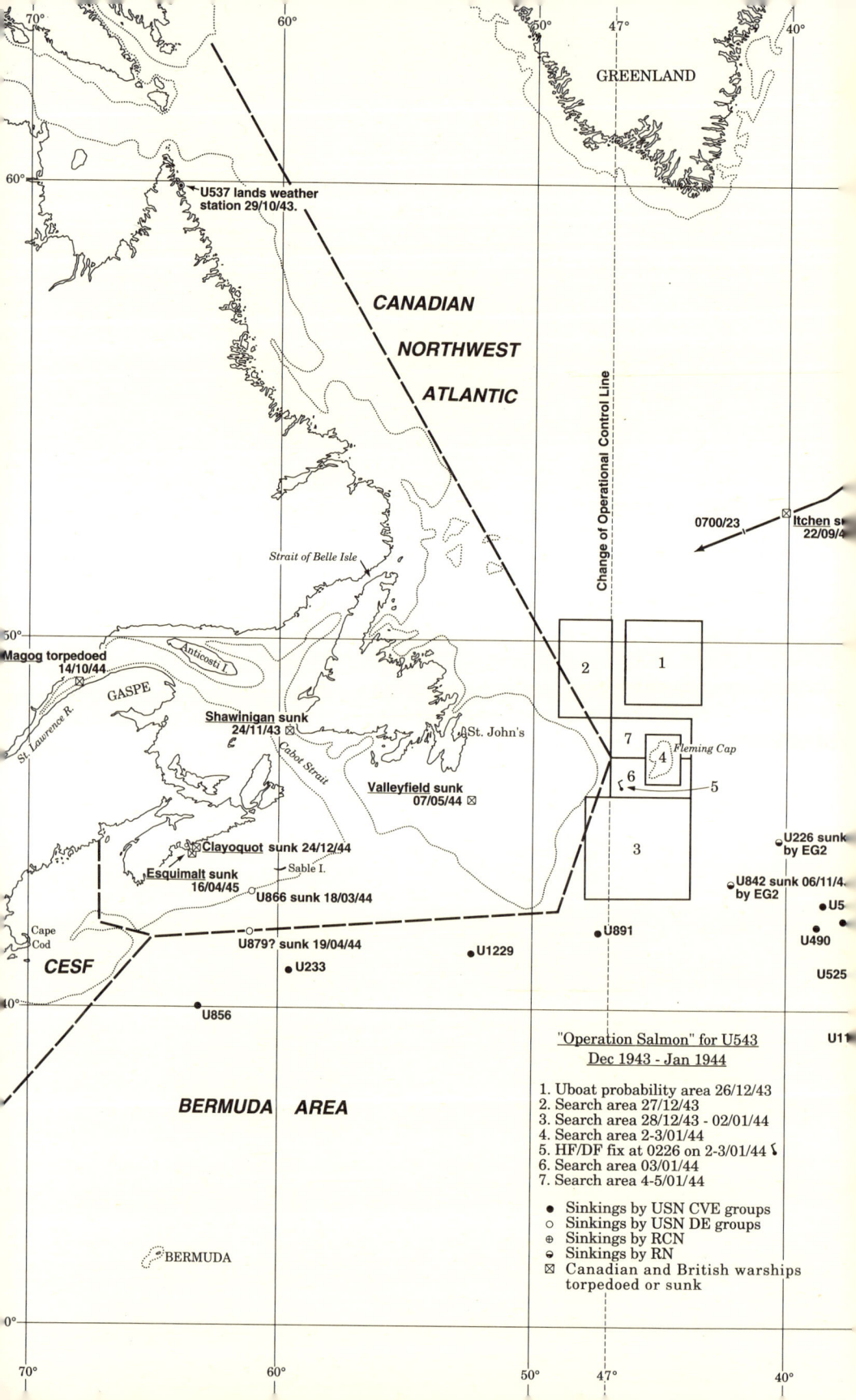

GREENLAND

CANADIAN NORTHWEST ATLANTIC

Change of Operational Control Line

U537 lands weather station 29/10/43

0700/23
Itchen s[unk]
22/09/4[3]

Strait of Belle Isle

Anticosti I.

Magog torpedoed 14/10/44
GASPE
St. Lawrence R.
Shawinigan sunk 24/11/43
Cabot Strait
St. John's
Valleyfield sunk 07/05/44
Fleming Cap
Clayoquot sunk 24/12/44
Sable I.
Esquimalt sunk 16/04/45
U866 sunk 18/03/44

U226 sunk by EG2
U842 sunk 06/11/4[3] by EG2
U5[
U490
U525

Cape Cod
CESF
U879? sunk 19/04/44
• U233
U1229
U891

• U856
U1[

"Operation Salmon" for U543
Dec 1943 - Jan 1944

1. Uboat probability area 26/12/43
2. Search area 27/12/43
3. Search area 28/12/43 - 02/01/44
4. Search area 2-3/01/44
5. HF/DF fix at 0226 on 2-3/01/44
6. Search area 03/01/44
7. Search area 4-5/01/44

- ● Sinkings by USN CVE groups
- ○ Sinkings by USN DE groups
- ⊕ Sinkings by RCN
- ⊖ Sinkings by RN
- ⊠ Canadian and British warships torpedoed or sunk

BERMUDA AREA

BERMUDA